WORLD SCIENTIFIC SERIES ON NONLINEAR SCIENCE

Editor: Leon O. Chua
University of California, Berkeley

Series A. MONOGRAPHS AND TREATISES*

*To view the complete list of the published volumes in the series, please visit:
http://www.worldscientific.com/series/wssnsa

WORLD SCIENTIFIC SERIES ON
NONLINEAR SCIENCE

Series A   Vol. 13

Series Editor: Leon O. Chua

CHAOS IN NONLINEAR OSCILLATORS
Controlling and Synchronization

M Lakshmanan
K Murali

Bharathidasan University,
Tiruchirapalli, India

World Scientific
Singapore • New Jersey • London • Hong Kong

Published by

World Scientific Publishing Co. Pte. Ltd.
5 Toh Tuck Link, Singapore 596224
USA office: 27 Warren Street, Suite 401-402, Hackensack, NJ 07601
UK office: 57 Shelton Street, Covent Garden, London WC2H 9HE

British Library Cataloguing-in-Publication Data
A catalogue record for this book is available from the British Library.

World Scientific Series on Nonlinear Science Series A — Vol. 13
CHAOS IN NONLINEAR OSCILLATORS
Controlling and Synchronization

ISBN-13 978-981-02-2143-0
ISBN-10 981-02-2143-6

PREFACE

It is now well accepted that chaos is an ubiquitous and robust nonlinear phenomenon frequently encountered in nature. During the past two decades or so the concept of chaos has permeated almost all branches of science and engineering. The field is growing into a stage where the initial surprises associated with the phenomenon are waning and new understandings are appearing, while actual controlling and harnessing of it are being contemplated. In these developments, the study of nonlinear oscillators has played a very important role in understanding the chaotic phenomenon. Many ubiquitous systems such as the Duffing oscillator, damped and driven pendulum, driven van der Pol oscillator, and so on have been treated as paradigms in chaos research. The study of such systems is based mostly on approximate analytic approaches and detailed numerical investigations. From another point of view, nonlinear electronic circuits complement these studies through analog simulations. Besides, many new nonlinear electronic circuits have been constructed, which are dynamical systems of interest on their own accord. Chua's diode and related circuits are foremost examples of nonlinear electronic circuits which act as veritable black boxes to study nonlinear phenomena. Thus the twin approaches of numerical analysis and circuit theoretic studies can complement each other in the investigation of bifurcation and chaos phenomena in nonlinear dynamical systems.

In recent times, one has witnessed considerable activity in the controlling of chaotic motions to desired regular orbits, through predetermined small perturbations. Various algorithms have been proposed and implemented successfully to avoid the harmful effects of chaos, when required, and bring back the system to desired regular states by minimal changes. But more surprisingly chaos can also be harnessed in a purposeful way leading to exciting technological applications. Against common beliefs, identical chaotic systems can be synchronized provided an appropriate coupling is introduced between them. This in turn leads to the possibility of spread-spectrum secure communications of both analog and digital signals.

The aim of this book is essentially to analyse the bifurcation and chaos phenomena in typical nonlinear oscillators, especially of damped and driven types, from both dynamical and circuit theoretical points of view, and then to introduce the concept of controlling and synchronization in them. Though many important books on chaotic dynamics have appeared in the recent literature stressing different aspects, the authors believe that the approach taken in this book and the topics covered deal with many aspects not readily discussed in other books on chaos.

Specifically, after giving a brief introduction to the topic of nonlinear dynamics in Chapter 1, we introduce the elementary notions on the dynamics of linear and nonlinear oscillators in Chapter 2. In Chapter 3, a brief introduction to linear and nonlinear circuit theory is provided and the relation to dynamical systems is explained. Bifurcation and chaos phenomena with specific reference to the Duffing oscillator are discussed in Chapters 4 and 5. The different types of attractors, bifurcations, and routes to chaos are discussed in detail for the double-well, single-well and double-hump Duffing oscillators by both numerical analysis and analog circuit simulation in Chapter 4. Complementing these studies, an analytic investigation of the Duffing oscillator is carried out in Chapter 5 through approximate (perturbation and linear stability) analyses, Melnikov criterion and analytic structure (Painlevé singularity structure) studies.

Chapter 6 deals with the bifurcation and chaos aspects of the Bonhoeffer-van der Pol (BVP) and Duffing–van der Pol oscillators, involving numerical, analytical and analog simulation studies. Chapter 7 is devoted to a study of bifurcation and chaos phenomena in nonlinear electronic circuits involving piecewise-linear Chua's diode by both experimental and numerical analyses. We consider the behaviour of Chua's oscillator, the autonomous Chua's circuit, the driven Chua's circuit, the simplest dissipative nonautonomous circuit, and the autonomous Duffing–van der Pol oscillator here.

The final part of the book, consisting of Chapters 8 and 9, deals with some very recent developments in chaotic dynamics, namely controlling of chaos and synchronization of chaotic systems. In Chapter 8 we give a brief account of the various algorithms suggested for controlling of chaos and apply these algorithms to the BVP oscillator as a test case. Some of the methods are also applied to the other oscillators mentioned above. Finally, in Chapter 9 we introduce the Pecora and Carroll method of chaos synchronization with and without cascading, as well as the alternative method of one-way coupling of identical chaotic systems. We then illustrate the possibility of transmitting analog and digital signals using synchronized chaotic signals as carriers in a secure way and apply it to the various oscillators discussed.

The book also contains three appendices on (i) perturbation methods, (ii) van der Pol oscillator, and (iii) some other standard oscillators. Also, a glossary of specialized terms is included.

In the absence of exact analytical methods, numerical studies alone cannot provide a complete picture of the dynamics in the parameter space. For most of the oscillators considered in this book, such phase diagrams given in the text cover only a portion of the parameter space. More extensive analysis is required to cover the entire space of the parameters. However, we do hope that the present book may motivate further work along this direction.

In our endeavour to write this book, we have received whole-hearted support from the members of the Nonlinear Dynamics Group at Bharathidasan University. We have freely used their research results in our discussions in the book. In addition, Dr. M. Daniel and Dr. S. Rajasekar helped us by providing critical comments on the manuscript. We thank them and other members of the group for their cooperation.

In the main body of the book, we have also used materials and figures from many published articles by different authors, which are referred to at appropriate places in the text. We thank the American Institute of Physics and Institute of Physics, U.K., for granting permission to reproduce some of the figures appeared in their journals. For a book of this nature, it is impossible to refer to all the related published literature. Though we have tried to give the relevant references which are familiar to us, we apologize to those authors whose work we did not mention due to either our ignorance or unfamiliarity.

Finally, we wish to record our acknowledgement of the continued support received from Bharathidasan University and the Department of Science and Technology, Government of India, for many years for our various activities in nonlinear dynamics. This support is the main inspiration for us to undertake the endeavour of writing this book.

April 1995   M. Lakshmanan
Tiruchirapalli   K. Murali

CONTENTS

CHAPTER 1

INTRODUCTION

1.1. General

Evolution of physical systems, subject to suitable (constraint-free) internal and external forces and appropriate initial conditions, is often expected to be completely and uniquely determined by Newton's equations of motion (Ref. [1]). For an N-particle system with masses m_i ($i = 1, 2, \ldots, N$) and forces $\mathbf{F}_i$ acting on them, the dynamics is in general described by the set of second order ordinary differential equations

$$m_i \frac{d^2 \mathbf{r}_i}{dt^2} = \mathbf{F}_i \left(t, \mathbf{r}_1, \mathbf{r}_2, \ldots, \mathbf{r}_n, \frac{d\mathbf{r}_1}{dt}, \frac{d\mathbf{r}_2}{dt}, \ldots, \frac{d\mathbf{r}_n}{dt} \right), \quad i = 1, 2, \ldots, n. \quad (1.1)$$

Here $\mathbf{r}_i$ is the position vector of the ith particle in an inertial frame of reference and Eq. (1.1) is subjected to the prescribed $6N$ initial conditions $\mathbf{r}_i(0)$, $\frac{d\mathbf{r}_i}{dt}|_{t=0}$. It is implicitly assumed here that the initial position and velocity vectors of each particle can be accurately and simultaneously provided. By solving the system of $3N$-second order coupled ordinary differential equations (1.1) along with the initial conditions, one can expect that the future of the system can be completely predicted with any required precision. Such a possibility, in fact, led Laplace to imagine that for a super-intelligence '*nothing could be uncertain and the future, as the past, would be present to its eyes*' (Ref. [2]).

1.2. Nonlinearity and Chaotic Motions

In spite of the impressive conceptual foundation, there are obvious limitations in Newton's description and so in Laplace's dictum:

1

(i) Presence of external random forces/fluctuations can always introduce a kind of indeterminacy, which is a statistical phenomenon.

(ii) Quantum effects can often lead to indeterminacy, dictated by the Heisenberg's uncertainty relations, due to our limitations in the simultaneous physical measurement of canonically conjugate dynamical variables such as position and momentum.

During recent times, it has been realized that a third kind of limitation can occur in Newton's description of evolution of even simple dynamical systems *when nonlinearity is present* (Refs. [2–10]) in a suitable form. It is true that in order to predict the future behaviour of a physical system accurately, leaving aside the limitations posed by statistical and quantum effects, one has only to solve the initial value problem of a system of deterministic differential equations of the form (1.1).

However, when nonlinear forces are present, the system can in general admit very complex motions and the associated equation of motion cannot in general be exactly integrated. As a result often one has to take recourse to numerical integration of the underlying differential equations. Then any small inaccuracy in the prescription of the initial state or round-off errors at any point or stage of the numerical calculation can build up exponentially fast to make the system deviate appreciably from the actual intended state in a finite time interval. One says that there is an *exponential divergence* of nearby trajectories. There is nothing much we can do about this indeterminism, because however accurate and fast calculating machines we are able to produce, there can still be some small error at some stage of the calculation which will multiply fast in a finite amount of time. This is a fact which we have to live with when nonlinearity is present in an appropriate form.

One might wonder whether the above effect is a mere mathematical or computational artifact or whether it has anything to do with the physical behaviour of the system at all. In fact, one knows now very well that an immediate physical realization of the above exponential divergence of nearby trajectories is the extreme *sensitiveness* of the behaviour of the system on initial conditions. This fact was after all anticipated by H. Poincaré in his celebrated analysis of celestial mechanics (Ref. [2]) itself. Any infinitesimal fluctuation at any time during the evolution of a system can in a finite time lead to a physically realizable effect, the so called 'butterfly effect' as termed by Lorenz (Refs. [2, 4]): "As small a perturbation as a butterfly fluttering its wings somewhere in the Amazons can in a few days time grow into a tornado in Texas".

The above type of complex behaviour admitted by appropriate nonlinear systems, exhibiting extreme sensitiveness to initial conditions, is termed as *chaotic motion* or simply *chaos*, which is a pure manifestation of nonlinearity. Of all possible nonlinear systems, especially of importance are dissipative and conservative systems. There are characteristic differences between the chaos exhibited by these two categories.

1.3. Dissipative and Conservative Nonlinear Systems

(i) *Dissipative systems*: The time evolution of these systems contracts volume in the phase-space (the abstract space of state variables) and consequently the trajectories approach asymptotically either a chaotic or a non-chaotic attractor. The latter may be a fixed point, a periodic limit cycle or a quasiperiodic attractor. These and the chaotic attractors are bounded regions of phase-space towards which the trajectory of the system, represented as a curve, converges in the course of long-time evolution (Refs. [5–10])). Bifurcation or qualitative changes of periodic attractors can occur leading to more complicated and chaotic structures, as a control parameter is varied.

The chaotic attractor is, typically, neither a point nor a curve but a geometrical structure having a self-similar and fractal (often multifractal) nature. Such chaotic attractors are called *strange attractors*. Many physically and biologically important nonlinear dissipative systems, both in low and high dimensions, exhibit strange attractors and chaotic motions. Typical examples are the various damped and driven nonlinear oscillators (Refs. [5–16]), the Lorenz system (Ref. [4]), the Brusselator model (Refs. [13, 14]), the Bonhoeffer–van der Pol oscillator (Ref. [17]), the piecewise linear electronic circuits (Refs. [18–20]), and so on.

(ii) *Conservative or Hamiltonian systems*: Nonlinear systems of *conservative* or *Hamiltonian type* also often exhibit chaotic motions (Refs. [21–23]). But here the phase-space volume is conserved and so no strange attractor is exhibited. Instead, chaotic orbits tend to visit all parts of a subspace of the phase-space uniformly. The dynamics of a nonintegrable conservative system is typically neither entirely regular nor entirely irregular, but the phase-space consists of a complicated mixture of regular and irregular components. In the regular region the motion is quasiperiodic and the orbits lie on tori while in the irregular regions the motion appears to be chaotic but they are not attractive in nature. Typical examples include coupled nonlinear oscillators, the Henon–Heiles system, the anisotropic Kepler problem, and so on. Similarly, the quantum manifestations of such Hamiltonian chaos, namely quantum chaos (Refs. [21–24]), are also of great physical relevance. However, this book does not deal with the Hamiltonian chaos aspects but concentrates only on dissipative systems.

It should be emphasized here that not every nonlinear dynamical system as a rule exhibits chaotic motions. Even very complicated nonlinear systems can sometimes exhibit very coherent and ordered structures such as solitons, dromions, instantons, and so on (Refs. [25, 26]). When a given nonlinear dynamical system will exhibit chaotic behaviour and when it will admit coherent and ordered behaviour are intricate mathematical problems, the understanding of which will constitute an important area of future investigations in the field. Some possible lines of thinking include the Painlevé singularity structure analysis (Refs. [27, 28]), investigation of generalized symmetries (Refs. [28–31]), Melnikov analysis (Refs. [10, 32]) and so on.

1.4. Bifurcations and Chaos-Controlling and Synchronization

In this book we will concentrate mainly on the chaotic motions exhibited by damped and driven nonlinear oscillator systems of interest in different fields of research and will illustrate the rich variety of bifurcations and chaos phenomenon exhibited by them. We will then also discuss how chaos can be controlled to regular motion by minimal efforts and finally the possible technological applications of it in secure communications through the concept of chaos synchronization. As a prelude to these developments we will first consider the oscillations of simple linear and nonlinear systems in the next Chapter.

CHAPTER 2

LINEAR AND NONLINEAR OSCILLATORS

The superposition principle which is valid for linear differential equations is no longer valid for nonlinear ones. A physical consequence is that the frequency of oscillation is in general amplitude-dependent in the case of nonlinear systems, while it is not so in the case of linear systems. Particularly, this can have dramatic consequences in the case of forced and damped nonlinear oscillators, leading to nonlinear resonance and jump (hysteresis) phenomenon for low strengths of nonlinearity parameters. Such behaviours can be analyzed using various perturbation methods. However, as the control parameter varies, the nonlinear systems can enter into more complex motions through different routes, where detailed numerical analysis and possible analog simulations using electronic circuits can be of much help. We will briefly introduce these ideas in the present and next Chapters, while more exhaustive studies will be taken up in the later Chapters. Before discussing the nature of nonlinear oscillations, we will first briefly discuss the salient features associated with a damped and driven linear oscillator in order to compare its properties with nonlinear oscillators.

2.1. Linear Oscillators and Predictability

Physical systems whose motion is described by linear differential equations are called linear systems. If they are associated with oscillatory behaviour, then they are designated as *linear oscillators* (Ref. [33]). The characteristic features of such linear systems are their insensitiveness to infinitesimal changes in initial conditions and the (at the most) constant separation of nearby trajectories in phase space. As a consequence the future behaviour becomes completely predictable. To illustrate these ideas, let us consider the simple example of a linear harmonic oscillator of unit mass, damped by a viscous drag force, and acted upon by an external periodic force. Such a model represents a very large number of physical systems ranging from forced oscillations in an LCR circuit to electron oscillations in an electromagnetic field (Ref. [33]). The oscillations are then described by an inhomogeneous, linear, second order differential equation of the form

$$\frac{d^2x}{dt^2} + \alpha\frac{dx}{dt} + \omega_0^2 x = f\sin\omega t\,, \tag{2.1}$$

where $x(t)$ is the amplitude of oscillation of a system of unit mass subjected to suitable initial conditions, say, $x(0) = A = $ constant, $\frac{dx}{dt}|_{t=0} = \dot{x}(0) = 0$. Here $\omega_0/2\pi$ corresponds to the natural frequency, α is the strength of the damping, while f and ω stand for the forcing amplitude and angular frequency, respectively, of the external force. We will first consider the special cases of (2.1) before looking at the full equation.

2.1.1. *Free Oscillations*

When the damping and external forcing are absent, $\alpha = 0$, $f = 0$, the system (2.1) is essentially a linear harmonic oscillator executing simple harmonic vibrations (Fig. 2.1(a)),

$$x(t) = A\cos\omega_0 t\,, \tag{2.2}$$

satisfying the initial conditions $x(0) = A$, $\dot{x}(0) = 0$. The phase trajectories are concentric ellipses in the phase $(x - \dot{x})$ space, characterized by the constant energy $E = \frac{\dot{x}^2}{2} + \frac{\omega_0^2 x^2}{2} \equiv \frac{1}{2}\omega_0^2 A^2$. In Fig. 2.1(b) one such ellipse is shown for a given A.

2.1.2. *Damped Oscillations*

Considering again (2.1), now with damping present and forcing absent, $\alpha \neq 0$, $f = 0$, we have three possibilities:

(i) underdamping : $\alpha < 2\omega_0$
(ii) overdamping : $\alpha > 2\omega_0$
(iii) critical damping : $\alpha = 2\omega_0$.

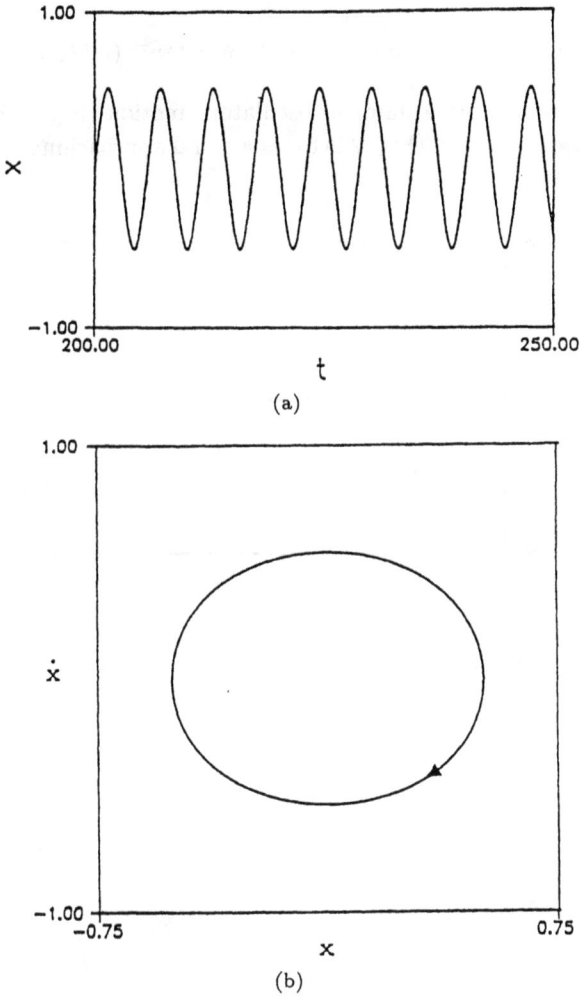

Fig. 2.1. (a) Solution curve $x(t)$ of Eq. (2.1) for $\alpha = 0.0$, $f = 0.0$, $\omega_0^2 = 1.0$, and $\omega = 1.0$. (b) Phase portrait in $(x - \dot{x})$ plane corresponding to (a).

In each of these cases, explicit solutions can be easily given. For example, in the physically more interesting case of underdamping, the solution to Eq. (2.1) for $f = 0$, $\alpha > 0$ can be given as

$$x(t) = \frac{A\sqrt{(\alpha^2/4) + C^2}}{C} e^{-(\alpha/2)t} \cos(Ct - \delta), \qquad (2.3a)$$

where

$$C = \sqrt{(\omega_0^2 - (\alpha^2/4))} \quad \text{and} \quad \delta = \tan^{-1}(\alpha/2C). \tag{2.3b}$$

Equation (2.3) represents a damped oscillatory motion (Fig. 2.2(a)) and the associated phase trajectory (Fig. 2.2(b)) is a spiral approaching the fixed point at the origin, which is a stable focus.

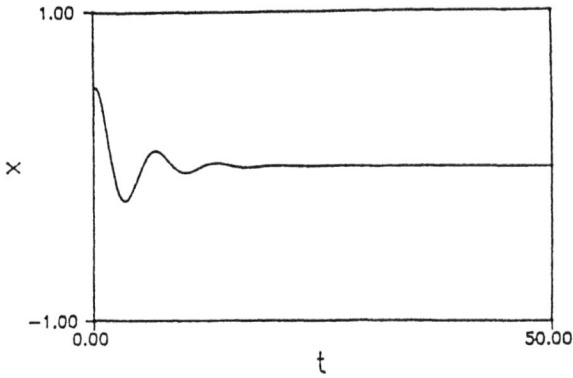

(a)

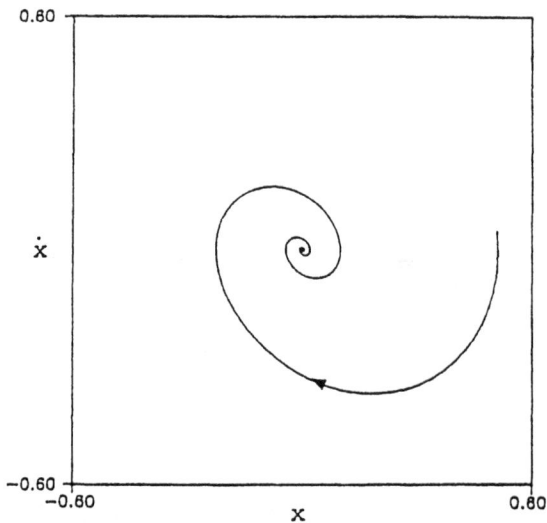

(b)

Fig. 2.2. (a) Solution curve $x(t)$ of Eq. (2.1) for $\alpha = 0.5$, $f = 0.0$, $\omega_0^2 = 1.0$, and $\omega = 1.0$. (b) Phase portrait of (a).

2.1.3. *Damped and Forced Oscillations: Resonance*

The full system (2.1) can be easily integrated to give

$$x(t) = \frac{A_t\sqrt{(\alpha^2/4) + C^2}}{C}e^{-(\alpha/2)t}\cos(Ct - \delta) + A_p\cos(\omega t - \gamma), \qquad (2.4a)$$

where

$$A_p = \frac{f}{\sqrt{(\omega_0^2 - \omega^2)^2 + \alpha^2\omega^2}} \text{ and } \gamma = \tan^{-1}\left[\frac{\alpha\omega}{(\omega_0^2 - \omega^2)}\right]. \qquad (2.4b)$$

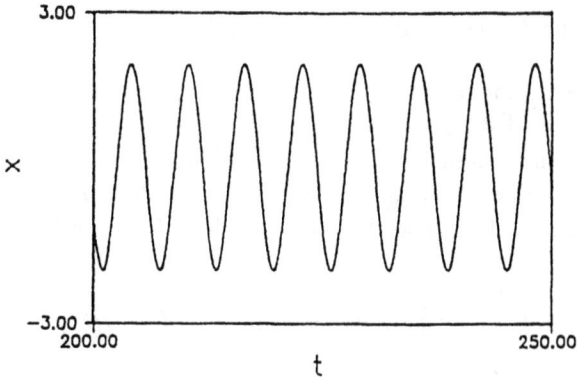

(a)

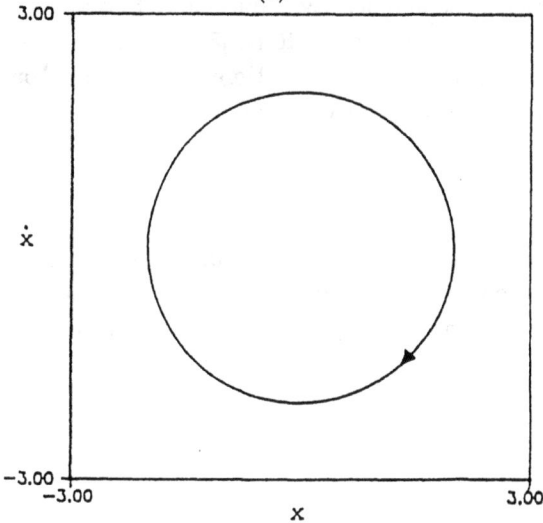

(b)

Fig. 2.3. (a) Solution curve $x(t)$ of Eq. (2.1) for $\alpha = 0.5$, $f = 1.0$, $\omega_0^2 = 1.0$, and $\omega = 1.0$. (b) Phase portrait of (a) (transients excluded).

Here the constant A_t is chosen so as to satisfy the initial conditions $x(0) = A$, $\dot{x}(0) = 0$: The first term in Eq. (2.4) is clearly a transient in the sense that it dies down to zero asymptotically. Thus for large t, the frequency of oscillation is effectively equal to the impressed frequency $(\omega/2\pi)$ and its amplitude is A_p. At the *resonance* value $\omega = \omega_0$, the amplitude of oscillation A_p takes a maximum value. The corresponding solution and the attractor in the phase-space, which is a limit cycle, are depicted in Figs. 2.3(a) and 2.3(b).

2.2. Damped and Driven Nonlinear Oscillators

As a prototype to understand the effect of nonlinearity on resonant linear oscillations described in the previous section, we ask the question as to how the dynamics of the system (2.1) gets modified when typical nonlinear spring forces are included. As a standard example, we include a cubic nonlinear force to the left-hand side of (2.1) so that the equation of motion becomes

$$\ddot{x} + \alpha\dot{x} + \omega_0^2 x + \beta x^3 = f \sin \omega t, \quad (\cdot = d/dt) \qquad (2.5)$$

where β is the strength of nonlinearity. Equation (2.5) is a ubiquitous nonlinear system called the Duffing oscillator (Refs. [5, 34]) about which we will have much more to say later in Chapters 3–4. Here we will confine ourselves to some of its elementary dynamical properties alone.

2.2.1. *Free Oscillations*

For the undamped and unforced case ($\alpha = 0, f = 0$), one again has bounded and periodic solutions but they are now described by Jacobian elliptic functions. For example when $\omega_0^2 > 0$, $\beta > 0$, we have the solution (Fig. 2.4(a))

$$x(t) = A \operatorname{cn}(\bar{\omega}t), \quad \bar{\omega} = \sqrt{(\omega_0^2 + \beta A^2)}, \qquad (2.6)$$

where cn is the Jacobian elliptic function of modulus $k = \left[\frac{\beta A^2}{\omega_0^2 + \beta A^2}\right]^{1/2}$. Again the phase trajectories are concentric curves with the equilibrium point at the origin being a center (as shown in Fig. 2.4(b) for a fixed A).

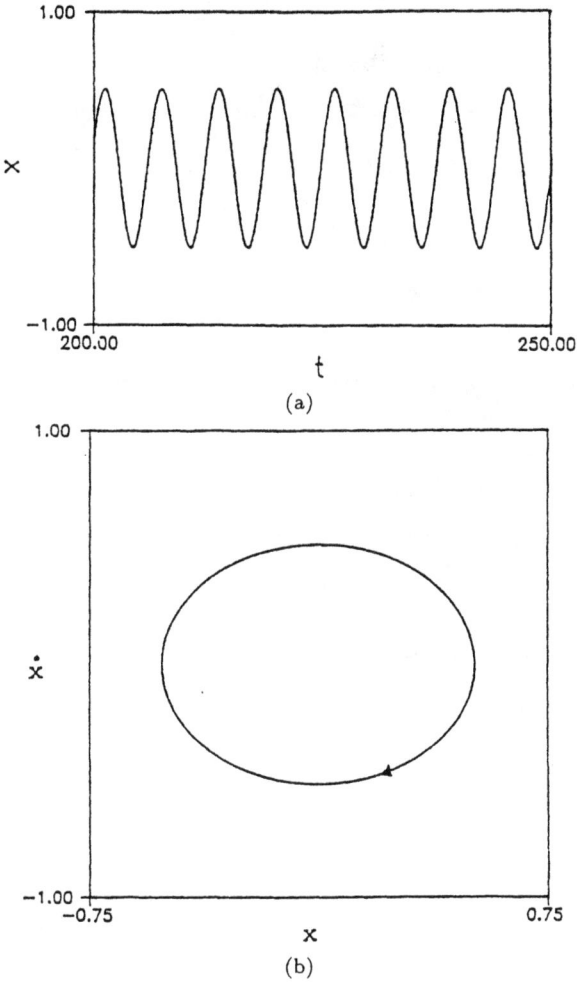

Fig. 2.4. (a) Solution curve $x(t)$ of Eq. (2.5) for $\alpha = 0.0$, $\beta = 1.0$, $f = 0.0$, $\omega_0^2 = 1.0$, and $\omega = 1.0$. (b) Phase portrait of (a).

2.2.2. *Damped Oscillations*

When $\alpha > 0$ and $f = 0$ in Eq. (2.5), i.e., in the physically interesting under-damped case $\alpha < 2\omega_0$, again we have damped oscillatory solution corresponding to an inwardly spiralling trajectory towards the equilibrium point at the origin, which is now a focus (Fig. 2.5). Though in the general case no explicit solution can be given, in the special case $\alpha = \pm(3/\sqrt{2})\omega_0$, and $\beta > 0$, one can give the exponentially decaying oscillatory solution (Ref. [35]) as

$$x(t) = (\omega_0/\sqrt{\beta})A\exp(-\omega_0 t/\sqrt{2})\text{cn}(Av; k),$$
$$v = -\sqrt{2}\exp(-\omega_0 t/\sqrt{2}) - v_0, \qquad (2.7)$$

where the modulus $k = 1/\sqrt{2}$ and A is a constant.

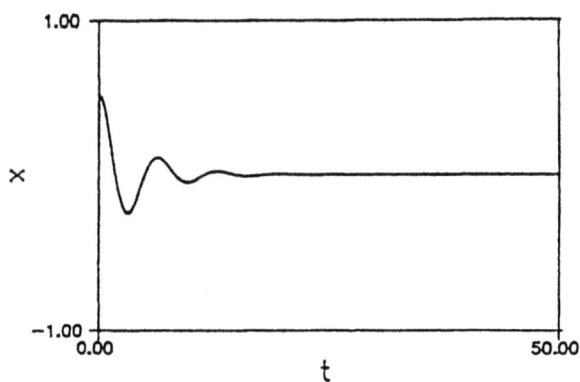

(a)

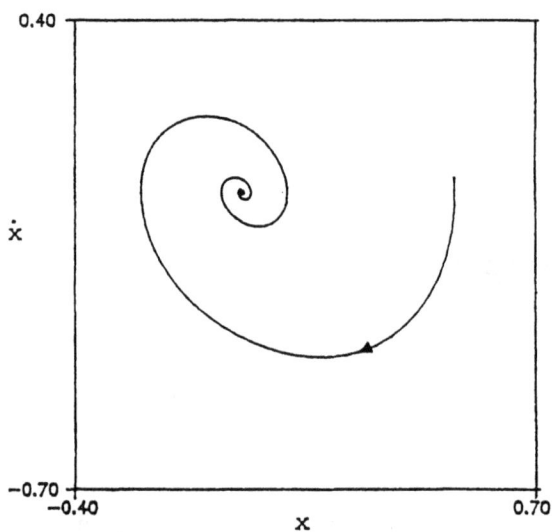

(b)

Fig. 2.5. (a) Solution curve $x(t)$ of Eq. (2.5) for $\alpha = 0.5$, $\beta = 1.0$, $f = 0.0$, $\omega_0^2 = 1.0$, and $\omega = 1.0$. (b) Phase portrait of (a).

2.2.3. *Forced Oscillations-Primary Resonance and Jump Phenomenon (Hysteresis)*

One of the most interesting aspects of nonlinear oscillators of the form (2.5) is that, even for very small β, its behaviour can be qualitatively different from that of the linear oscillator discussed in Sec. 2.1. This can be seen very easily as shown below (Ref. [33]).

Following our discussion on resonant linear oscillators in Sec. 2.1, we can assume the solution of Eq. (2.5) for $\beta \ll 1$ and $\omega \approx \omega_0$ (*primary resonance*) in the lowest order to be of the form

$$x(t) = A\cos(\omega t + \delta),\tag{2.8}$$

where δ is a phase constant to be fixed. Using Eq. (2.8) in Eq. (2.5) and equating the coefficients of $\cos\omega t$ and $\sin\omega t$, neglecting higher harmonic terms, we can obtain the *frequency-response equation* in the form

$$[(\omega_0^2 - \omega^2)A + \frac{3}{4}\beta A^3]^2 + (\alpha A\omega)^2 = f^2.\tag{2.9}$$

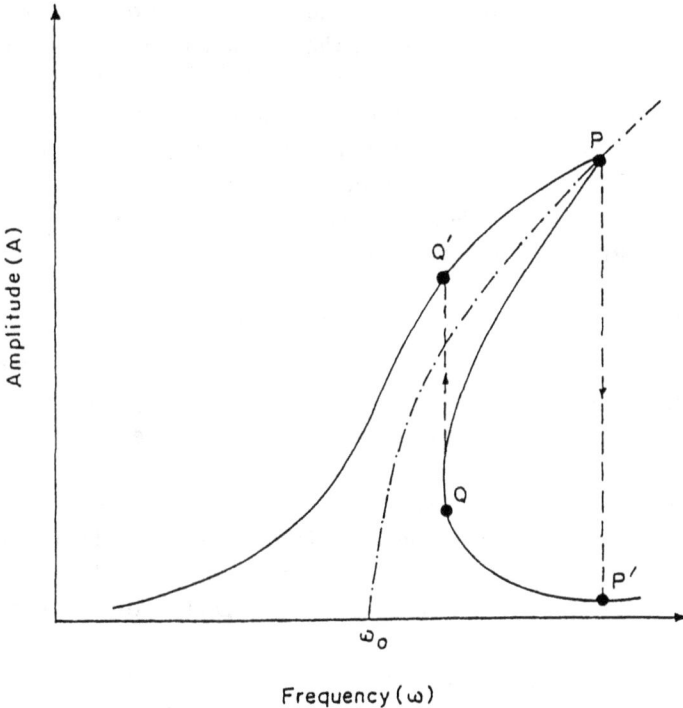

Fig. 2.6. Resonance curve of the nonlinear oscillator (2.5) for $\beta > 0$.

Equation (2.9) defines the *resonance curve* in the nonlinear oscillator case (2.5). A typical form of it is shown in Fig. 2.6 for $\beta > 0$.

We note that the resonance curve is not symmetrical as in the linear case (Eq. 2.4(b)), but leans to the right. (For $\beta < 0$, the resonance curve in Fig. 2.6 will lean to the left). When the frequency of the periodic driving force is gradually increased while keeping its magnitude constant, the amplitude of oscillation will be found to increase steadily and reach a maximum value P. Further increase of frequency results in a markedly different behaviour as the amplitude of oscillation makes a discontinuous jump from P to P'. Similarly, an amplitude jump from Q to Q' will occur if the frequency is gradually lowered from a value beyond P'. One should however note that this *jump or hysteresis phenomenon* does not imply that these jumps are truly discontinuous but the amplitude changes over a finite time.

2.2.4. *Secondary Resonances (Subharmonic and Superharmonic)*

Apart from the jump phenomenon and primary resonance another characteristic behaviour of nonlinear systems is the secondary resonances (Ref. [36]). For example, when $w \approx 3\omega_0$, $\beta \ll 1$, the solution to Eq. (2.5) can be approximated by $x(t) = a\cos((\omega t/3) + \delta) + (f/(\omega^2 - \omega_0^2))\cos\omega t$. The first term corresponds to a free oscillation of one-third *subharmonic resonance*. Similarly for $\omega \approx \omega_0/3$, $\beta \ll 1$, one can obtain the approximate solution $x(t) = a\cos(3\omega t + \delta) + (f/(\omega^2 - \omega_0^2))\cos\omega t$, in which the first term corresponds to a *superharmonic resonance*. These and other subharmonic and superharmonic resonances constitute the *secondary resonances* phenomena.

2.3. Nonlinear Oscillations and Bifurcations

The dynamics of the nonlinear oscillators of the form (2.5) is too complicated to be described in terms of resonant oscillations of the type mentioned above alone. The equation of motion is in general not solvable exactly. Qualitative and quantitative ideas (Refs. [36, 37]) on the types of oscillations and their stability for small strength of the nonlinearity parameter can be obtained by making use of one of the several perturbation methods available in the literature. Some of the most often used methods are briefly indicated in Appendix A. In Chapter 5, we apply one such method to the Duffing oscillator to discuss its dynamics for small β.

However, a complete picture of the dynamics can be obtained essentially by a straightforward and detailed numerical analysis using any one of the available standard numerical algorithms, like the fourth order Runge–Kutta integration method. The result is that one obtains a rich variety of *bifurcation* phenomena, namely the successive qualitative (and quantitative) changes in the nature of oscillations as the nonlinearity parameter (or another control parameter for a fixed nonlinearity parameter) is varied. Detailed analysis of the Duffing oscillator system will be undertaken in Chapters 3–4 and for other oscillators in the remainder of the book.

The above-mentioned numerical analysis can be used in different ways to extract information about the dynamics (Refs. [5, 9, 37]). They include the following.

1. *Trajectory plot*: The actual plot of the motion $x(t)$ versus t as shown in Figs. 2.1(a)–2.5(a).

2. *Phase portrait*: It is the two-dimensional projection of the trajectories in the $(x - \dot{x})$ plane of the three-dimensional '*phase-space*' $(x, \dot{x}, t)$ (as shown in Figs. 2.1(b)–2.5(b)) of the Duffing and similar oscillators. Periodic and complex motions can be distinguished visually from these plots. For higher dimensional systems, it is again any two-dimensional projection of the associated phase-space.

3. *Poincaré map*: It is the stroboscopic (snap-shot) portrait of the phase-space trajectories at every period $T = (2\pi/\omega)$ of the impressed external periodic force. For autonomous systems, it is an appropriate Poincaré plane. It will contain much valuable information about the dynamics. In this plot, periodic solutions of period NT will correspond to a finite number of points N. On the other hand, complex and chaotic solutions will have self-similar fractal structures possessing non-integer Hausdorff dimensions typically.

4. *Power spectrum*: The power spectrum or power density

$$P(\hat{\omega}) = \left| \frac{1}{2\pi} \int_{-\infty}^{\infty} x(t) e^{-i\hat{\omega}t} \right|^2$$

of the signal $x(t)$ as a function of frequency $\nu = (\hat{\omega}/2\pi)$ can detect the presence of harmonics and subharmonics as suitable peaks while a continuous broadband spectrum of it indicates the presence of chaotic motions.

5. *Bifurcation diagram*: It is the plot of the Poincaré points of the solution $x(t)$ for every period $(T = 2\pi/\omega)$ of the external forcing signal versus the control parameter. This diagram clearly depicts the associated period doubling and other bifurcations from periodic to chaotic motions as the control parameter is varied.

6. *Lyapunov spectrum*: It is the plot of the maximal Lyapunov exponent versus the bifurcation control parameter of a nonlinear system. Lyapunov exponents are numbers that measure the exponential attraction or separation in time of two adjacent trajectories in phase-space with different initial conditions. A positive Lyapunov exponent indicates chaotic motion in a dynamical system.

Besides these, one can also use correlation functions, basin of attractions, and other measures to characterize regular and chaotic motions (Refs. [9, 37]).

2.4. Dynamical Systems and Chaos

The Duffing oscillator equation (2.5) can also be equivalently written as a system of three coupled first order odes

$$\dot{x} = y, \tag{2.10a}$$

$$\dot{y} = -\alpha y - \omega_0^2 x - \beta x^3 + f \sin z, \tag{2.10b}$$

$$\dot{z} = \omega. \tag{2.10c}$$

Similarly, any evolution equation of a dynamical system of the form (1.1) can always be written as a first order vector differential equation in an $\bar{n}$-dimensional space:

$$\dot{X}(t) = \frac{dX}{dt} = F(X(t), \mu), \quad X(0) = X_0, \tag{2.11}$$

where $X(t) = (x_1(t), x_2(t), \ldots, x_{\bar{n}}(t))^T \in R^{\bar{n}}$ is the state vector defining the trajectory in the state space (phase-space) and $F(X(t)) = (F_1(X(t)), F_2(X(t)), \ldots, F_{\bar{n}}(X(t)))^T$ is an $\bar{n}$-dimensional map or vector field. Here X_0 is the initial condition and μ stands for a control parameter.

When looking at the time evolution of the state variables of Eqs. (2.11) one usually distinguishes between transient behaviour, which disappears after certain time, and permanent features which persist in time. The latter is often referred to as asymptotic behaviour. One often encounters the following behaviours in the analysis of *dissipative* nonlinear systems.

1) *Fixed points*: An equilibrium point or fixed point of Eq. (2.11) is a state X_Q at which the vector field F is zero. Thus, $F(X_Q) = 0$ and $X_Q(t) = X_Q$, a trajectory starting from an equilibrium point remains indefinitely at that point.

2) *Periodic response*: All solutions converge to a periodic waveform having the natural frequency ω_n for the autonomous case and ω of the input signal frequency in the nonautonomous case.

3) *Subharmonic response*: All solutions converge to a periodic waveform whose frequency Ω is a submultiple of the input signal frequency ω; that is $\Omega = (\omega/n)$, where n is a positive integer. For the autonomous case $\Omega = (\omega_n/n)$.

4) *Superharmonic response*: All solutions converge to a periodic waveform whose frequency Ω is a multiple of the input signal frequency, $\Omega = nw$ ($n > 1$, integer), or $\Omega = n\omega_n$ for the autonomous case.

5) *Almost periodic (quasiperiodic) response*: All solution waveforms are made up of periodic components whose fundamental frequencies are incommensurable.

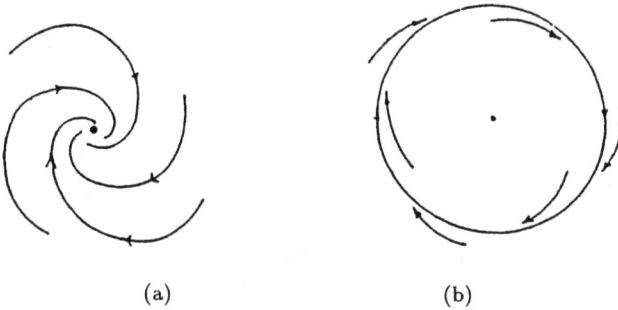

(a)   (b)

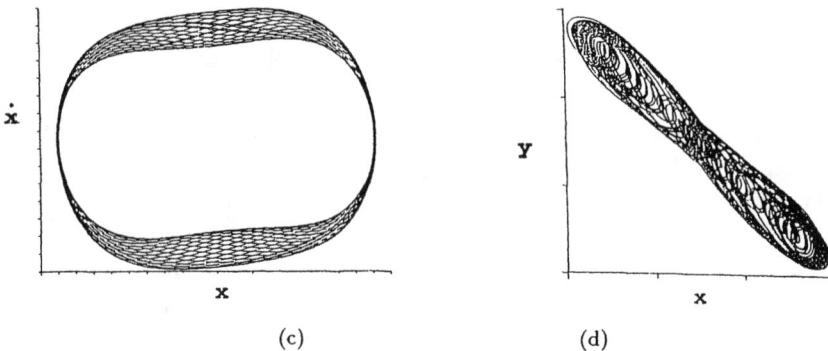

(c)   (d)

Fig. 2.7. (a) Trajectories ending at a stablefocus.(b) Stable limit-cycle attractor of period T and angular frequency $\omega = 2\pi/T$. (c) A quasiperiodic attractor of the Duffing–van der Pol oscillator (cf. Chapter 6). (d) Chaotic attractor of the Murali–Lakshmanan–Chua circuit (cf. Chapter 7).

6) *Chaotic response*: These are responses which depend sensitively on the initial conditions as mentioned in Chapter 1. These motions are not even periodic and close initial states lead to time evolutions that become more and more distant from each other in finite time exponentially, within an enclosed region of the phase-space, namely the strange attractor.

Typical forms of some of these responses in the $(x - \dot{x})$ phase-space are shown in Fig. 2.7.

2.5. Routes to Chaos

In the past two decades or so, a large number of analytical, numerical and experimental studies have been carried out on different nonlinear systems with an effort to understand the various features associated with the occurrence of chaotic behaviour. These include the Duffing oscillator, van der Pol oscillator, Duffing-van der Pol oscillator, Bonhoeffer–van der Pol oscillator, damped and driven pendulum, Lorenz system, Brusselator, and so on (Refs. [5–17, 37–41]). One essentially tries to vary one or more of the control parameters in the system so that the parameter ranges for which regular and periodic behaviour occurs and the regimes for which chaotic behaviour occurs can be identified. In many nonlinear dissipative dynamical systems, including damped and driven oscillators, chaotic motion is found to set in mainly through one of the three predominant routes which are all familiar in the chaos literature (Refs. [5, 8, 37, 38, 41]).

1. *Feigenbaum scenario(period-doubling route)*: In this route the transition to chaos proceeds in the following way (Refs. [5, 8, 9, 13, 41–43]). Let a nonlinear dynamical system has a stable periodic solution with period T for a range of values of a control parameter μ on which it depends. If μ is varied, it turns out that

(i) When it reaches a certain value μ_1, there is a bifurcation at which the periodic orbit loses its stability and a new stable solution with period $2T$ is produced. This is followed by an infinite sequence of similar period doubling bifurcations at $\mu = \mu_k, k = 2, 3, \ldots, \infty$, at which solutions with periods $2^{k-1}T$ lose their stability while new stable ones with period $2^k T$ are created.

(ii) The period doubling sequence accumulates at a certain critical value $\mu = \mu_\infty$, following which chaos sets in.

(iii) Associated with this period doubling bifurcation phenomena there is a universal constant δ given by $\delta_k = [(\mu_{k+1} - \mu_k)/(\mu_{k+2} - \mu_{k+1})]$. In the limit $k \to \infty$ the value of δ approaches the universal value 4.6692....

(iv) For $\mu > \mu_\infty$, the system may exhibit chaotic solution followed by periodic windows, period doubling bifurcations of periodic windows, chaos, etc.

2. *Ruelle–Takens–Newhouse scenario* (*Quasiperiodic route*): In this route when the control parameter of the system is changed, the initial stationary state becomes unstable and undergoes a Hopf bifurcation. Further change of the control parameter makes the system undergo one more Hopf bifurcation so that a doubly periodic orbit occurs. The precursor to chaotic motion is the presence of two simultaneous periodic oscillations. When the frequencies of these oscillations are incommensurate with each other the motion is said to be quasiperiodic. This quasiperiodic motion then bifurcates to a chaotic motion as the control parameter is further changed (Refs. [5, 9, 13, 37, 41–44]).

3. *Pomeau–Manneville scenario* (*Intermittency route*): Over a certain range of parameter the dynamical system has a periodic orbit (laminar phase). As the parameter is changed beyond a critical value, some irregular short chaotic bursts appear among the long regular motions. As the value of the parameter changes further, the chaotic bursts appear more frequently and the average time between two consecutive bursts shortens. Eventually the system moves into a chaotic regime (Refs. [5, 9, 13, 37–45]).

These are the three prominent routes to chaos in dissipative systems. Besides, there are many other not so common routes, such as period-adding sequence bifurcations (Ref. [46]), equal-periodic bifurcations (Ref. [47]), and so on.

We shall encounter many of these bifurcations in our further study in this book. Before looking at actual systems, we will briefly consider the possibility of analog-simulating these systems through nonlinear electronic circuits and also constructing new nonlinear circuits in the next Chapter.

CHAPTER 3

ELECTRONIC CIRCUITS AS OSCILLATORS AND ANALOG SIMULATION OF DYNAMICAL SYSTEMS

Even though numerical analysis can help to bring out a detailed picture of the dynamics of a system such as (2.5), it requires much computer power and enormous time to scan the entire parameter space, particularly if more than one control parameters are involved, in order to understand the rich variety of bifurcations and chaotic orbits. In this connection, *analog simulation* studies of nonlinear oscillators of the form (2.5) through appropriate electronic circuits are often helpful in a dramatic way for a quick scan of the parameter space and also for avoiding long transients to reach the steady states as in numerical studies. These analog circuits normally use operational amplifiers (op-amps) and multipliers. The principal advantage of an electronic analog simulation circuit lies in its fast response as the control parameters are varied. The construction of such circuits is fairly easy for many nonlinear differential equations (Refs. [48–54]).

From another point of view, in recent times a variety of nonlinear electronic circuits consisting of either real nonlinear physical devices (Refs. [48–60]) such as nonlinear diodes, capacitors, inductors and resistors or devices constructed with ingenious piecewise-linear circuit elements have been utilized as veritable black boxes to explore different properties of chaotic dynamics. These circuits are unique in being easy to build, easy to analyze and easy to model. They often provide a convenient framework to understand the various mechanisms underlying the onset of chaos and for possible technological applications.

In the case of nonlinear circuits, currents and voltages across various branches of the circuit play the role of dynamical variables. On the other hand, in the case of analog simulation circuits which are used to mimic the dynamics of nonlinear odes, it is mainly the node voltages of the operational amplifier circuit modules that play the corresponding role. In either case the study of nonlinear circuits has become an important and active area of study of nonlinear dynamics. With the great advances in nonlinear circuit theory over the past thirty years or so (Refs. [49, 55, 56]), it is now possible to understand highly complex nonlinear behaviours with simple models and minor extensions of linear circuit theory. In this Chapter, we will give an introduction to these aspects.

3.1. Linear and Nonlinear Circuit Elements

To start with, we briefly discuss the properties of various circuit elements, namely *resistors*, *capacitors* and *inductors*. They can be either linear or nonlinear depending upon their characteristic curves, namely the v–i (voltage–current), v–q (voltage–charge) and i–ϕ (current–magnetic flux) curves respectively.

(a) *Resistors*: A two-terminal resistor whose current $i(t)$ and voltage $v(t)$ falls on some fixed (characteristic) curve in the v–i plane represented by the equation $f_R(v, i) = 0$ at any time t is called a time-invariant resistor. If the measured v–i characteristic curve is a straight line passing through the origin as shown in Fig. 3.1(a) then the resistor is said to be *linear* and it satisfies the Ohm's law

$$v(t) = Ri(t). \tag{3.1}$$

If a resistor is characterized by a v–i curve other than a straight line through the origin, it is called a *nonlinear resistor*. The characteristic curve of a typical nonlinear resistor is given in Fig. 3.1(b).

(b) *Capacitors*: A two-terminal circuit element whose charge $q(t)$ and voltage $v(t)$ falls on some fixed (characteristic) curve in the q–v plane represented by the equation $f_C(q, v) = 0$ at any time t is called a time-invariant capacitor. If the measured q–v characteristic curve is a straight line passing through the origin as shown in Fig. 3.2(a), then the capacitor is said to be *linear* and it satisfies the current–voltage relation $i = C\frac{dv}{dt}$. In all other cases, the capacitor is *nonlinear*. The characteristic curve of a typical nonlinear capacitor is given in Fig. 3.2(b).

(i)

(a)

(i)

(ii)

(b)

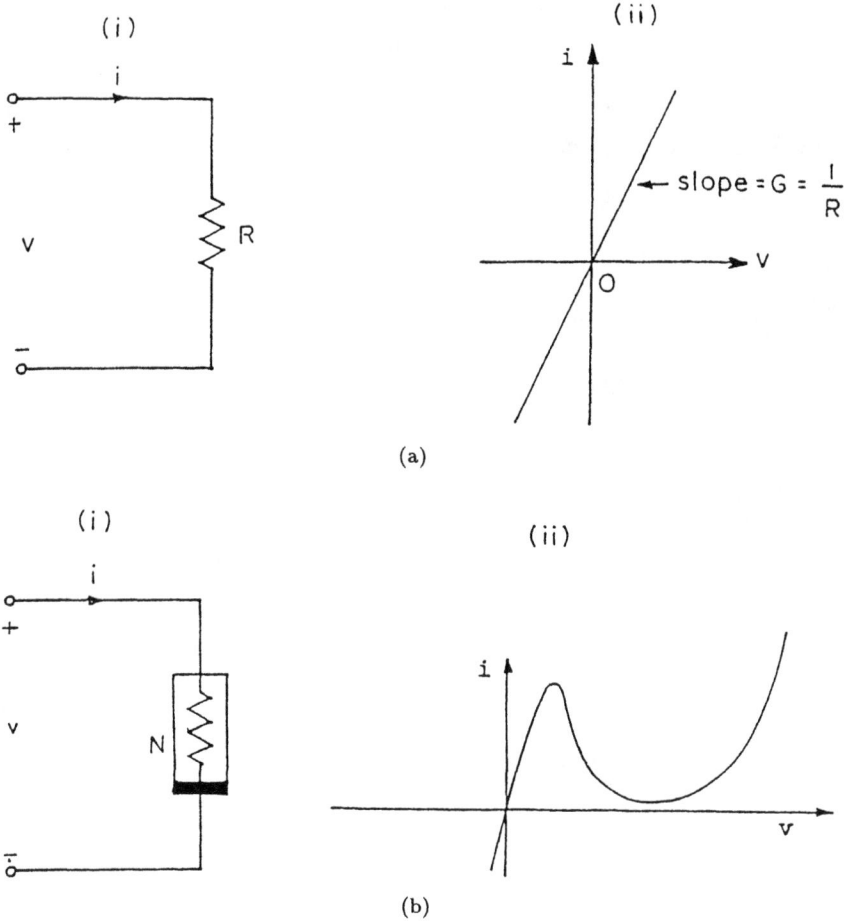

Fig. 3.1. (a): (i) Symbol of a two-terminal linear resistor; (ii) Characteristic curve of the linear resistor. (b): (i) Symbol of a two-terminal nonlinear resistor; (ii) Characteristic curve of a typical nonlinear resistor.

(c) *Inductors*: A two-terminal circuit element whose flux $\phi(t)$ and current $i(t)$ falls on some fixed (characteristic) curve in the $\phi\text{-}i$ plane represented by the equation $f_L(\phi, i) = 0$ at any time t is called a time-invariant inductor. If the measured $\phi\text{-}i$ characteristic curve is a straight line passing through the origin as shown in Fig. 3.3(a), then the inductor is said to be *linear* and it satisfies the voltage–current relation

$$v = L\frac{di}{dt}.$$
(3.2)

In all other cases, the inductor is *nonlinear*. The characteristic curve of a nonlinear inductor is given in Fig. 3.3(b).

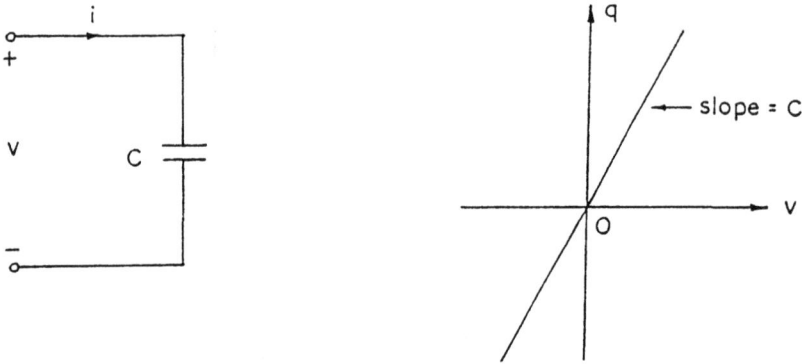

(a)

(i)   (ii)

(b)

Fig. 3.2. (a): (i) Symbol of a two-terminal linear capacitor; (ii) Characteristic curve of the linear capacitor. (b): (i) Symbol of a two-terminal nonlinear capacitor; (ii) Characteristic curve of a typical nonlinear capacitor.

(i)

(ii)

(a)

(i)

(ii)

(b)

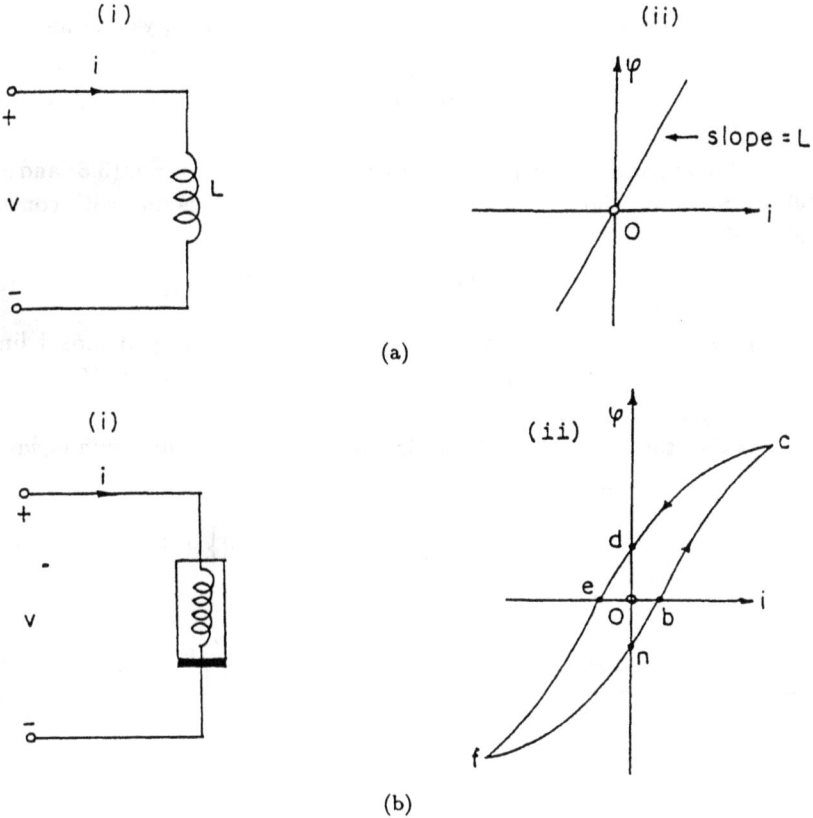

Fig. 3.3. (a): (i) Symbol of a two-terminal linear inductor; (ii) Characteristic curve of the linear inductor. (b): (i) Symbol of a two-terminal nonlinear inductor; (ii) Characteristic curve of a typical nonlinear inductor.

3.2. Linear Circuits

As noted above, linear circuits are those which contain linear elements (linear resistors, capacitors and inductors) only. In this subsection, we will consider some ubiquitous linear circuits.

A. *The resonant linear RLC circuit: Analog simulation of damped forced oscillator*: In the case of linear oscillatory behaviour described in Chapter 2, the well-known RLC resonant circuit plays the role of a paradigm. It can also be thought of as an analog simulation of the damped forced linear oscillator (2.1). It consists of a linear resistor, a linear inductor, a linear capacitor and a time-dependent voltage source $f(t) = F_s \sin(\omega_s t)$ as shown in Fig. 3.4(a).

Applying Kirchoff's voltage law to the circuit of Fig. 3.4(a), we obtain

$$L\frac{di_L}{dt} + Ri_L + v = F_s \sin\omega_s t, \qquad (3.3)$$

with $i_L(0) = i_{L0}$ and $v(0) = v_0$. Substituting $i_L = C\frac{dv}{dt}$ into Eq. (3.3) and simplifying, we obtain the second order linear inhomogeneous ode with constant coefficients

$$\frac{d^2v}{dt^2} + (R/L)\frac{dv}{dt} + (1/LC)v = (F_s/LC)\sin\omega_s t. \qquad (3.4)$$

Equation (3.4) is identical in form with the resonant damped forced linear oscillator (2.1) with the identification, $v \to x$, $(R/L) = \alpha$, $(1/LC) = \omega_0^2$, $(F_s/LC) = f$ and $\omega_s = \omega$.

Now using the solution (2.4) for Eq. (3.4), one can easily write down the solution to the voltage v across the capacitor as

$$v(t) = v_0 e^{-(R/2L)t}\left\{\cos[\sqrt{((1/LC) - (R/2L)^2)}t - \theta]\right\} + F_p \sin(\omega_s t - \nu), \qquad (3.5a)$$

where

$$F_p = \frac{(F_s/LC)}{\{((1/LC) - \omega_s^2)^2 + (R\omega_s/L)^2\}^{(1/2)}}, \quad \nu = \tan^{-1}\left\{\frac{R\omega_s}{L[(1/LC) - \omega_s^2]}\right\}. \qquad (3.5b)$$

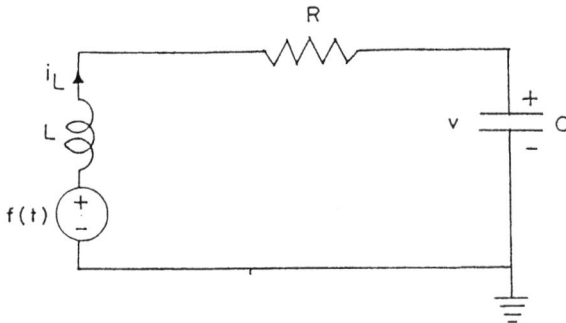

(a)

Fig. 3.4. (a) A forced LCR circuit (It can also be thought of as an analog simulation circuit of Eq. (2.1)). Here, $f(t) = F_s \sin\omega_s t$. (b): (i) Wave form v of circuit of Fig. 3.4(a) for $R = 500$ Ω, $L = 18$ mH, $C = 10$ nF, $F_s = 0.1$ V, and frequency $f_s = 8890$ Hz after discarding the transients (*asymptotic solution*); (ii) Trajectory plot in (v–i_L) plane. (c) Trajectory plot in (v–i_L) plane of the experimental circuit of Fig. 3.4(a).

b (i)

b (ii)

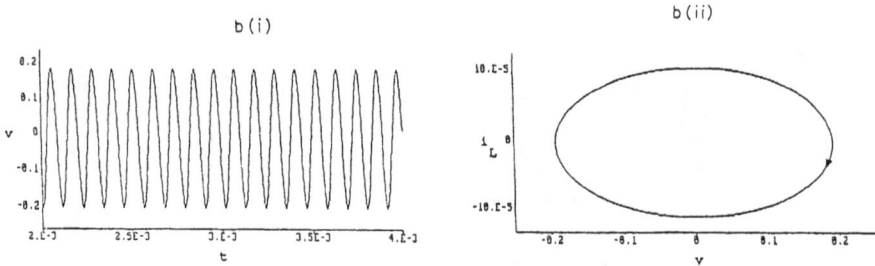

(b)

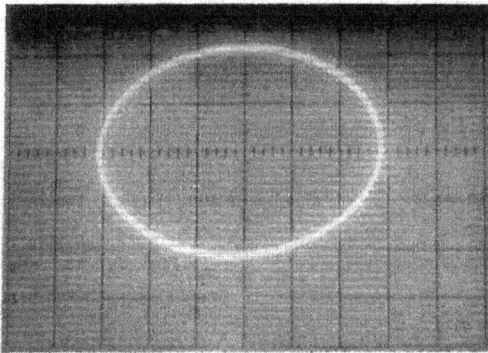

(c)

Fig. 3.4. (*Continued*).

Obviously, the asymptotic orbit of the circuit (Fig. 3.4(a)) is a closed orbit (ellipse), corresponding to the sinusoidal steady state of the circuit. All the waveforms v and i_L converge to a periodic waveform having the same frequency resonating with the input signal of frequency ω_s, as shown in Fig. 3.4(b), and it mimics exactly the phase portrait of Fig. 2.3. This is the most regular asymptotic behaviour one can expect from a linear circuit with a sinusoidal source. The actual experimental circuit result in the form of a phase-portrait in the $(v\text{--}i_L)$-plane is shown in Fig. 3.4(c).

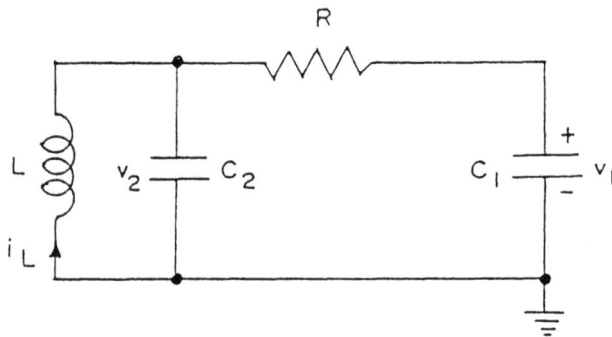

Fig. 3.5. A third-order linear circuit.

B. *Higher order linear circuits*: Interestingly, if one further includes an additional linear passive capacitor C_2 in parallel with the inductor L as depicted in Fig. 3.5 but without a signal generating source, then we have the third-order autonomous linear circuit represented by the state equations

$$C_1\frac{dv_1}{dt} = (1/R)(v_2 - v_1),\tag{3.6a}$$

$$C_2\frac{dv_2}{dt} = (1/R)(v_1 - v_2) + i_L,\tag{3.6b}$$

$$L\frac{di_L}{dt} = -v_2,\tag{3.6c}$$

where v_1, v_2 and i_L are the voltage across the capacitor C_1, the voltage across the capacitor C_2 and the current through the inductor L respectively, with appropriate initial values. All the circuit elements are assumed to be positive. System (3.6) is equivalent to a third-order linear differential equation, which can again be solved explicitly. For a specific choice of parameters, namely $C_1 = 10$ nF, $C_2 = 100$ nF, $R = 1740\,\Omega$ and $L = 18$ mH, the trajectories approach an equilibrium point as indicated in Fig. 3.6. Similarly for a different set of parametric choice, one can observe a periodic solution as well. One can check analogous phenomena even for still higher order linear circuits. We can in general conclude that at the most one expects either a fixed point solution or a periodic steady-state oscillation from a linear circuit for chosen nominal circuit parameters.

(a)

(b)

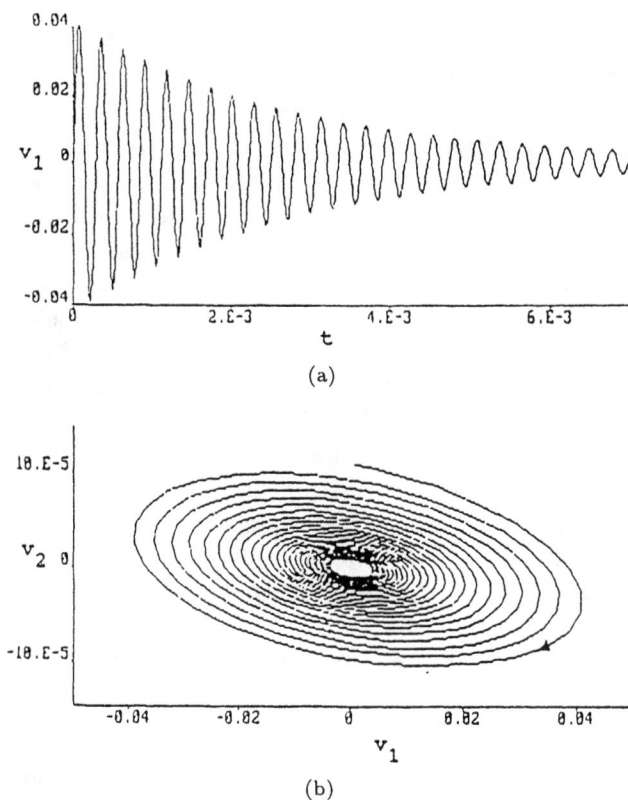

Fig. 3.6. (a) Wave form v_1 of the circuit of Fig. 3.5 for $R = 1740\ \Omega$, $L = 18$ mH, $C_1 = 10$ nF, $C_2 = 100$ nF, approaching the equilibrium as time progresses. (b) Trajectory plot in the (v_1-v_2) plane.

3.3. Nonlinear Circuits: Asymptotic Behaviour (Refs. [49, 55, 56])

A circuit is said to be nonlinear if it contains at least one nonlinear circuit element like a nonlinear resistor, nonlinear capacitor or nonlinear inductor. Another basic inventory in nonlinear circuit analysis is the use of piecewise-linear circuit elements designed ingeniously for specific needs, whose characteristic curves are piecewise-linear. These elements include piecewise-linear resistors, capacitors and inductors respectively (Refs. [49, 55–57]).

In the literature a large number of nonlinear circuits have been widely discussed, especially in the study of chaotic phenomena. These nonlinear circuits include both driven nonlinear RLC circuits and undriven nonlinear RLC networks. There are at least three groups of studies here. In the first group, a number of authors have reported circuits with typical nonlinear elements such as nonlinear capacitors (varactor diode, junction diode) (Refs. [50, 58–60]), nonlinear inductors (saturable core inductor, Josephson junctions, ferroresonant power systems) (Refs. [61–64]) and nonlinear resistors (tunnel diode, thyristor, dead-zone conductor, serially connected Zener diodes, neon bulb, etc.) (Refs. [50, 65–72]). In the second group, circuits with ingeniously devised piecewise-linear elements like piecewise-linear resistors, capacitors and inductors have been discussed (Refs. [48, 57, 73–84]). These studies comprise both low-order and high-order circuits. In the third group, nonlinear circuits including phase-locked loops (Refs. [85, 86]), digital filters (Ref. [87]), cellular neural networks (Ref. [88]), RC-ladder phase-shift networks (Ref. [89]), DC–DC converter (Ref. [90]), etc., have been extensively discussed.

In all these cases, the dynamics changes radically when nonlinear circuit elements are admitted to the linear circuits discussed in the previous section. Apart from the "normal" behaviour, many qualitatively very different time evolutions are possible which are direct counterparts of the type of motions mentioned in Chapter 2, Sec. 2.4 for nonlinear dynamical systems.

3.4. Simplest Circuits with a Nonlinear Resistor (Chua's Diode)

Of all the possible nonlinear circuit elements nonlinear resistors are easy to build and model. In this connection, *Chua's diode* (Refs. [73–76]) is a simple nonlinear resistor with piecewise-linear characteristics and is widely used by circuit theorists. In the following we shall discuss some simple circuits which contain this nonlinear resistor along with additional linear circuit elements.

(a) *Autonomous case*: The most natural extension of linear circuit theory to the world of nonlinear circuits is through piecewise-linear circuit modelling. Thus one can modify the series RLC circuit to a nonlinear circuit, by placing a piecewise-linear (and so nonlinear) resistor N in parallel with the capacitor C. First let us consider the case in which the third-order autonomous circuit of Fig. 3.5 is generalized by including the nonlinear resistor N, which can be represented by its characteristic curve as shown in Fig. 3.7, so that we have an autonomous third-order nonlinear circuit of Fig. 3.8. This is one of the simplest and most widely studied real nonlinear dynamical systems, and it is called *Chua's autonomous circuit* (Refs. [73–76]). The single voltage-controlled two-terminal nonlinear resistor N is called *Chua's diode* (Refs. [73–76]). For this remarkable circuit the presence of chaos has been reported experimentally,

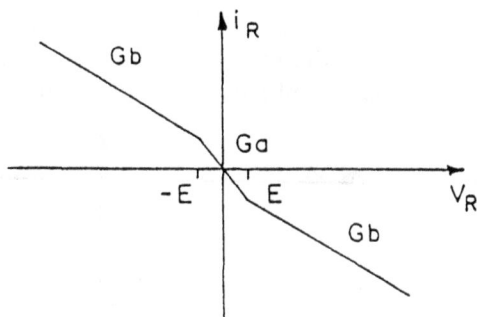

Fig. 3.7. A typical three-segment piecewise-linear characteristic curve of a nonlinear resistor (N).

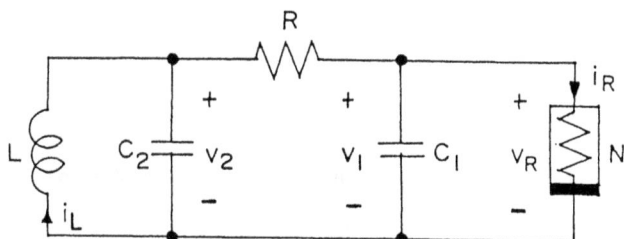

Fig. 3.8. Chua's circuit consisting of a linear inductor L, two linear capacitors (C_1, C_2), a linear resistor R, and a voltage-controlled nonlinear resistor N, namely, Chua's diode.

confirmed numerically, and proven mathematically. The circuit is readily constructed at low cost using standard electronic components and exhibits a rich
variety of bifurcations and chaos behaviour. The circuit equations of Fig. 3.8
are represented by the set of three coupled differential equations

$$C_1 \frac{dv_1}{dt} = (1/R)(v_2 - v_1) - f(v_1), \tag{3.7a}$$

$$C_2 \frac{dv_2}{dt} = (1/R)(v_1 - v_2) + i_L, \tag{3.7b}$$

$$L \frac{di_L}{dt} = -v_2, \tag{3.7c}$$

where $f(v_1)$ represents the functional form of the piecewise-linear nature of N.
The chaotic dynamics of (3.7) is discussed in detail later in Chapter 7.

  (b) *Non-autonomous case:* Now let us include the nonlinear resistor N of
Fig. 3.7 discussed above to our familiar linear forced RLC circuit of Fig. 3.4(a).

The modified circuit is shown in Fig. 3.9. This simplest non-autonomous circuit has been first reported by Murali, Lakshmanan and Chua (Refs. [77–79]) recently. The governing equations of motion of this circuit are

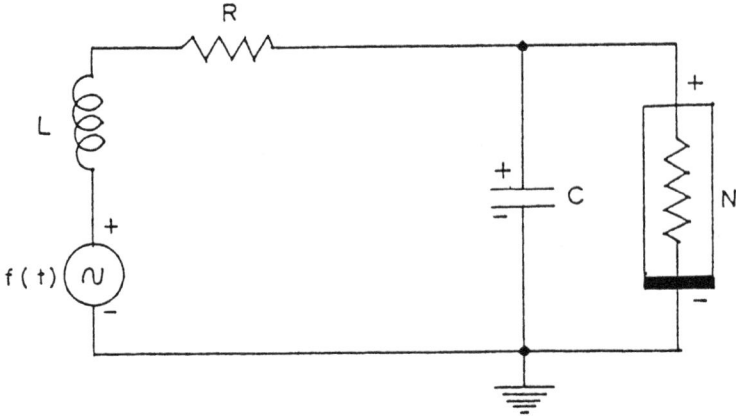

Fig. 3.9. Circuit diagram of the simplest dissipative nonautonomous circuit.

$$C\frac{dv}{dt} = i_L - f(v)\,, \tag{3.8a}$$

$$L\frac{di_L}{dt} = -v - Ri_L + F_s \sin(\omega_s t)\,, \tag{3.8b}$$

and its complete chaotic dynamics is again discussed in detail later in Chapter 7.

3.5. Analog Circuit Simulations of Nonlinear Odes

Apart from the construction of ingenious nonlinear circuits to study chaotic dynamics, one can also utilize suitable analog simulation circuits to mimic the dynamics of certain nonlinear differential equations. These circuits are modelled with suitable operational amplifier modules and multipliers. Usually operational amplifiers (Op-amps) are used as integrators, differentiators, sign-changers, adders and so on (Refs. [51–54, 91]). The mathematical operations are generally obtained either by using individually an op-amp with resistor feedback or an op-amp with capacitor feedback. These op-amp mathematical modules are a better choice because of the following advantages.

(i) The high input impedance allows one to add many input signals.

(ii) The low output impedance allows one to connect these devices to several others without changing the characteristic time of the single modules (especially integrators and differentiators).

A typical scale changer and adding integrator with resistor and capacitor feedback respectively are shown in Figs. 3.10(a) and (b). In Fig. 3.10(a), the

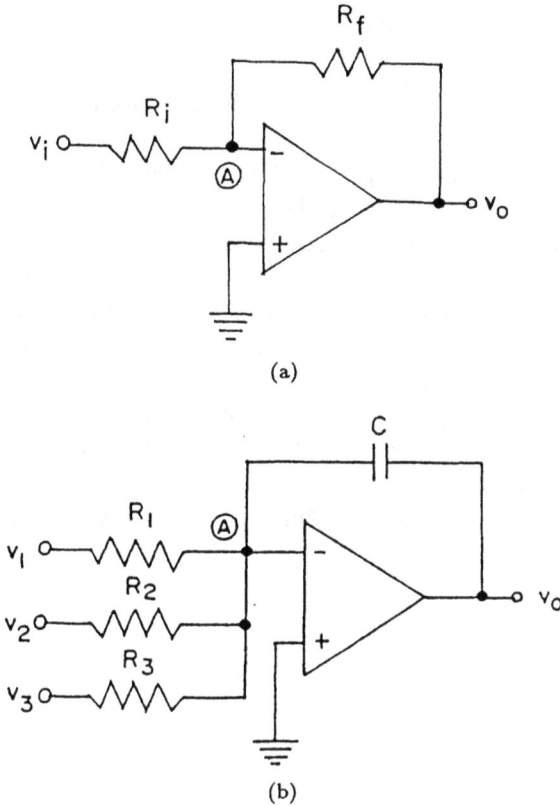

(a)

(b)

Fig. 3.10. (a) Circuit diagram of an op-amp scale changer. (b) Circuit diagram of an op-amp adding integrator.

resistor R_i is called input resistor and the resistor R_f is called feedback resistor. Applying Kirchoff's current law at node A we obtain

$$v_i/R_i = -v_0/R_f ,$$ (3.9)

then

$$v_0 = -(R_f/R_i)v_i .$$ (3.10)

It may be observed that the output voltage (v_0) equals the negative of the input voltage (v_i) multiplied by a constant. By choosing an appropriate ratio of R_f to R_i, one may fix the multiplication constant.

For Fig. 3.10(b), by writing the current equation at node A, we obtain

$$v_1/R_1 + v_2/R_2 + v_3/R_3 = -C(dv_0/dt). \qquad (3.11)$$

The above equation may be integrated to give

$$v_0(t) = -(1/C) \int_0^t (v_1/R_1 + v_2/R_2 + v_3/R_3)dt' - v_0(0) \qquad (3.9)$$

and thus we have an adding integrator. Similar circuits for other mathematical operations can be constructed using the op-amps.

Because of the considerable development of electronic devices today, it turns out to be easy and inexpensive to perform product and division operations using multiplier and divider chips. A typical multiplier/divider chip is the AD532 or AD534 made by the Analog Devices. Further, due to the differential nature of the inputs, we are in a position to assemble 'minimum component devices' using op-amp modules and multipliers so as to simulate nonlinear differential equations (Refs. [51–54, 91]). One such analog simulation circuit constructed with adding integrators, scale changers and multipliers helped us to study the chaotic dynamics of the Duffing oscillator, which will be discussed later in Chapter 4.

CHAPTER 4

DUFFING OSCILLATOR:
BIFURCATION AND CHAOS

The Duffing oscillator equation (2.5) is a ubiquitous nonlinear differential equation, which makes its presence in many physical, engineering and even biological problems (Refs. [5, 7, 9, 10, 52, 92–97]). Originally the model was introduced by the German electrical engineer Duffing in 1918 (Ref. [34]). The Duffing equation*

$$\frac{d^2x}{dt^2} + \alpha\frac{dx}{dt} + \omega_0^2 x + \beta x^3 = f(t), \alpha > 0; \ f(t) = f\sin\omega t \ (\text{or } f\cos\omega t), \quad (4.1)$$

can be thought of as the equation of motion for a particle of unit mass in the potential well

$$V(x) = \frac{1}{2}\omega_0^2 x^2 + \frac{\beta}{4}x^4 \qquad (4.2)$$

subjected to a viscous drag force of strength α and driven by an external periodic signal of period $T = (2\pi/\omega)$ and strength f. However we can distinguish three types of potential wells of physical relevance here:

(i) $\omega_0^2 < 0, \beta > 0$: A double-well with potential minima at $x = \pm\sqrt{(|\omega_0^2|/\beta)}$ and a local maximum at $x = 0$.

(ii) $\omega_0^2 > 0, \beta > 0$: A single-well with a potential minimum at the equilibrium point $x = 0$.

(iii) $\omega_0^2 > 0, \beta < 0$: A double-hump potential well with a local minimum at $x = 0$ and maxima at $x = \pm\sqrt{(\omega_0^2/|\beta|)}$.

---

*Traditionally one uses $f(t) = f\sin\omega t$ for experimental systems and $f(t) = f\cos\omega t$ for numerical investigations as a matter of convenience.

Each one of the above three cases has become a classical central model to describe inherently nonlinear phenomena, exhibiting a rich and baffling variety of regular (periodic) and complex (chaotic) motions which can coexist or exist in neighbouring parametric regimes. We will discuss each of these cases separately.

4.1. Scaling Property and Three-dimensional Control Parameter Space

In the analysis of Eq. (4.1), one would typically start from the behaviour of the linear oscillator (that is, $\beta = 0$), where resonant oscillations occur (when $\omega_0^2 > 0$) and then investigate the behaviour of the nonlinear oscillator as β increases slowly for ranges of other parameters. Equation (4.1) contains five external control parameters, namely $\alpha, \omega_0^2, \beta, f$ and ω. However, one can introduce two scaling parameters a and b by the rescaling $t \rightarrow at$, $x \rightarrow bx$. As a result one can fix two of the control parameters by choosing a and b appropriately, while the remaining three parameters will be the free control parameters. It is customary to fix the values of the linear and nonlinear restoring force strengths ω_0^2 and β, while the rescaled damping constant α, the strength f and frequency ω of the external forcing are varied over the required ranges. This is because it is much easier to vary α, f and ω than ω_0^2 and β from an experimental point of view. In the following we will concentrate essentially on the $(\alpha\text{--}f\text{--}\omega)$ parameter space alone by choosing suitable values for ω_0^2 and β, whenever necessary. Naturally no loss of generality will occur in the qualitative understanding of the dynamics due to this choice of the parameter space.

As mentioned in the earlier Chapters, most of our understanding of the dynamics of nonlinear oscillators such as (4.1) is based on numerical and analog studies. While the most ideal procedure will be to draw a three-dimensional $(\alpha\text{--}\beta\text{--}f)$ phase diagram, practical considerations confine one to concentrate on

(i) the $(f\text{--}\omega)$ phase diagram by fixing the damping parameter α or sometimes even the $(\alpha\text{--}\omega)$ or $(f\text{--}\alpha)$ phase diagrams and

(ii) the bifurcation diagram for either f or ω while fixing the other two parameters.

Our further studies on the Duffing oscillator will essentially concern with the bifurcations in the f parameter space to a large extent and on the $(f\text{--}\omega)$ phase diagram to a lesser extent.

4.2. The Double-Well Oscillator

The Duffing oscillator equation (4.1) for the double-well (or twin-well) potential ($\omega_0^2 < 0, \beta > 0$) has been studied by Holmes and Moon and coworkers since 1979 (Refs. [10, 93, 94]) and was derived as a mathematical model for a buckled beam or plasma oscillations, among others. The simplest experimental device for it is a particle placed in a twin-well potential with a base vibrating with periodic motion (See Fig. 4.1). The potential well of this model is shown in Fig. 4.2. When the amplitude of excitation is large enough, the particle can escape from one of the potential wells into the other in a highly complex and random-like fashion. Another interesting mechanical model is the oscillations of a buckled beam under the action of a periodic force as shown in Fig. 4.3 (Ref. [7]).

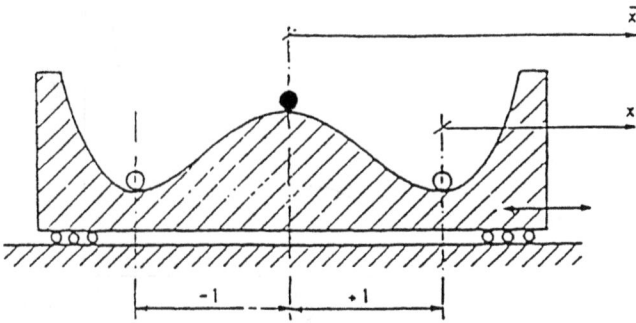

Fig. 4.1. A physical model of double-well potential oscillator ($|\omega_0^2| = 1.0, \beta = 1.0$).

We notice that the three equilibrium points of system (4.1) for $f = 0$ for the present double-well case ($\omega_0^2 < 0$) correspond to

$$\omega_0^2 x + \beta x^3 = 0, \tag{4.3a}$$

so that we have the stable fixed points

$$x_{1,2}^{(s)} = \mp\sqrt{(|\omega_0^2|/\beta)}, \tag{4.3b}$$

and an unstable fixed point which is a saddle at

$$x_0^{(u)} = 0. \tag{4.3c}$$

Then the oscillations about the two stable equilibrium points are governed essentially by the equation

$$\frac{d^2x}{dt^2} + \alpha\frac{dx}{dt} + 2|\omega_0^2|x \mp 3\sqrt{|\omega_0^2|\beta}\,x^2 + \beta x^3 = f(t)\,. \qquad (4.4)$$

The above equation is obtained by shifting x in Eq. (4.1) to $x = \bar{x} \mp \sqrt{(|\omega_0^2|/\beta)}$ and then dropping the bars.

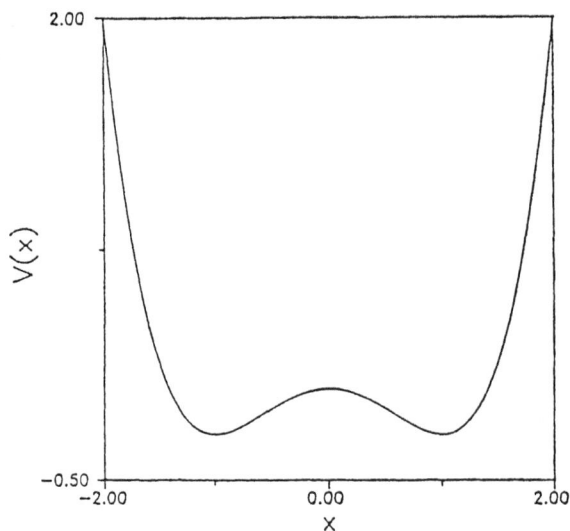

Fig. 4.2. The double-well potential ($\omega_0^2 < 0$).

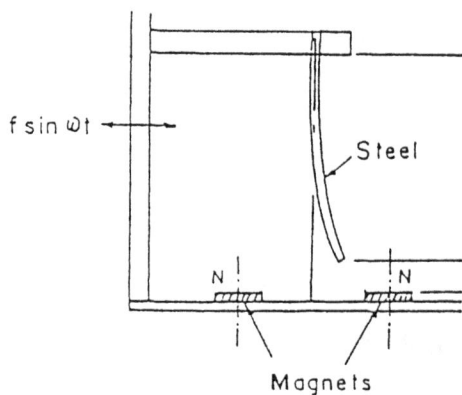

Fig. 4.3. A mechanical model (*buckling beam*) having a double-well potential.

4.2.1. *Period-Doubling Scenario in f-space*

We consider the dynamics of the double-well Duffing oscillator (4.1) with $\omega_0^2 < 0, \beta > 0$ and $f(t) = f \sin \omega t$ by scanning the drive amplitude f upward from 0. The restoring force parameters are fixed as $\omega_0^2 = -1.0$ and $\beta = 1.0$. The other two control parameters are fixed at the values $\alpha = 0.5$ and $\omega = 1.0$ for our computer studies (Ref. [53]). From the nature of the results of the numerical investigations obtained by solving Eq. (4.1) for the above parameters, using the standard fourth-order Runge–Kutta algorithm, we infer the following picture. The results are substantiated by (i) phase-portrait, (ii) power spectrum, and (iii) Poincaré surface of section, wherever required.

To start with, we observe that for $f = 0$ the system asymptotically approaches the stable fixed point (focus) (the left equilibrium point $x_1 = -1.0$ in the present case) as in Fig. 4.4(a) corresponding to purely damped oscillation. (It might be noted that one can also start with a different initial condition so as to asymptotically reach the other stable fixed point $(x = 1.0)$ in the right well). We realize from the theory of nonlinear oscillations that the system can, as f is increased, exhibit two types of periodic motions:

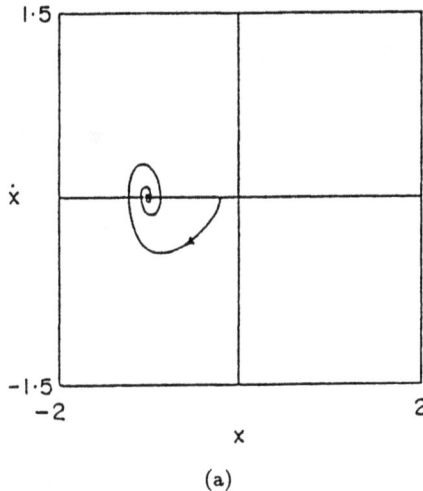

(a)

Fig. 4.4. (a) Phase-portrait denoting the trajectory approaching the fixed point $(x < 0)$ for $f = 0.0$. (b) (i) Period-$T(= 2\pi/\omega)$ limit cycle for $f = 0.33$; (ii) Power spectrum of signal x. (c) (i) Period-$2T$ limit cycle for $f = 0.35$; (ii) Power spectrum of signal x. (d) (i) Period-$4T$ limit cycle for $f = 0.357$; (ii) Power spectrum of signal x. (e) (i) One-band chaos for $f = 0.37$; (ii) Power spectrum of signal x. (f) (i) Double-band chaos for $f = 0.42$; (ii) Power spectrum of signal x; (iii) Poincaré map of (i). (g) Period-$3T$ window for $f = 0.664$. (h) Period-T boundary for $f = 0.85$.

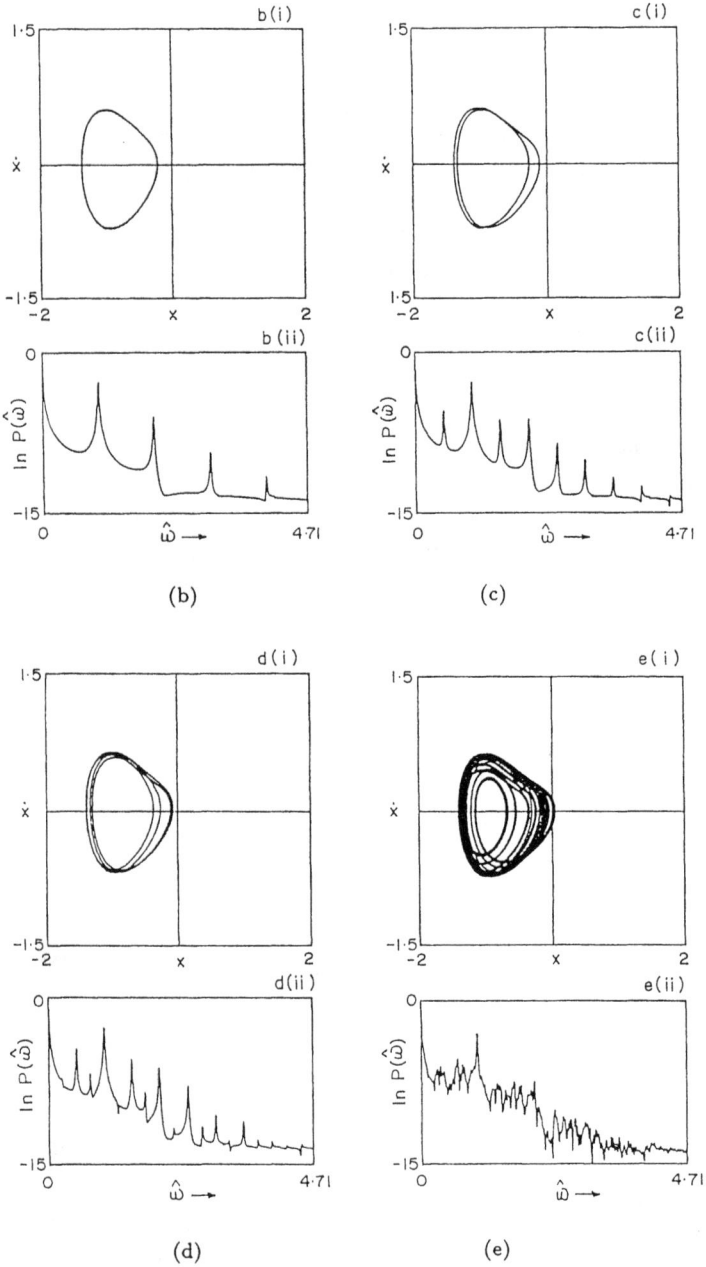

Fig. 4.4. (*Continued*).

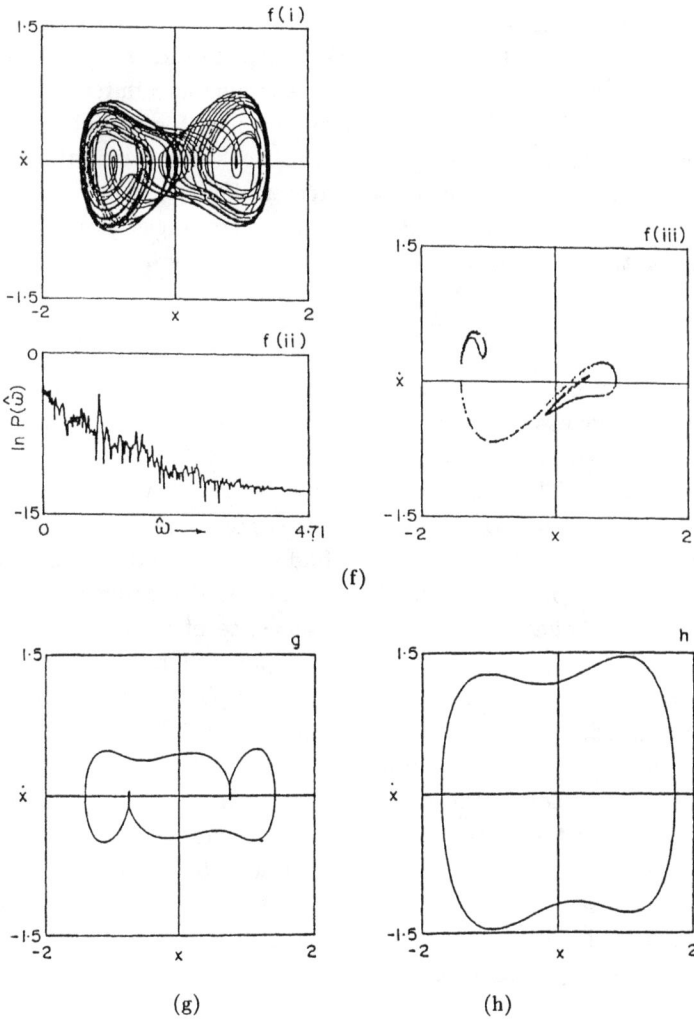

Fig. 4.4. (*Continued*).

(i) *Small orbits (SOs) oscillations* around one of the two stable equilibria $x_{1,2}$ given by Eq. (4.3b).

(ii) *Large orbits (LOs), large amplitude oscillations* that encircle all the three equilibrium points x_1, x_2 and x_0.

A. *Small orbit oscillations*: As f is increased slightly from 0, a stable period $(T = 2\pi/\omega)$ limit cycle about the left equilibrium point $x_1(= -1.0)$ occurs, which persists up to $f = 0.34$. Figure 4.4(b) depicts such a limit cycle. As f is increased to a critical value 0.35, the phase trajectory bifurcates into a new limit cycle of period $2T$, again about x_1 as shown in Fig. 4.4(c). The system then undergoes further period-doubling bifurcations as shown in the rest of the Figs. 4.4 when f is smoothly increased further. This cascade of bifurcations accumulates at $f \equiv f_c = 0.361$ and the sequence converges geometrically at a rate known as the Feigenbaum's ratio, given by

$$\lim_{n \to \infty} \frac{f_{n+1} - f_n}{f_{n+2} - f_{n+1}} \equiv \delta = 4.6692\ldots, \tag{4.5}$$

where f_n is the value of f at which the $2^{n-1}T$ period solution bifurcates into the $2^n T$ period solution. From our numerical results above, the calculated value of δ turns out to be $\delta = 4.77674$, which is in fair agreement with the value of Eq. (4.5). This convergence rate δ tells us how quickly successive bifurcations occur as the control or bifurcation parameter f is varied.

Beyond the critical value $f = f_c$, one finds that nonperiodic chaotic orbits occur which correspond to an infinite number of stretching and folding of trajectories in the phase-space. The consequence of such a stretching and folding is that initially nearby trajectories exponentially separate in a finite time interval. As mentioned in Chapter 1, the phase trajectory of such a nonperiodic chaotic orbit exhibiting exponential instability is called a strange attractor and it has in general a noninteger, or *fractal*, dimension which is a measure of the complexity and self-similarity of the attractor. Figure 4.4(e) exhibits a typical one-band (left well) chaotic attractor for $f = 0.37$. Also, the nature of the spectrum of this attractor is seen to be continuous and broad, which indicates the aperiodicity of the solution $x(t)$. When f is increased a little above 0.37 the onset of *crisis* occurs, which is used to denote a sudden change in the chaotic state when the control parameter f is changed.

B. *Hopping oscillations*: With the onset of crisis, for $f > 0.37$, the chaotic motion which was originally confined to a limited range of $x(t)$ (namely, the left well in the present case) may suddenly expand to a broad range of $x(t)$, say covering both the wells. Now the orbit begins to migrate into the right valley of the phase plane and exhibits a hopping cross-well chaotic state as shown in Fig. 4.4(f) for $f = 0.42$, which we might call a double-band chaos. Then the previous chaotic state in Fig. 4.4(e) may be designated as one-band or single-band chaos.

As the bifurcation parameter f is continuously increased further beyond 0.42, the state keeps on hopping first and then is locked into some subharmonics typically termed as windows. This state exists in a small interval of the driving amplitude f and, beyond that, further period-doubling bifurcations to chaos occurs, followed by reverse bifurcations (see below), further windows and chaos as f is slowly varied as shown in the bifurcation diagram (Fig. 4.5). These are further discussed in the following.

C. *Large orbit oscillations*: For sufficiently larger values of f, the chaotic state suddenly disappears and becomes stable with a symmetrical phase portrait having a period $mT(m > 0)$ and extending over the two valleys, encompassing all the three equilibrium points x_1, x_0 and x_2. This implies that the central

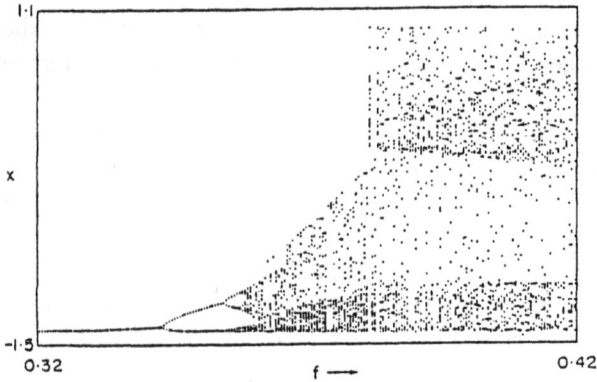

(a)

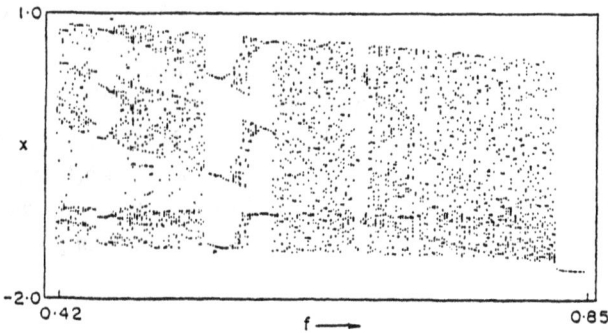

(b)

Fig. 4.5. Bifurcation diagrams in $(f–x)$ plane computed through numerical simulation.

barrier of the potential well (Fig. 4.1(a)) plays no significant effect on the motion of the system for large amplitude excitation. Figures 4.4(g) and 4.4(h) represent large amplitude, period-$3T$ window and period-T boundary respectively at $f = 0.664$ and $f = 0.85$.

The above results can be put in a nutshell in the form of a bifurcation diagram (Fig. 4.5) in the $(f-x)$ plane obtained from the numerical analysis, where the x value is the Poincaré points of the solution $x(t)$ for every period-(2π) of the external frequency.

D. *Antimonotonicity*: Recently, Yorke and coworkers (Ref. [98]) have shown that antimonotonicity — inevitable reversals of period-doubling cascades — is a fundamental phenomenon for a large class of nonlinear systems. One finds that periodic orbits are not only created but also destroyed when one increases the control parameter monotonically (smoothly) in any neighbourhood of a homoclinic tangency value. This phenomenon is now called *antimonotonicity* (Refs. [98, 99]).

Now we wish to point out that for the present double-well Duffing oscillator problem antimonotonicity can be observed in two different ways by looking at the bifurcation diagram, Fig. 4.5, for the range $0.32 \leq f \leq 0.85$. The first bifurcation pattern is shown in Fig. 4.6(a)(i). As f is increased, chaotic behaviour is followed by a complete sequence of reverse period-doubling cascades of a period-m orbit; then, a complete period-doubling sequence of period-m orbit is followed by chaotic oscillations. The pattern is called a *reverse period-m bubble*. Such a reverse period-5 bubble in the range $0.43 \leq f \leq 0.48$ is shown in Fig. 4.6(a)(ii).

The second bifurcation pattern is shown schematically in Fig. 4.6(b)(i). As f is increased a tangent bifurcation of periodic orbit with period-m is followed by complete sequence of period-doubling bifurcations; then, after an interval with chaotic behaviour, there is a reverse period-doubling sequence, ending in a periodic orbit with period m. This pattern is called a *period-m bubble*. An example of this pattern for $m = 3$ is shown in Fig. 4.6(b)(ii) in the range $0.53 \leq f \leq 0.62$. A detailed numerical study shows that the two patterns of bifurcations diagrams are indeed typical for certain range of parameters in the Duffing oscillator with a double well.

E. *Further periodic windows*: Finally, there also appear a period-4 window and a period-2 window within the chaotic regimes as f is slowly varied, as shown in the bifurcation diagrams Fig. 4.6(c) ($0.49 \leq f \leq 0.51$) and Fig. 4.6(d) ($0.62 \leq f \leq 0.71$) respectively.

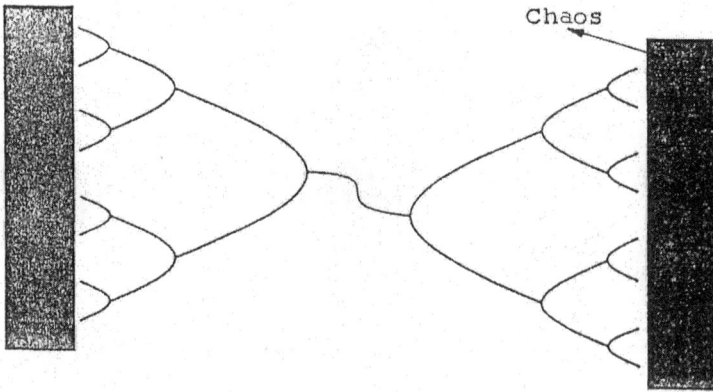

(a) (i)

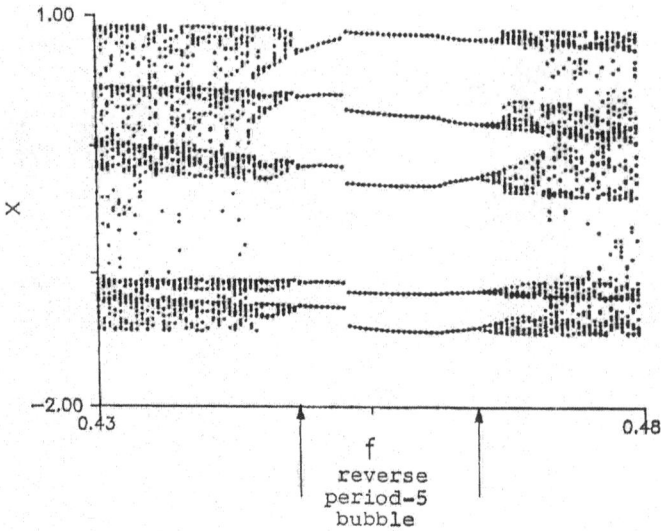

(a) (ii)

Fig. 4.6. (a): (i) Schematic diagram of a reverse bubble. (b): (i) Schematic diagram of a bubble. (a): (ii) Bifurcation diagram in the range 0.43–0.48 of the parameter f. Note the bifurcation pattern of Fig. 4.6(a): (i), henceforth it is called a reverse period-5 bubble. (b): (ii) $0.53 \leq f \leq 0.62$; due to the bifurcation pattern of Fig. 4.6(b): (i), henceforth it is called a period-3 bubble. (c) Bifurcation diagram in (f-x) plane ($0.49 \leq f \leq 0.5$). (d) Bifurcation diagram in (f-x) plane ($0.62 \leq f \leq 0.71$).

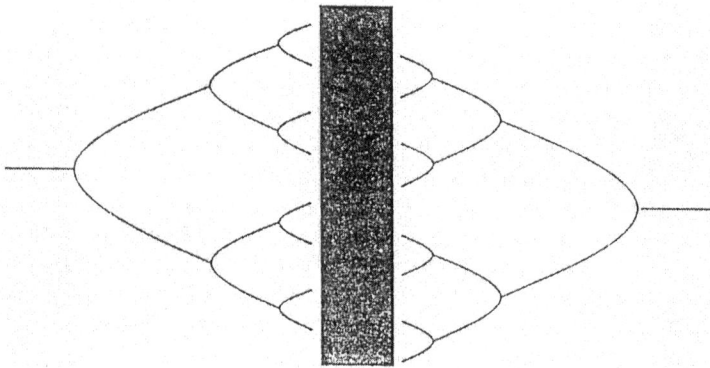

(b) (i)

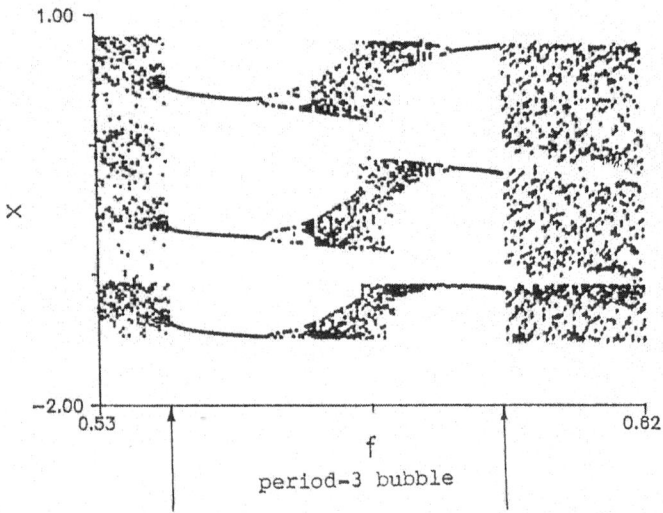

(b) (ii)

Fig. 4.6. (*Continued*).

In Table 4.1, we give a succinct account of the different types of attractors admitted by the double-well Duffing oscillator in the above parametric regime. Experimental confirmation of the above results through the use of analog circuits is discussed later in Sec. 4.6.

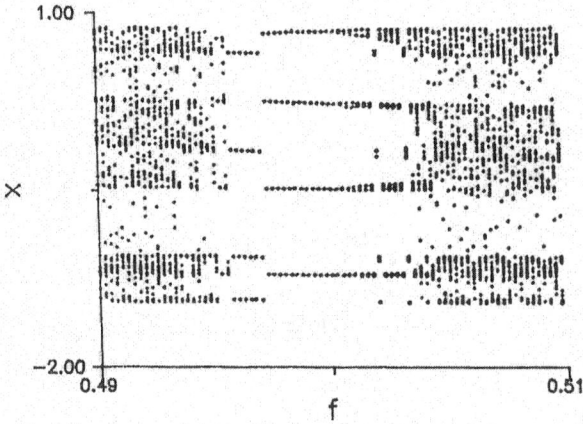

(c)

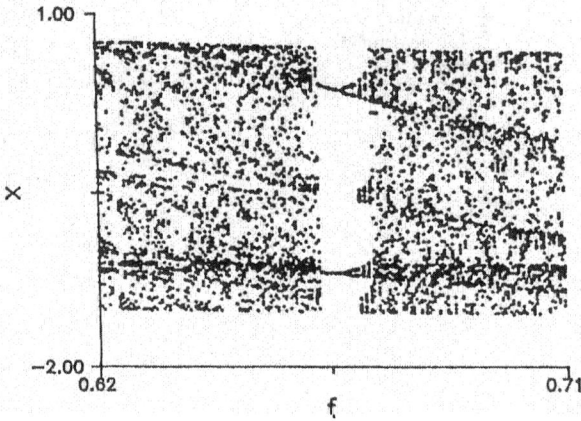

(d)

Fig. 4.6. (*Continued*).

Finally there is one more important bifurcation which can occur in the above type of oscillations which is the *intermittency bifurcations* (or *interior catastrophe, or explosion, or interior crisis*). A typical case occurs for $\omega_0^2 = -1.0, \beta = 1.0, \omega = 1.0, \alpha = 0.25$ and $f = 0.266$ (see Ref. [9], p. 276). We do not consider such bifurcations in detail in this book.

Table 4.1. Summary of bifurcation phenomena of Eq. (4.1) with $\alpha = 0.5, \omega_0^2 = -1.0, \beta = 1.0$ and $\omega = 1.0$.

Value of f	Nature of solution	Attractor in the phase space in the $(x$–$\dot{x})$ plane (simulation results)	
		Numerical Fig. 4.4	Analog Fig. 4.23
$f = 0$	Damped oscillation to the stable focus at $x < 0$	(a)	
$0 < f < 0.34$	Period-T oscillation	(b)	
$0.34 \le f < 0.358$	Period-$2T$ oscillation	(c)	
$0.358 \le f < 0.362$	Period-$4T$ oscillation	(d)	
$0.362 \le f < 0.37$	Further period doublings		
$0.37 \le f < 0.4$	One-band chaos	(e)	
$0.4 \le f = 0.42$	Double-band chaos	(f)	
$0.42 < f$	Chaos, windows, reverse period-doublings, chaos, boundary crisis, etc. (See bifurcation diagram of Fig. 4.5)	(g), (h)	

4.2.2. *Phase Diagram in $(f$–$\omega)$ Space*

Szemplinska-Stupnicka and Rudowski (Ref. [96]) have recently made a detailed analysis of the double-well Duffing oscillator equation of the form (4.1) with $f(t) = f \cos \omega t$ $(\omega_0^2 < 0, \beta > 0)$ in the $(f$–$\omega)$ space. We give a brief summary of their computer-based results, within the frequency zone $0.25 < \omega < 1.1$ at fixed damping $\alpha = 0.1$, and $\omega_0^2 = -0.5$ and $\beta = 0.5$. Figure 4.7 depicts the $(f$–$\omega)$ phase diagram, where small orbit (SO), symmetric large orbit (SLO), unsymmetric large orbit (ULO), and cross-well chaotic or regular stable attractors exist. The small orbit motions occur within the whole $(f$–$\omega)$ plane except for two V-shaped regions: One with cusp at $\omega = 0.8$ and $f = F_2$ at the principal resonance region, and the other with cusp at $\omega = 0.4$ and $f \approx 0.14$, that is at superharmonic resonance zone. Inside the two V-shaped regions the system can exhibit cross-well chaotic (or regular) motion. Also in some regions two different attractors coexist, while in others single steady-state motion can be observed. In the latter case, the attractors are globally stable.

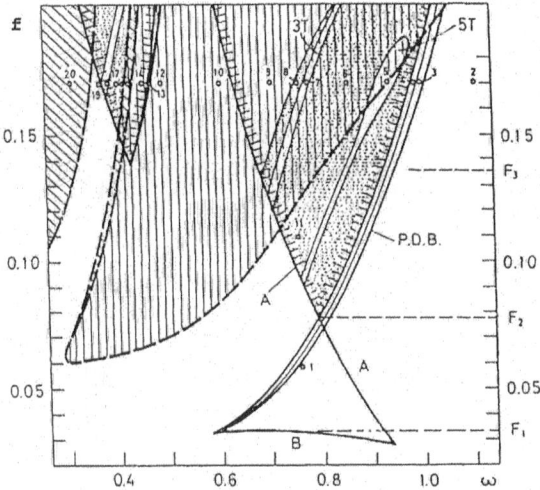

Fig. 4.7. The $(f–\omega)$ phase diagram indicating regions of different steady states exhibited by the double-well potential oscillator for $\alpha = 0.1, \omega_0^2 = -0.5, \beta = 0.5$ in Eq. (4.1) with $f(t) = f\cos\omega t$. |||||: LO symmetric; \\\\\\: LO unsymmetric; shaded area: cross-well chaotic motion; cross-thatched V areas: SO occurs outside V-shaped regions. Adapted from Ref. [96].

A. *Typical orbits*: The forms of the various single and coexisting states denoted in Fig. 4.7 by numbers $1, 2, \ldots, 20$ are then shown in Figs. 4.8. Regular attractors are illustrated by their phase portraits and chaotic attractors by Poincaré maps. They correspond to the following:

(i) *Point 1 (low f value, $f < F_2$)*: Resonant, large amplitude or nonresonant low-amplitude small-orbit motion.
(ii) *Points 2–10 (large f, here $f = 0.17 > F_3$)*: Principal resonance region for decreasing driving frequency.

  2–4 : Period-doubling bifurcations of SO
   5 : cross-well periodic motion (periodic window)
   6 : cross-well chaotic attractor coexisting with symmetric LO
   7 : $3T$ periodic cross-well motion coexisting with LO attractors
   8 : Cross-well chaotic attractor coexisting with symmetric LO
   9 : unique symmetric LO attractor
  10 : coexistence of SO and LO
  11 : cross-well unique chaotic attractor.

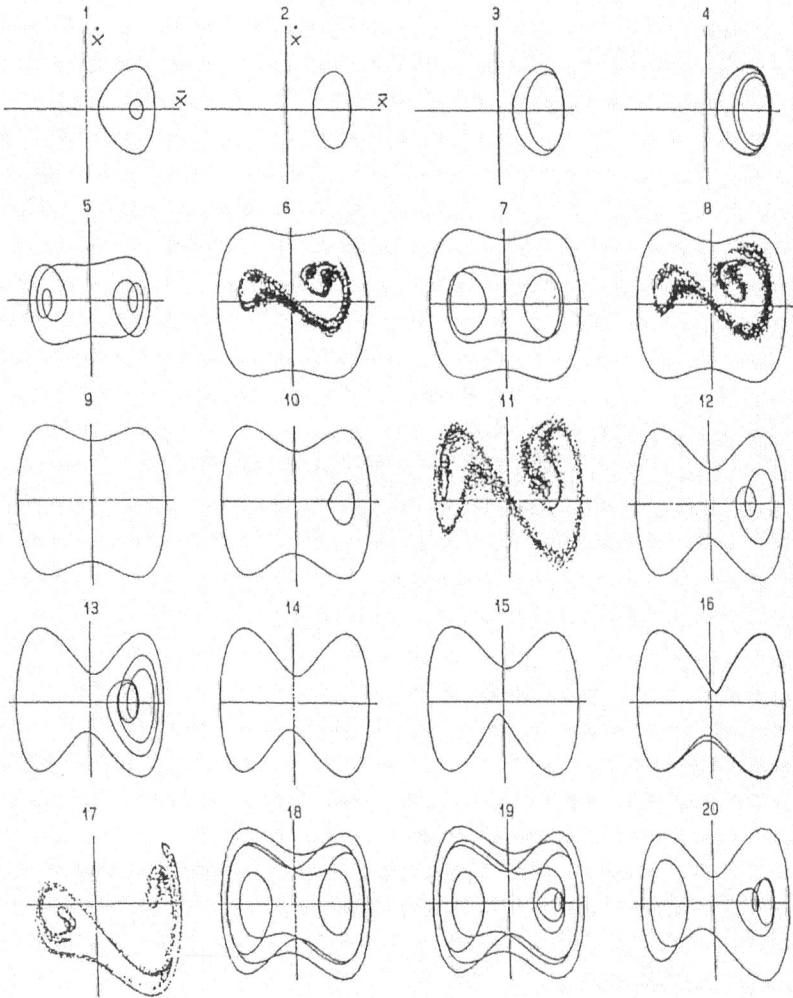

Fig. 4.8. Various types of steady-state attractors: $\alpha = 0.1$, (1) $f = 0.06, \omega = 0.74; f = 0.17$; (2) $\omega = 1.1$; (3) $\omega = 1.0$; (4) $\omega = 0.982$; (5) $\omega = 0.93$; (6) $\omega = 0.85$; (7) $\omega = 0.79$; (8) $\omega = 0.75$; (9) $\omega = 0.70$; (10) $\omega = 0.6$; (11) $f = 0.11, \omega = 0.75; f = 0.17$; (12) $\omega = 0.48$; (13) $\omega = 0.45$; (14) $\omega = 0.44$; (15) $\omega = 0.41$; (16) $\omega = 0.4045$; (17) $\omega = 0.40$; (18) $\omega = 0.3845$; (19) $\omega = 0.38$; (20) $\omega = 0.30$ (see Ref. [96]).

(iii) *Points 12–20 (Zone of superharmonic resonance)*

    12 : coexistence of LO and SO but with multifrequency response

    13 : unique symmetric LO obtained after bifurcation of small
        T-periodic orbit into $2T$-periodic orbit which then disappears

    14 : symmetry-breaking bifurcations

  15/16 : period-doubling bifurcation of LO

    17 : chaotic cross-well motion

    18 : regular cross-well motion

    19 : coexistence of regular cross-well motion with T-periodic SO
        attractor

    20 : superharmonic resonance of LO motion.

Frequency spectrum: Figure 4.9 gives the frequency (power) spectrum of some of the steady-state motions. From Figure 4.9(a) we can infer that the SO at point 2 involves a large fundamental harmonic, a constant term, and a very small second harmonic. Figure 4.9(b) clearly shows that the first period-doubling bifurcation has occurred, while the broad-band continuous spectrum of Fig. 4.9(c) represents the cross-well chaotic attractor. Finally Fig. 4.9(d) shows that the LO is highly regular and very close to the harmonic function of time with frequency ω.

B. *Resonance curves*: For small-orbit motion in the neighborhood of the principal resonance, that is at $w \approx 1$, we notice that the resonance curve $x_{max} \equiv x_{max}(\omega) = A(\omega)$ is bent to the left as may be inferred from Eq. (4.4) for motion in the right well. At sufficiently low values of the forcing term the response is approximately a harmonic function of time with frequency ω, and it shows the resonance behaviour (Fig. 4.10(a)) as discussed in Chapter 2, but now the resonance curve is left-leaning.

When f exceeds a certain critical value F_1, $F_1 < f < F_2$, the T-periodic nearly harmonic solution bifurcates into a $2T$-periodic solution at the top of the resonant branch of the resonance curve, but does not develop a complete period-doubling cascade and "jumps" down to the nonresonant branch so that hysteresis behaviour still occurs (Fig. 4.10(b)). On further increasing the parameter f, $f > F_2$, the first period doubling is followed by the classic Feigenbaum period-doubling cascade and further complex bifurcations. Finally the system escapes from the potential well $x = 0$ (which now, in the transformed state variable, corresponds to the right well) and falls into the other (left) well and cross-well chaotic motion develops (Fig. 4.10(c)). It is also probable that just before the escape, single-well chaos as in Sec. 4.2.1-A occurs.

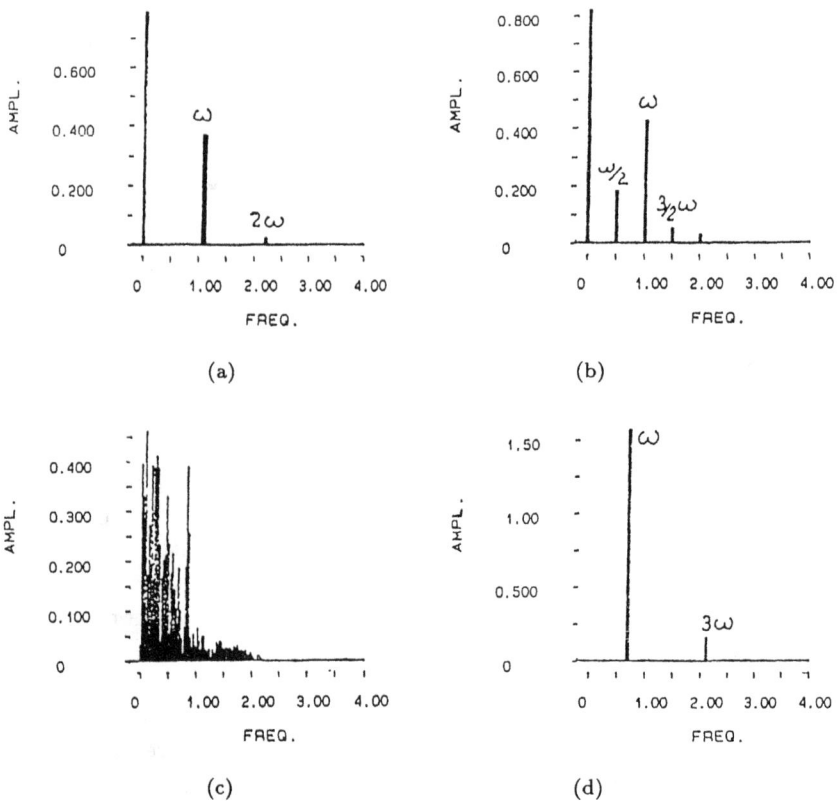

Fig. 4.9. Fourier spectra of various steady-state attractors, $\alpha = 0.1$: (a) point 2, (b) point 3, (c) point 6, (d) point 9 (Ref. [96]).

As the higher frequency boundary of the cross-well chaos is related to the resonant branch of the resonance curve, it is preceded by the universal period-doubling cascade. All the enormously complex bifurcations occur, however, within a narrow frequency zone, denoted by $\Delta\omega$ in Fig. 4.10(c). The lower frequency boundary corresponds to a saddle-node bifurcation (Refs. [32, 100]), where a sudden change to/from T-periodic SO from/to chaotic attractor occurs and the two different steady states are separated by transient motion only. Figure 4.10(c) also depicts a resonance curve of LO motion, where the LO T-periodic response coexists with cross-well chaos, or with nonresonant T-periodic SO. Depending on the initial conditions the system exhibits one or the other steady state. For more details for still higher values of f, we refer the reader to the paper of Szemplinska-Stupnicka and Rudowski (Ref. [96]).

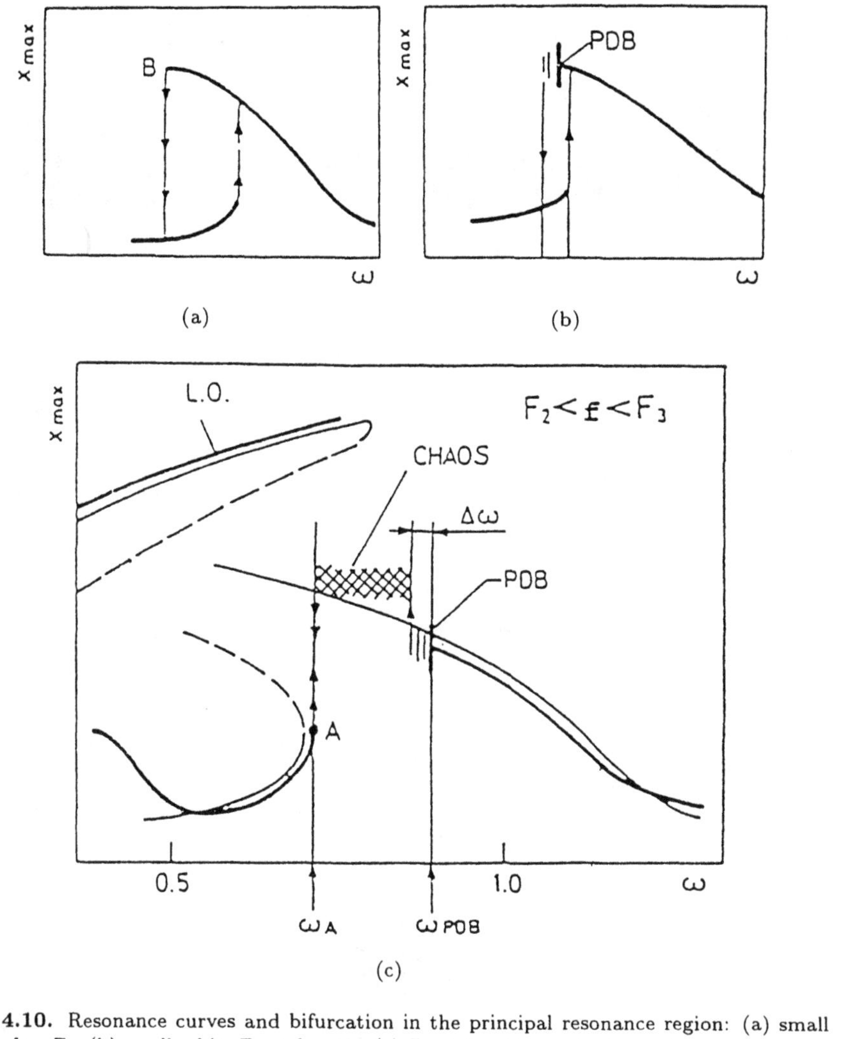

(a)

(b)

(c)

Fig. 4.10. Resonance curves and bifurcation in the principal resonance region: (a) small orbit, $f > F_1$. (b) small orbit, $F_1 < f < F_2$. (c) $F_2 < f < F_3$-small orbit and coexisting large orbit motion. Thick line: computer simulations (stable branches only), thin line: theoretical resonance curve (Ref. [96]).

4.3. Single-Well Duffing Oscillator (Refs. [5, 7, 38, 95, 97, 101])

When $\omega_0^2 > 0$ and $\beta > 0$ in Eq. (4.1), the Duffing oscillator has only a single-well symmetric potential (Fig. 4.11(a)). A mechanical model of it is given in Fig. 4.11(b), which is self-explanatory. We can choose $\omega_0^2 = 1, \beta = 1$ and $f(t) = f \cos \omega t$ without loss of generality, and then the equation of motion

becomes

$$\frac{d^2x}{dt^2} + \alpha\frac{dx}{dt} + x + x^3 = f\cos\omega t. \qquad (4.6)$$

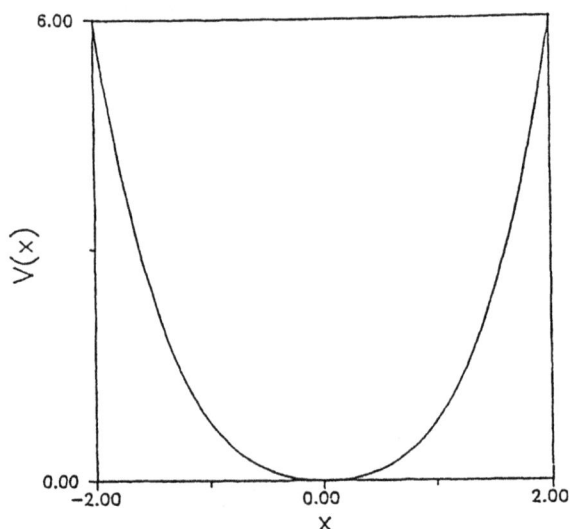

(a)

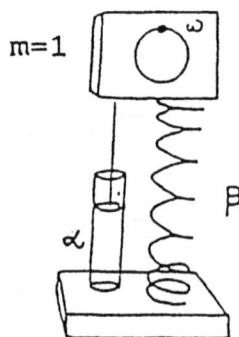

(b)

Fig. 4.11. (a) Single-well potential ($\omega_0^2 \geq 0, \beta > 0$). (b) Mechanical model having a single-well potential.

Fixing $\alpha = 0.1$, one can study the dynamics of (4.6) in the (f–ω) plane. Here we give a very brief account of the salient features following the work of Olson and Olsson (Ref. [101]):

A. *Very low f* ($\ll 1$): The excitation is nearly that of the linear system and the anharmonic effects are easily calculated using any standard perturbation theory.

B. *Low f* ($0.5 < f < 5.0$): *Resonance and symmetry breaking* Already near $f = 0.5$, the hysteresis effect at the primary resonance is quite evident and harmonic resonances with peaks at $\omega_{res} = (2n+1)^{-1}$ are starting to appear, as briefly mentioned in Chapter 2. In Fig. 4.12, we show the maximum displacement (amplitude) of the response for $f = 3$ for a range of frequencies sweeping both up and down. The characteristic features are

(i) hysteresis in the primary resonance region (for comparison lowest order perturbation curve is also included in Fig. (4.12)),
(ii) harmonic resonance peaks (labeled by their Fourier components) and
(iii) the dynamical symmetry breaking in the region $S_2 S_2'(0.88 < \omega < 1.05)$, where the second harmonic (with frequency 2ω) begins and ends. Here the response is not symmetric even though the potential is symmetric.

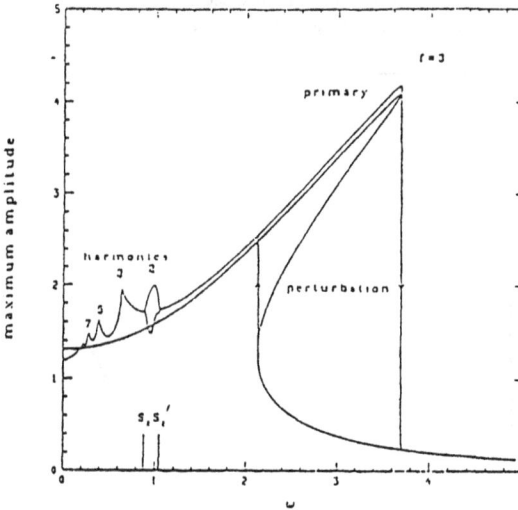

Fig. 4.12. Maximum displacement of the single-well Duffing oscillator (Eq. (4.6)) with $f = 3.0$. The driving frequency was varied both up and down in small increments. Mechanical hysteresis is observed with sudden jumps between the two stable attractors. The lowest-order perturbative result is shown for comparison. Harmonic resonances are labeled by their Fourier components, and the dynamical symmetry-breaking region $S_2 S_2'$ is indicated (Ref. [101]).

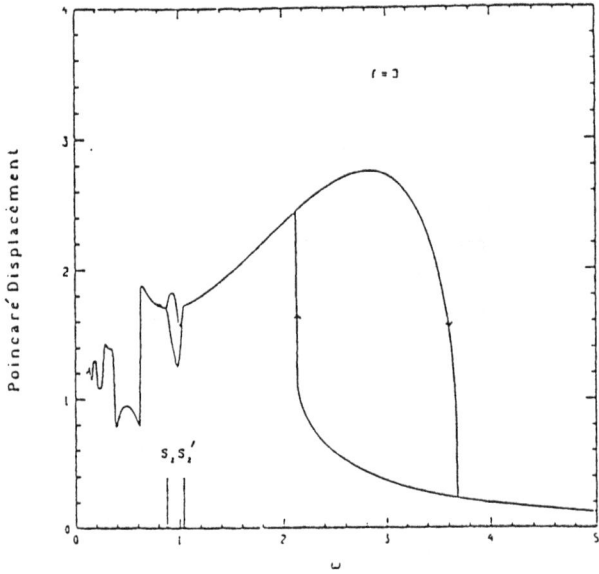

Fig. 4.13. Displacement component of the Poincaré section at zero driver phase and $f = 3.0$. All subsequent Poincaré plots are also at zero driving phase (Ref. [101]).

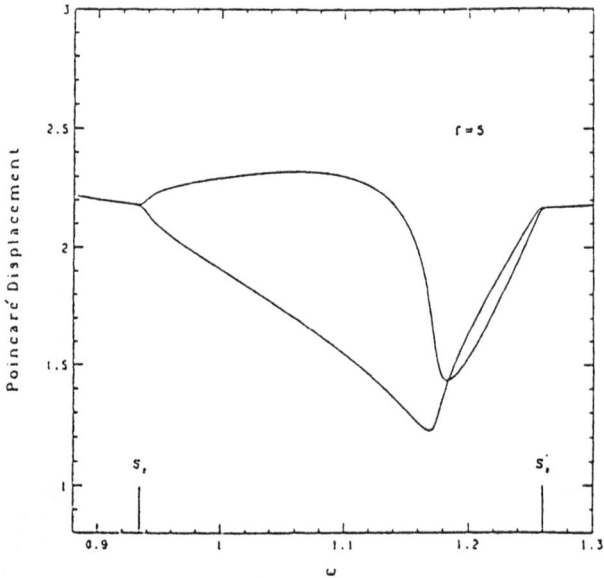

Fig. 4.14. $f = 5.0$. Poincaré displacement in the $S_2 S_2'$ region. The two attractors are the two symmetry-breaking cases. They correspond to the same orbit (Ref. [101]).

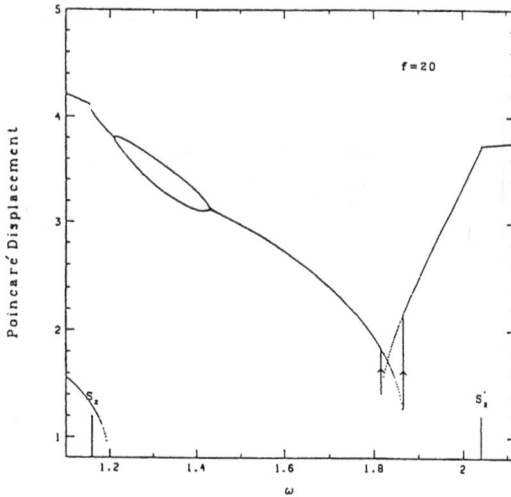

Fig. 4.15. $f = 20.0$. Poincaré displacement. A period-doubling bifurcation appears between $1.2 < \omega < 1.4$. The hysteresis near $\omega = 1.8$ has become prominent. Again, only one symmetry-breaking attractor is shown. The other branch, as in Fig. 4.14, is the other side of the same orbit (Ref. [101]).

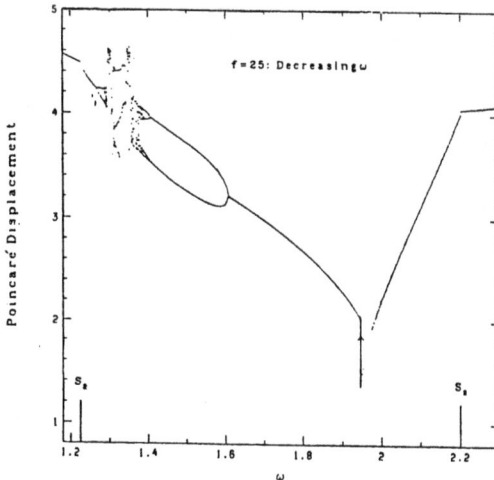

Fig. 4.16. $f = 25.0$. Poincaré displacement. The plot is based only on a downward frequency sweep. A complete cascade of period-doubling bifurcations and a reverse sequence occurs. A region of chaos occurs between these cascades. Near $\omega = 1.33$, a three-cycle appears (Ref. [101]).

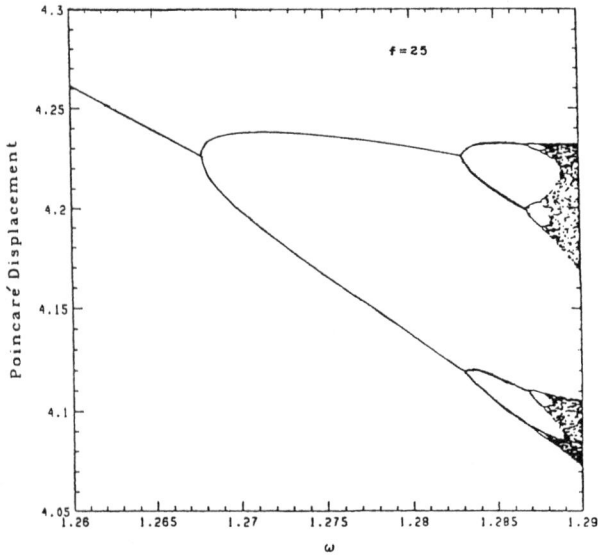

Fig. 4.17. $f = 25.0$. Poincaré displacement detail of the bifurcation cascade (Ref. [101]).

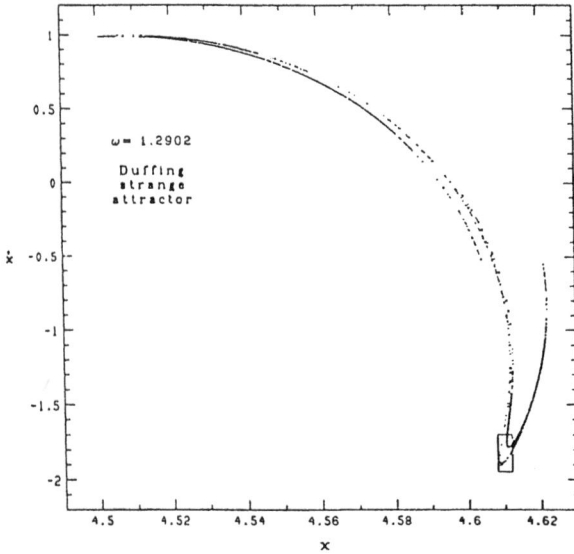

Fig. 4.18. $f = 25.0$. Poincaré strange attractor at $\omega = 1.2902$ (Ref. [101]).

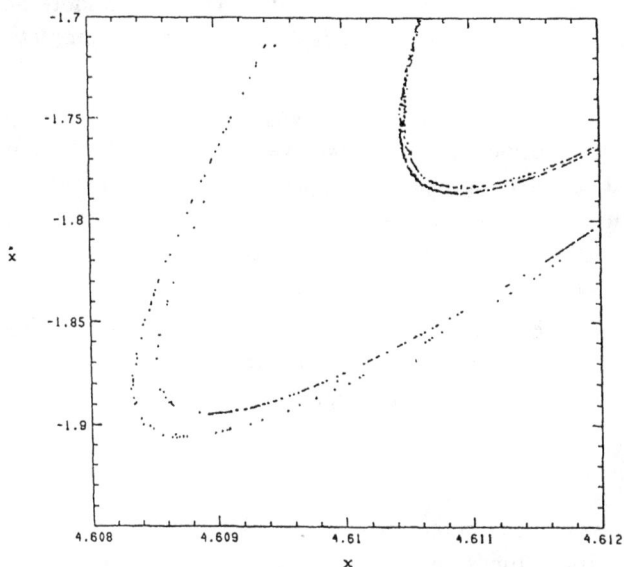

Fig. 4.19. Blowup of the strange attractor of Fig.4.18 contained in the rectangle (Ref. [101]).

Figure 4.13 is the Poincaré section for $f = 3$, which can be compared with the previous figure.

C. *Medium f* ($5 < f < 15$): In the region $5 < f < 15$, the symmetry-breaking orbits will exhibit increasing hysteresis and more exaggerated frequency dependence of the orbit. Figure 4.14 is a magnified Poincaré view of the $S_2 S_2'$ region for $f = 5$. Within the $S_2 S_2'$ region two curves are shown, indicating that the attractor is asymmetric. Even though the Poincaré section is taken at zero phases of the drive, there are two equivalent orbits corresponding to the two ways that the symmetry can be broken.

D. *Large f* ($15 < f < 25$): When the driving amplitude is increased to $f = 20$, the first true period-doubling bifurcation occurs. The Poincaré displacement at $f = 20$ is shown in Fig. 4.15. In this figure only one symmetry-breaking attractor is shown. The other branch is, as in Fig. 4.14, the other side of the same orbit. A further increase in the driving amplitude to $f = 25$ produces a complete cascade of bifurcations and a chaotic region as depicted in Fig. 4.16. Again, we have suppressed the branch corresponding to the alternate symmetry breaking except in the region $\omega \approx 1.3$, where the chaotic

regions of the two branches overlap. Figure 4.17 shows a more detailed view
of the initial part of the bifurcation region, revealing a complete cascade of
period doublings.

Slightly beyond the accumulation point of the bifurcation sequence lies a
region of chaotic motion. At $\omega = 1.2902$ we show in Fig. 4.18 the full Poincaré
map, again at zero driving phase. It apparently consists of curves in the $(x–\dot{x})$
plane, but upon closer examination these curves are found to have internal
structures. A closer view of the portion of the attractor within the rectangle
in Fig. 4.18 is shown in Fig. 4.19.

In the present case at $f = 25$, with a further increase in frequency, the
attractor collapses back in a reverse bifurcation sequence and again becomes
simple attractor near $\omega = 1.6$ (See Fig. 4.16).

4.4. Double-Hump Potential

As the third type of the potential V of Eq. (4.2), we now consider the double-
hump Duffing oscillator (Refs. [102, 103]) with $\omega_0^2 > 0$ and $\beta < 0$ in Eq. (4.1):

$$\frac{d^2x}{dt^2} + \alpha\frac{dx}{dt} + \omega_0^2 x - |\beta|x^3 = f\cos\omega t, \quad \omega_0^2 > 0. \tag{4.7}$$

The typical form of the associated $V = (1/2)\omega_0^2 x^2 - (|\beta|/4)x^4$ is shown in
Fig. 4.20. Equation (4.7) has also considerable physical interest and is of relevance in condensed matter physics related to problems such as weakly pinned
charge density waves in anisotropic solids, and superionic conductors. For the
typical case of $\alpha = 0.4, \omega_0^2 = 1.0, |\beta| = 4.0$ and $f = 0.114$, the following picture
emerges from analog and numerical studies (Refs. [102, 103]).

Figure 4.21 shows the amplitude-frequency phase diagram. For decreasing
ω, the chaotic region is preceded by a splitting of the one inversion-symmetric
attractor into two mutually inversion-symmetric attractors (right and left).
The splitting is due to the symmetry rule that forbids inversion-symmetric
attractors from having an even period. Both the attractors undergo period-
doubling cascade to chaos where windows of both even and odd periods occur.
Both the chaotic attractors broaden so that they merge at $\omega \approx 0.5268$ to
form again one attractor, the general form of which is inversion-symmetric (an
interior crisis). After this, due to the symmetry rule only windows with odd
periods occur and their period-doubling are inhibited. The jump occurs when
the chaotic attractor collides with the unstable attractor (a boundary crisis).

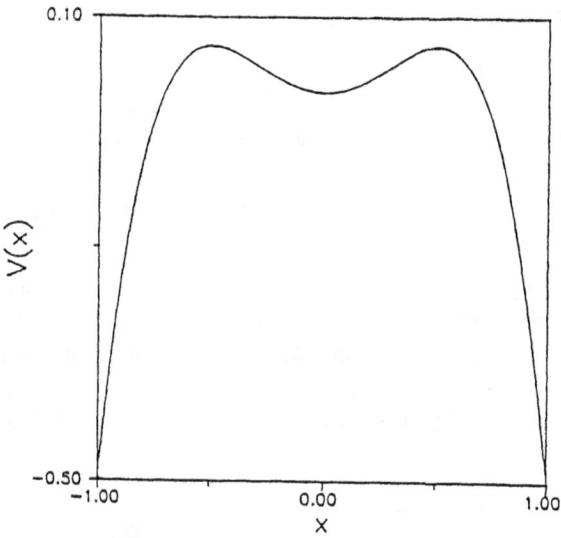

Fig. 4.20. The double-hump potential.

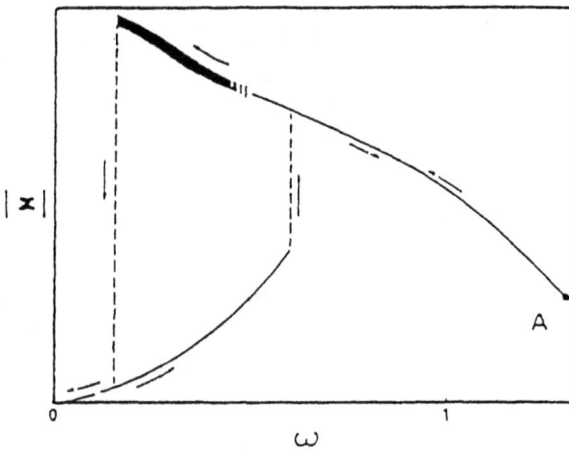

Fig. 4.21. Schematic phase diagram for the anharmonic oscillator of Eq. (4.7). The thin solid lines denote periodic solutions, whereas the thick solid one corresponds to the chaotic state. The short vertical lines denote the set of cascading bifurcations (Ref. [102]).

4.5. Analog Simulation and Experimental Verification

The electronic analog simulator of Eq. (4.1) can be easily constructed, as discussed in Chapter 3, using conventional operational amplifiers and four-quadrant multipliers. Replacing now $\dot{x}(t)(\equiv \frac{dx}{dt})$ and $x(t)$ by

$$\dot{x}(t) = \int \ddot{x}\, dt, \quad x(t) = \int \dot{x}\, dt, \tag{4.8}$$

the equivalent schematic simulation circuit (Ref. [53]) is shown in Fig. 4.22. In this figure the blocks IC_1 and IC_2 represent two integrators, IC_3 is an inverter (sign-changer) and M_1 and M_2 are four-quadrant multipliers. The operational amplifiers used in IC_1, IC_2 and IC_3 are op-amp μA741C and the two multipliers M_1 and M_2 are Analog Devices multiplier AD532 or AD534.

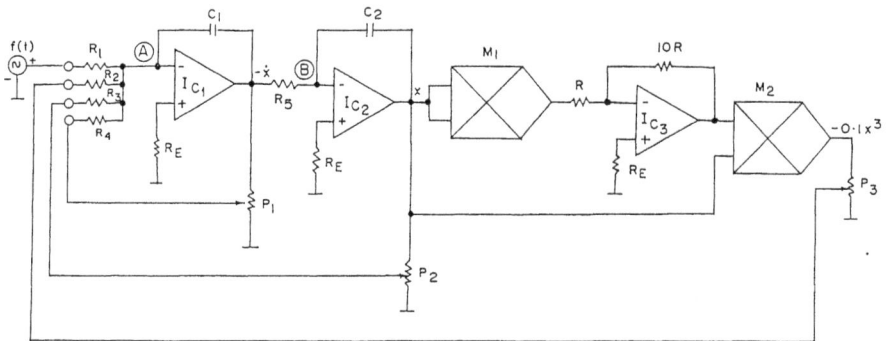

Fig. 4.22. Analog simulation circuit of the driven Duffing oscillator (double-well potential).

Let the output voltage of IC_1 be v_1 and that of IC_2 be v_2. Then, applying Kirchoff's current law at junctions A and B, we obtain the following equations.

Junction A:

$$-C_1\frac{dv_1}{dt} = \frac{f(t)}{R_1} - \frac{0.1p_3 v_2^3}{R_2} + \frac{p_2 v_2}{R_3} + \frac{p_1 v_1}{R_4}, \tag{4.9}$$

where $f(t) = f\sin\omega' t$.

Junction B:

$$-C_2\frac{dv_2}{dt} = \frac{v_1}{R_5}. \tag{4.10}$$

Combining (4.9) and (4.10), we obtain

$$C_1 C_2 R_4 R_5 \frac{d^2 v_2}{dt^2} = \frac{R_4 f(t)}{R_1} - \frac{0.1 R_4 P_3 v_2^3}{R_2} - P_1 R_5 C_2 \frac{dv_2}{dt} + \frac{P_2 R_4 v_2}{R_3}. \quad (4.11)$$

Now choosing for the various resistors and capacitors the values $R_1 = R_3 = R_4 = R_5 = R_6 = 100\text{k}\Omega, R_E = R_2 = R_7 = 10\text{k}\Omega, C_1 = C_2 = 0.01\mu\text{F}$, redefining v_2 as $x(t)$ and rescaling $t \to \tau = R_1 C_1 t$, the Duffing equation for the double-well case of Eq. (4.1) follows, provided the parameters are redefined as

$$\tau = t, \quad p_1 = \alpha, \quad p_2 = |\omega_0|^2, \quad p_3 = \beta, \quad R_1 C_1 \omega' = \omega. \quad (4.12)$$

In the actual experiments, we have fixed the frequency of the external sinusoidal signal at $\frac{\omega'}{2\pi} = 160\text{Hz}$, so that the corresponding redefined frequency $\omega = R_1 C_1 \omega' = (100\text{K}\Omega) \times (0.01\mu\text{F}) \times (2\pi \times 160\text{Hz}) \approx 1.0$. This corresponds to the value of the parameter ω chosen in our numerical analysis in Sec. 4.2. Choosing $\omega_0^2 = -1, p_3 = \beta = 1.0$ and $p_1 = \alpha = 0.5$, the circuit shown in Fig. 4.22 will immediately give the waveforms x and $\dot{x}$ and they can be traced by an oscilloscope to show the trajectory plot in the $(x{-}t)$ plane and the phase-portrait in the $(x{-}\dot{x})$ plane. The experimental results are given

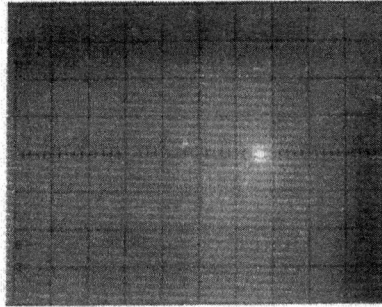

(a)

Fig. 4.23. (a) Phase-portrait denoting the fixed point $(x_1 < 0)$ for $f = 0.0$. (b): (i) Period-$T(= 2\pi/\omega)$ limit cycle for $f = 0.33$; (ii) Lower trace: wave form $x(t)$ of (i); upper trace: external periodic signal. (c): (i) Period-$2T$ limit cycle for $f = 0.35$; (ii) Lower trace: wave form $x(t)$ of (i); upper trace: external periodic signal. (d): (i) Period-$4T$ limit cycle for $f = 0.357$; (ii) Lower trace: wave form $x(t)$ of (i); upper trace: external periodic signal. (e): (i) One-band chaos for $f = 0.37$; (ii) Lower trace: wave form $x(t)$ of (i); upper trace: external periodic signal. (f): (i) Double-band chaos for $f = 0.42$; (ii) Lower trace: wave form $x(t)$ of (i); upper trace: external periodic signal. (g) Period-$3T$ window for $f = 0.664$. (h) Period-T boundary for $f = 0.85$.

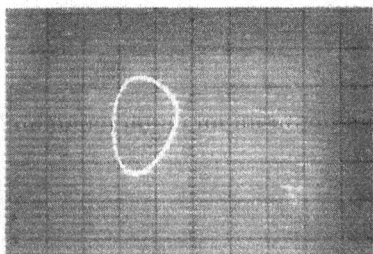 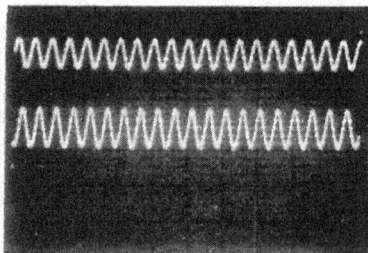

(b) (i)   (b) (ii)

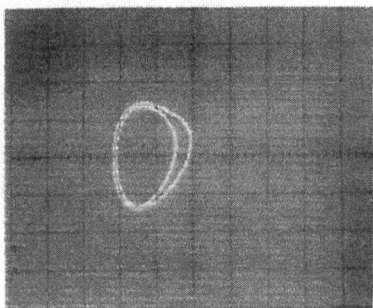 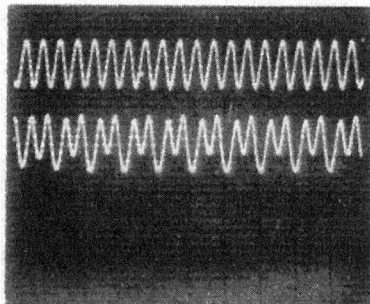

(c) (i)   (c) (ii)

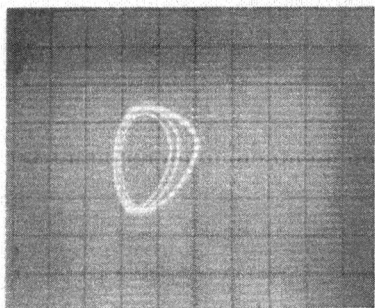 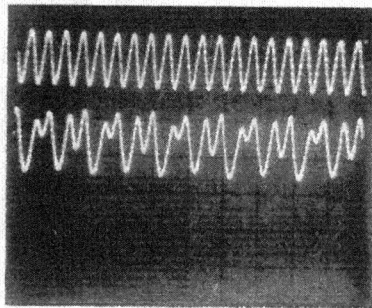

(d) (i)   (d) (ii)

Fig. 4.23. (*Continued*).

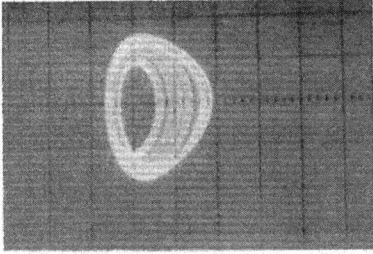

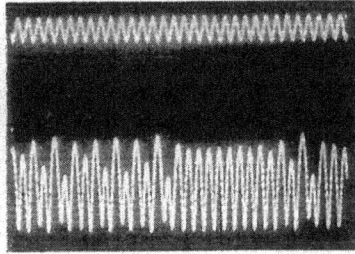

(e) (i)   (e) (ii)

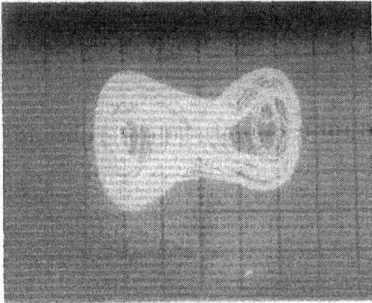

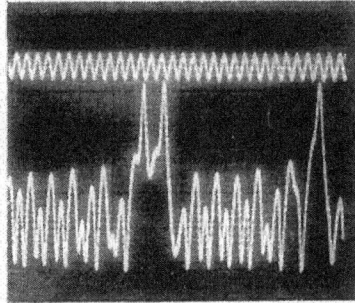

(f) (i)   (f) (ii)

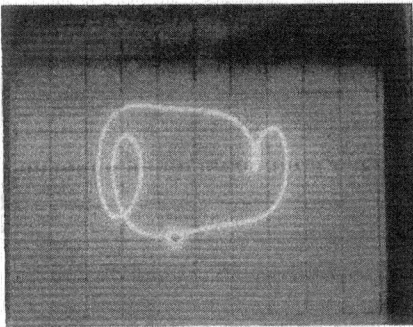

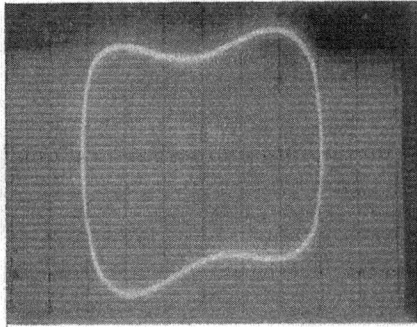

(g)   (h)

Fig. 4.23. (*Continued*).

in Figs. 4.23 which are direct confirmations of the numerical results of Figs. 4.4, thereby complementing each other.

With simple changes in the above circuitry the single-well and double-hump Duffing oscillators can also be analogously simulated and the numerical results confirmed by the experimental results. In fact, in Chapter 4, we will discuss such an analog simulation for the Duffing–van der Pol oscillator which is a generalization of the double-hump Duffing oscillator.

CHAPTER 5

DUFFING OSCILLATOR:
ANALYTIC APPROACHES

In Chapters 3 and 4, we have seen how complex the dynamics of one of the simplest of the nonlinear oscillators is, namely the Duffing oscillator, through detailed numerical analysis and analog simulation studies. As noted, even though no exact analytical treatment can be given to realize all the complex behaviours of the system, useful information can be obtained through approximation methods. In particular, the various perturbation methods (see Appendix A) that are available in the theory of nonlinear oscillations (Ref. [36]) can be profitably employed to obtain the T-periodic solutions and to study their stability properties. One such method, namely the multiple scale method, is employed in Sec. 5.1 to obtain the large T and small T periodic orbits of the double-well Duffing oscillator (4.1) with $\omega_0^2 < 0$. Further, we point out that this approximate analysis can bring out useful criteria for cross-well chaos. In Sec. 5.2 we compare the approximate criteria with that of another well-known method, namely the Melnikov method (Ref. [10]), to estimate the onset of chaos. Finally in Sec. 5.3, we introduce an altogether new analytic method, namely the Painlevé singularity structure analysis of nonintegrable and chaotic systems (Ref. [28]). We show that very complicated clustering of singularities with multisheet structure can arise in the complex t-plane in the chaotic regions of the Duffing oscillator, in contrast to the regular nature of integrable systems.

5.1. Theoretical Analysis of Large-orbit *T*-periodic Solution for the Double-well Duffing Oscillator (Ref. [96])

In Chapter 4, Sec. 4.2.2, we noticed that for the double-well Duffing oscillator (4.1) with $\omega_0^2 < 0$ and $\beta > 0$, the large-orbit (LO) *T*-periodic motion within a wide range of driving frequency is close to a harmonic function of time with frequency ω. One would then expect a first-order approximate solution and its stability analysis might give a good estimation of the system parameter domain in which LOs occur. With this aim in mind, one may seek (Ref. [96]) a *T*-periodic solution of Eq. (4.1) with $(\omega_0^2 < 0, \beta > 0)$,

$$\frac{d^2x}{dt^2} + \alpha\frac{dx}{dt} - |\omega_0^2|x + \beta x^3 = f\cos\omega t\,, \qquad (5.1)$$

which is close to a harmonic function of time as

$$x^{(0)} = A^{(0)}\cos(\omega t + \phi)\,, \qquad (5.2)$$

where $A^{(0)}$ and ϕ are to be determined. Thus we have to transform Eq. (5.1) into the form

$$\ddot{x} + \omega^2 x + \mu f(x, \dot{x}, \omega t) = 0, \quad (\cdot = d/dt) \qquad (5.3)$$

where μ is a small parameter. Since in (5.1) the linearized system (with $\beta = 0$) has a natural frequency which is imaginary, one can look for sufficiently large amplitude oscillations with real frequencies (for $\beta \neq 0$). Therefore setting $f = \alpha = 0$ in Eq. (5.1), we assume the harmonic solution

$$x = a\cos\Omega t\,, \qquad (5.4)$$

where

$$\Omega^2(a) = -|\omega_0^2| + (3/4)\beta a^2 > 0 \quad \text{if } a^2 > (4/3\beta)|\omega_0^2|\,. \qquad (5.5)$$

The condition (5.5) can be obtained by applying the harmonic balance method. Then, one can rewrite Eq. (5.1) as

$$\ddot{x} + \omega^2 x + \mu(\bar{\alpha}\dot{x} + x\sigma - \bar{\Omega}^2 x - |\bar{\omega}_0^2|x + \bar{\beta}x^3 - \bar{f}\cos\omega t) = 0\,, \qquad (5.6)$$

where

$$\mu\bar{\alpha} = \alpha, \ \mu\sigma = \Omega^2 - \omega^2, \ \mu\bar{f} = f, \ \mu|\bar{\omega}_0^2| = |\omega_0^2| = \frac{1}{2}, \ \mu\bar{\beta} = \beta = \frac{1}{2}, \ \mu\bar{\Omega}^2 = \Omega^2\,. \qquad (5.7)$$

Here both the parameters $|\omega_0^2|$ and β are chosen as 0.5 to compare with the numerical results of Sec. 4.2.2.

5.1.1. *Multiple Scale Method and Perturbative Solution for LO*

Now we look for a periodic solution of Eq. (5.6) in the form

$$x(t,\mu) = x_0(T_0,T_1,\ldots) + \mu x_1(T_0,T_1,\ldots) + \ldots, \tag{5.8}$$

where the multiple time scales $T_0 = t, T_1 = \mu t, \ldots$, and

$$\frac{d}{dt} = \sum_{n=0}^{\infty} \mu^n D_n, \quad D_n = \frac{\partial}{\partial T_n}, \tag{5.9}$$

from which (d^2/dt^2) can also be calculated. Then using (5.8) in Eq. (5.6) in the $0(\mu)$ approximation we have

$$D_0^2 x_0 + \omega^2 x_0 = 0, \quad (D_0 = \partial/\partial T_0) \tag{5.10}$$

$$\begin{aligned} D_0^2 x_1 + \omega^2 x_1 = &-2D_0 D_1 x_0 - \bar\alpha D_0 x_0 \\ &-(\sigma - \bar\Omega^2 - |\bar\omega_0^2|)x_0 \\ &-\bar\beta x_0^3 + \bar f \cos\omega t. \quad (D_1 = \partial/\partial T_1) \end{aligned} \tag{5.11}$$

The general solution of Eq. (5.10) can be written as

$$x_0 = A_1(T_1)\exp i\omega T_0 + \text{c.c}, \tag{5.12}$$

where A_1 is complex in general. Then Eq. (5.11) becomes

$$\begin{aligned} D_0^2 x_1 + \omega^2 x_1 = &-[2i\omega(D_1 A_1 + \frac{\bar\alpha}{2}A_1)]\exp(i\omega T_0) \\ &-(\sigma - \bar\Omega^2 - |\bar\omega_0^2|)A_1 \exp(i\omega T_0) \\ &-\bar\beta[A_1^3 \exp(3i\omega T_0) + 3A_1^2 A_1^* \exp(i\omega T_0)] \\ &+(\bar f/2)\exp(i\omega T_0) + \text{c.c.} \end{aligned} \tag{5.13}$$

In order to eliminate secular terms in the solution x_1 of Eq. (5.13), we require that

$$-[2i\omega(D_1 A_1 + \frac{\bar\alpha}{2}A_1) + (\sigma - \bar\Omega^2 - |\bar\omega_0^2|)A_1 + 3\bar\beta A_1^2 A_1^*] + \frac{\bar f}{2} = 0. \tag{5.14}$$

Making the substitution

$$A_1(T_1) = \frac{1}{2}a(T_1)\exp i\phi(T_1) \tag{5.15}$$

in Eq. (5.14) and equating the real and imaginary parts, one obtains the system of coupled differential equations (after using the relation $\Omega^2 \approx -|\omega_0|^2 + \frac{3}{4}\beta a^2$),

$$a' + \frac{\bar{\alpha}}{a} = -\frac{\bar{f}}{2\omega}\sin\phi, \tag{5.16a}$$

$$a\phi' - \frac{\sigma}{2\omega} = -\frac{\bar{f}}{2\omega}\cos\phi. \qquad (' = d/dT_1) \tag{5.16b}$$

For steady-state conditions, $a' = \phi' = 0$ and it follows from (5.16) that

$$a = \frac{f}{\sqrt{(\Omega^2 - \omega^2)^2 + \alpha^2\omega^2}}, \quad \tan\phi = \frac{-\alpha\omega}{(\Omega^2(a) - \omega^2)}, \tag{5.17a}$$

where

$$\Omega^2 = -|\omega_0^2| + \frac{3}{4}\beta a^2 = -\frac{1}{2} + \frac{3}{8}a^2. \tag{5.17b}$$

The first of the equations (5.17a) indeed defines the frequency-amplitude resonance curve in the first order as depicted in Fig. 5.1.

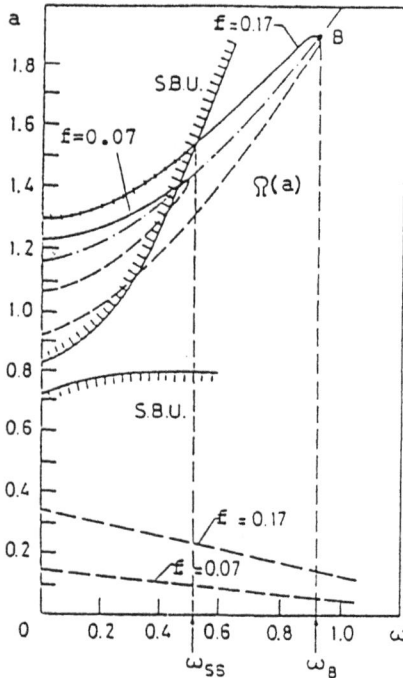

Fig. 5.1. Resonance curves and unstable regions in LO approximate solution: curved line with tic marks: symmetry breaking instability; — — — — : branches unstable in the sense of criterion (5.23); —|—|—|— : branches unstable in the sense of criterion (5.33); _____: stable portion of resonance curves (adapted from Ref. [96]).

Now using the above results and the secularity condition (5.14), Eq. (5.13) can be solved to obtain the first-order correction to the solution of the equation (5.6) as

$$\mu x_1(t) = A_3 \cos 3(\omega t + \phi), \quad A_3 = \frac{\mu \bar{\beta} a^3}{32\omega^2} = \frac{a^3}{64\omega^2}. \quad (5.18)$$

Finally, a refined first-order approximate solution, which describes LO periodic motion, is obtained as

$$x(t) = a \cos(\omega t + \phi) + \frac{a^3}{64\omega^2} \cos 3(\omega t + \phi) = x(t + T). \quad (5.19)$$

5.1.2. Stability of LO Solution

A. *Soft mode instability:*

To examine the stability of the solution (5.19), we may look at a specific form of instability which manifests itself by an exponential growth with time of the harmonic components that are involved in $x(t)$. This can be done by adding small disturbances to the amplitude and phase of the solution (5.19), so that the disturbed solution is written as

$$\tilde{x}(t) = (a + \delta a) \cos(\omega t + \phi + \delta\phi) + \frac{(a + \delta a)^3}{64\omega^2} \cos 3(\omega t + \phi + \delta\phi). \quad (5.20)$$

Using Eq. (5.20) in Eq. (5.1) or (5.6), we obtain the linearized eigenvalue problem in terms of δa and $\delta\phi$:

$$\begin{pmatrix} (\dot{\delta a}) \\ (\dot{\delta\phi}) \end{pmatrix} = \begin{bmatrix} -\alpha/2 & -(a/2\omega)(\Omega^2 - \omega^2) \\ (1/2\omega a)(\Omega^2 - \omega^2) & -\alpha/2 \end{bmatrix} \begin{pmatrix} \delta a \\ \delta\phi \end{pmatrix}. \quad (\cdot = d/dt)$$

$$(5.21)$$

The linear stability of the solution (5.19) then depends upon the eigenvalues of the 2 × 2 coefficient matrix in the right-hand side of Eq. (5.21).

One can easily check (see for example, Ref. [36], p. 172) that the stability limit of the solution (5.19) occurs at those points of the resonance curves $a = a(\omega)$ and $\phi = \phi(\omega)$ given by Eq. (5.17) that have the vertical tangent

$$\frac{da}{d\omega} = \frac{d\phi}{d\omega} = \infty. \quad (5.22)$$

One of these points is shown as B in Fig. 5.1.

Moreover, from the approximate theory of nonlinear oscillations (Refs. [36, 104]) of *first-order or soft-mode instabilities*, it is also known that the unstable branches of the resonance curve are those for which $\frac{da}{d\omega}$ and $\omega - \Omega(a)$ have the same sign:

$$\frac{da}{d\omega} > 0, \quad \omega - \Omega(a) > 0 \tag{5.23a}$$

or

$$\frac{da}{d\omega} < 0, \quad \omega - \Omega(a) < 0. \tag{5.23b}$$

These are also indicated in Fig. 5.1.

B. *Symmetry-breaking instability*:

Considering more general possible instabilities, let us study the effect of small disturbance to $x(t)$ in the form

$$\tilde{x}(t) = x(t) + \delta x(t). \tag{5.24}$$

The linear variational equation for $\delta x(t)$ is then

$$\delta\ddot{x} + \alpha\delta\dot{x} + \left.\frac{\partial W}{\partial x}\right|_{x(t)} \delta x = 0, \quad W = -\frac{1}{2}x + \frac{1}{2}x^3. \tag{5.25}$$

Using the approximate solution (5.19) and expanding $\partial W/\partial x$ into a Fourier series, one can obtain the equation

$$\delta\ddot{x} + \alpha\delta\dot{x} + \left(\lambda_0 + \sum_{n=2,4,6} \lambda_n \cos n\omega t\right)\delta x = 0, \tag{5.26}$$

where

$$\lambda_0 = -\frac{1}{2} + \frac{3}{4}a^2 + \frac{3}{4}a_3^2, \quad \lambda_2 = \frac{3}{4}a^2 + \frac{3}{2}aa_3, \tag{5.27a}$$

$$\lambda_4 = \frac{3}{2}aa_3, \quad \lambda_6 = \frac{3}{4}a_3^2, \tag{5.27b}$$

$$a_3 = \left(\frac{a^3}{64\omega^2}\right). \tag{5.27c}$$

As the time-dependent coefficients in Eq. (5.26) are even harmonics with period $T' = T/2$ by virtue of Floquet's theorem, one can seek particular solutions of the form

$$\delta x(t) = e^{\varepsilon t}\phi(t), \tag{5.28}$$

where ε is real and positive in *unstable* regions and $\phi(t)$ is a periodic function of time with the period $2T' = T$ or $T' = \frac{T}{2}$. Looking for a suitable Fourier series representation for $\phi(t)$, and following the standard procedure (Ref. [96]), one can easily check that while there no period-doubling instability occurs, a buildup of the even harmonics can result leading to *symmetry-breaking* instability.

To examine such a symmetry-breaking instability of the symmetric solution (5.19), we assume a solution of the form

$$\delta x(t) = e_1^{\varepsilon_1 t}(b_0 + b_{21}\cos 2\omega t + b_{22}\sin 2\omega t). \tag{5.29}$$

Using this in (5.26) and applying the harmonic balance method (that is, equating the constant term, coefficients of $\cos 2\omega t$ and of $\sin 2\omega t$ separately to zero), one can obtain a set of linear algebraic equations for b_0, b_{21} and b_{22}. Equating to zero the characteristic determinant, rescaling ε_1 by $\varepsilon_1' - \frac{\alpha}{2}$ and then dropping the prime, we obtain

$$\Delta(\varepsilon_1^2) = \begin{vmatrix} \lambda_0 + \varepsilon_1^2 - \frac{\alpha^2}{4} & \lambda_2/2 & 0 \\ \lambda_2 & -4\omega^2 + \lambda_0 - \frac{\alpha^2}{4} + \varepsilon_1^2 + \frac{\lambda_4}{2} & 4\omega\varepsilon_1 \\ 0 & -4\omega\varepsilon_1 & -4\omega^2 + \lambda_0 - \frac{\alpha^2}{4} + \varepsilon_1^2 - \frac{\lambda_4}{2} \end{vmatrix} = 0. \tag{5.30}$$

At the stability limit, where $\varepsilon_1^2 - (\frac{\alpha^2}{4}) = 0$, the above determinant simplifies to a quadratic polynomial of ω^2:

$$\omega_{SB}^4 - 2B\omega_{SB}^2 + C = 0, \quad B \equiv B(a), \quad C \equiv C(a). \tag{5.31}$$

Equation (5.31) gives us the desired relationship between ω and the amplitude a to be satisfied at the symmetry-breaking instability limit:

$$\omega_{SB} = \omega_{SB}(a). \tag{5.32}$$

By expanding the determinant (5.30) in powers of small values of $\varepsilon_1^2 - (\frac{\alpha^2}{4})$, one can readily establish that the unstable region corresponds to

$$\varepsilon_1^2 - \left(\frac{\alpha^2}{4}\right) > 0, \quad \Delta(\alpha^2) < 0. \tag{5.33}$$

The above analysis indicates that the resonant branch of $a = a(\omega)$, which seems to be stable in the sense of the criterion (5.23), is unstable in the sense of the symmetry-breaking instability criterion (5.33) for $\omega < \omega_{SB}$ (as shown in Fig. 5.1). Figure 5.2 depicts both the stability limits in the $(f-\omega)$ plane: the first-order stability limit, which coincides with point B in Fig. 5.1 and the symmetry-breaking stability limit defined by Eq. (5.31), against the background of numerical results (compare with Fig. 4.7). One observes that the theoretical $\omega_B(f)$ values are very close to the true boundary of existence of large orbit motion. On the other hand, the theoretical symmetry-breaking stability limit ω_{SB} overestimates the values of driving frequency, for which symmetry-breaking instability really occurs. It is also clear that the above theoretical analysis cannot give reasonable results in the region of f, ω parameters, where the two stability limits approach each other.

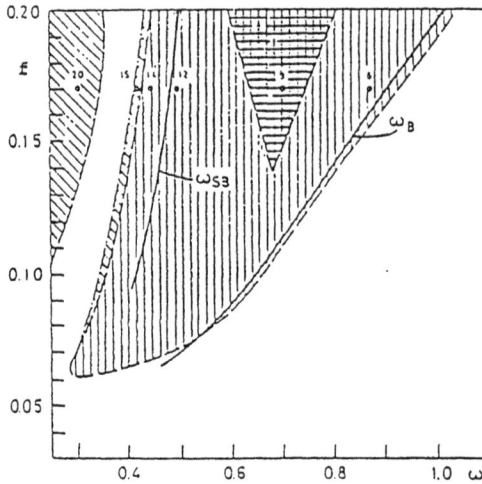

Fig. 5.2. Regions of LO attractor: computer simulation and theoretical stability limits: $\alpha = 0.1$ (Ref. [96]).

5.2. SO Motion in the Double-well Duffing Oscillator and Its Theoretical Analysis (Ref. [96])

Numerical analysis of Eq. (5.1), as discussed in Sec. 4.2, clearly shows that the small orbit (SO) motion in the neighbourhood of the principal resonance is also very close to harmonic functions of time with frequency ω, even for high values of the forcing parameter $(f > F_2)$. This is especially valid for values of the driving which are outside the zone defined by ω_A and ω_{PDB} (see Fig. 4.10), where ω_A is the cyclic fold bifurcation point and ω_{PDB} is the first period-doubling bifurcation.

Consequently, we may consider a T-periodic solution of Eq. (5.1) in the right well which is approximately of the form

$$x^{(0)} = A^{(0)} \cos(\omega t + \phi) \tag{5.34}$$

and deduce the instability regions, and obtain approximate criteria for steady-state cross-well chaotic motion, as we have done for the LO motion in the previous section.

In order to apply the multiple-scale perturbation method, we transform Eq. (5.1) into the form

$$\ddot{x} + \omega^2 x + \mu \bar{\omega}_1 x^2 + \mu^2 (\bar{\alpha}\dot{x} + \bar{\omega}_3 x^3 + \sigma x - \bar{f}\cos \omega t) = 0, \qquad (5.35a)$$

where

$$\mu \bar{\omega}_1 = \frac{3}{2}, \ \mu^2 \bar{\omega}_3 = \frac{1}{2}, \ \mu^2 \bar{\alpha} = \alpha, \ \sigma \mu^2 = 1 - \omega^2, \ \mu^2 \bar{f} = f, \ \mu \ll 1. \qquad (5.35b)$$

Now using the multiple-scale perturbation method used in Sec. 5.1.1, the $0(\mu)$ solution can be seen to satisfy the equation

$$D_0^2 x_1 + \omega^2 x_1 = -2i\omega D_1 A \exp(i\omega T_0) - \bar{\omega}_1 [A^2 \exp(2i\omega T_0) + 2AA^*] + c.c.,$$
$$D_0 = \partial/\partial T_0, \quad D_1 = \partial/\partial T_1, \qquad (5.36)$$

where we have used the zeroth order solution in the form

$$x_0 = A(T_1, T_2) \exp(i\omega T_0) + c.c.. \qquad (5.36a)$$

Eliminating the secular terms, the correct perturbative solution becomes

$$x_1 = \bar{\omega}_1 [-2AA^* + (A^2/3) \exp(2i\omega T_0)] + c.c. \qquad (5.37)$$

Similarly the $0(\mu^2)$ equation, on using (5.37), becomes

$$D_0^2 x_2 + \omega^2 x_2 = -[2i\omega(D_2 A + \bar{\alpha}A/2) + \sigma A + (3\bar{\omega}_3 - 10\bar{\omega}_1^2/3)A^2 A^*$$
$$- \bar{f}/2]e^{i\omega T_0}$$
$$-(2/3)\bar{\omega}_1^2 A^3 \exp(3i\omega T_0) - A\bar{\omega}_3 \exp(3i\omega T_0). \qquad (5.38)$$

Again eliminating the secular terms and using the steady-state conditions, we obtain

$$a' = -\bar{\alpha}a/2 - (\bar{f}/2\omega)\sin \phi, \qquad (5.39a)$$
$$\phi' = \sigma a/2w + (3\bar{\omega}_3 - 10\bar{\omega}_1^2/3)(a^3/8\omega) - (\bar{f}/2w)\cos \phi, \quad ('= d/dT_2)(5.39b)$$

where we have used the substitution

$$A = (1/2)a\exp(i\phi). \qquad (5.39c)$$

Consequently the amplitude and phase solutions become

$$a = \frac{f}{\sqrt{\Omega^2(a) - \omega^2 + \alpha^2\omega^2}}, \qquad \tan\phi = \frac{-\alpha\omega}{\Omega^2(a) - \omega^2}, \qquad (5.40a)$$

where

$$\Omega^2(a) = 1 - ka^2, \quad k = \frac{5}{6}\mu^2\bar{\omega}_1^2 - \frac{3}{4}\mu^2\bar{\omega}_3 = \frac{3}{2}. \qquad (5.40b)$$

Finally the second-order approximate solution for SO motion near the principal resonance $\omega \approx 1$ becomes

$$x(t) = a\cos(\omega t + \phi) - \frac{3}{4}a^2 + \frac{1}{4}a^2\cos 2(\omega t + \phi) + (a^3/16)\cos 3(\omega t + \phi). \qquad (5.41)$$

A. *First-order instability*: The natural frequency $\Omega(a)$ is decreasing with the amplitude, and consequently the resonance curves $a = a(\omega)$ are bent to the left as shown in Fig. 5.3. They preserve the classic shape and possess the peak amplitude at the point B for which the condition similar to (5.22) holds: $da/d\omega = \infty$, unless $f > F_2$. The theoretical limit value of the forcing parameter for point B to exist at small damping is

$$F_1 \approx 1/\sqrt{4k} \approx 1/\sqrt{6}. \qquad (5.42)$$

For $f > F_1$, the resonance curve $a \equiv a(\omega)$ looks like those for an undamped system.

The above analysis demonstrates that the first-order stability limits as given by the criteria (5.23) exist on the nonresonant branch only (point A in Fig. 5.3, and the whole resonant branch seems to be "stable" as per (5.23)).

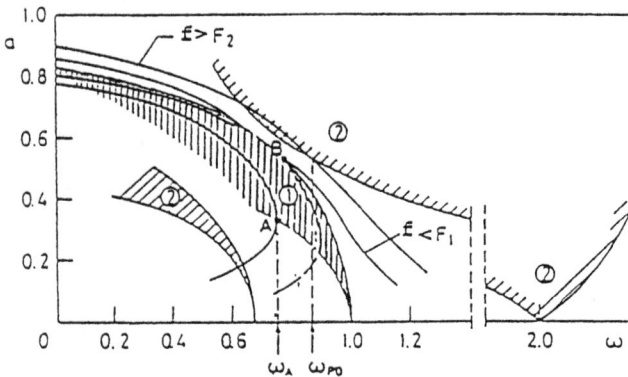

Fig. 5.3. Resonance curves for $\alpha = 0.1$ and two types of unstable regions of SO solution for $\alpha = 0$: (1) first-order instability; (2) period-doubling instability (Ref. [96]).

B. *Period-doubling instability:* To examine other forms of instability of the T-periodic solution (5.41), we consider the Hill's type variational equation (cf. Eq. (5.26))

$$\delta\ddot{x} + \alpha\delta\dot{x} + \left(\lambda_0 + \sum_{n=1}^{4}\lambda_n\cos n\omega t\right)\delta x = 0, \quad \lambda_i = \lambda_i(a). \tag{5.43}$$

It is clear that the periods of the periodic coefficients in Eq. (5.43) is equal to the period of $x(t)$: $T_1 = T$. Again an application of Floquet's theorem now ensures that a period-doubling instability corresponding to a growth of the harmonic components of period $2T$ of $\phi(t)$ in the perturbation,

$$\delta x(t) = e^{(\varepsilon_1 - \frac{\alpha}{2})t}\phi(t),$$

can occur. To examine this instability, one can assume a two-term solution for $\delta x(t)$

$$\delta x(t) = e^{[\varepsilon_1 - (\frac{\alpha}{2})]t}\sum_{n=1,3,5}\left(b_{nc}\cos\frac{n\omega t}{2} + b_{ns}\sin\frac{n\omega t}{2}\right). \tag{5.44}$$

Following the procedure of the previous section, we arrive at a fourth-order equation for the stability limit ω_{PD}:

$$\omega_{PD}^{8} + b_6\,\omega_{PD}^{6} + b_4\,\omega_{PD}^{4} + b_2\,\omega_{PD}^{2} + b_0 = 0, \quad b_i = b_i(a). \tag{5.45}$$

Figure 5.3 depicts the resonance curves $a \equiv a(\omega)$ and the two types of unstable regions: the first-order instability defined by Eq. (5.23) and period-doubling unstable regions given by (5.45). The two-term perturbative solution (5.44) gives rise to two period-doubling unstable regions: one of which is evaluated from the ω-axis at $\omega = 2$, and the other at $\omega = \frac{2}{3}$. At low amplitudes they correspond to the $\frac{1}{2}$ subharmonic resonance and $\frac{3}{2}$ supersubharmonic resonance, respectively. One also observes that the period-doubling instability also visits the principal resonance region and the stability boundary crosses the resonant branch of $a \equiv a(\omega)$, if the forcing parameter f exceeds certain critical value $f > F_2$.

One thus concludes from the approximate theory that the T-periodic solution (5.41) considered is unstable within the region of the driving frequency $\omega_A < \omega < \omega_{PD}$ and no other SO stable solution is possible. Figure 5.4 depicts the theoretical stability boundary defined by the frequencies ω_A and ω_{PD} in the $(f-\omega)$ plane, which is then compared with the numerical results. The above approximate analysis is seen to give surprisingly good estimation of the system critical parameter values, values for which cross-well chaos really occurs.

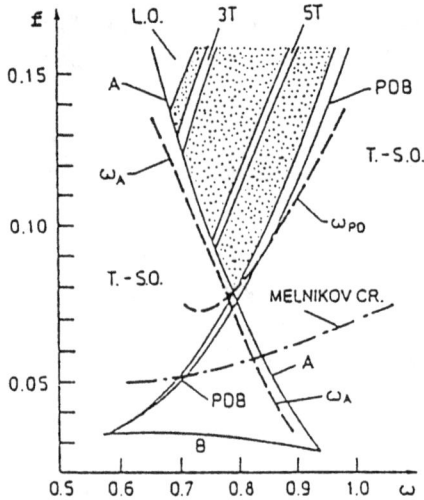

Fig. 5.4. Regions of SO attractor: computer simulations and theoretical stability limit. Shaded area: cross-well chaos (Ref. [96]).

5.3. The Melnikov Criterion

Another useful analytic method which often leads to a lower threshold criterion for the onset of chaotic motion is the Melnikov method, which has been well discussed in the literature (Refs. [10, 32]). In this method one investigates typical ways in which a fixed point of an unperturbed system breaks down and the onset of complicated motion when a weak perturbation sets in. An interesting case is the so-called Smale-horseshoe chaos, or simply horseshoe chaos, corresponding to the transverse intersection of the stable and unstable manifolds of a saddle-point (in the Poincaré-surface of section). The Melnikov method, which analytically determines the regions of the parameter space where horseshoe chaos occurs, is essentially to calculate the Melnikov integral.

In general, the existence of the horseshoe chaos does not imply that the trajectories will be asymptotically chaotic. However, orbits created by the horseshoe mechanism display an extreme sensitive dependence on initial conditions and possibly exhibit either a chaotic transient before settling to stable orbits or a strange attractor (Refs. [10, 32]). In many dynamical systems, namely the Duffing oscillator, the periodically forced pendulum, the periodically forced Morse oscillator, etc., the onset of chaos has been found to occur near the Melnikov threshold curve, and so it is often considered as a lower threshold for the onset of chaos.

We will therefore evaluate the threshold condition for occurrence of horse-shoe chaos for the double-well Duffing oscillator and compare it with the predictions of the analytic perturbation theory discussed in the previous Sec. 5.2 for SO cross-well chaos (Fig. 5.4). As the method is well known, we will give only a brief outline of it. We consider essentially a system of two odes (for convenience) of the form

$$\frac{d\vec{X}}{dt} = h_0(\vec{X}) + \varepsilon h_1(\vec{X}, t),\tag{5.46}$$

where $\vec{X} = (x_1, x_2)^T$, $h_0 = (f_0, g_0)^T$ and $h_1 = (f_1, g_1)^T$, satisfying the following conditions:

a) When $\varepsilon = 0$, the system has two equilibrium points $\vec{X}_0$ and $\vec{X}_s$, which are centre and saddle, respectively.

b) $(f_0(\vec{X}), g_0(\vec{X}))$ and $(f_1(\vec{X}, t), g_1(\vec{X}, t))$ are analytical in $\vec{X}$ in a sufficiently large neighbourhood of $\vec{X}_0$.

c) For $\varepsilon = 0$, (5.46) possesses a homoclinic orbit $\vec{X}_h(t) = (x_{1h}(t), x_{2h}(t))^T$, which is an orbit connecting the saddle to itself.

d) The functions f_1 and g_1 are periodic in t, that is, $f_1(t + 2\pi) = f_1(t)$ and $g_1(t + 2\pi) = g_1(t)$.

Then it has been shown [see (Ref. [10]) for details] that the stable and unstable manifolds intersect transversely if the Melnikov integral

$$M(t_0) = \int_{-\infty}^{\infty} \left\{ h_0[\vec{X}_h(t - t_0)] \wedge h_1[\vec{X}_h(t - t_0), t] \right\}$$

$$\times \exp\left\{ -\int_0^{t-t_0} \text{Trace } D_x h_0[\vec{X}_h(s)]ds \right\} dt\tag{5.47}$$

has a simple zero, independent of ε, and if $\frac{dM}{dt} \neq 0$ at $t = t_0$.

Rewriting now the Duffing oscillator equation (5.1) with a double-well in the form (with $|\omega_0^2| = \frac{1}{2}$ and $\beta = \frac{1}{2}$)

$$\frac{d^2x}{dt^2} - \frac{1}{2}x + \frac{1}{2}x^3 = \varepsilon\left[-\alpha\frac{dx}{dt} + f\cos\omega t \right],\tag{5.48}$$

or

$$\frac{dx}{dt} = y,\tag{5.49a}$$

$$\frac{dy}{dt} = \frac{1}{2}x - \frac{1}{2}x^3 + \varepsilon[-\alpha y + f\cos\omega t), \quad \varepsilon \ll 1,\tag{5.49b}$$

and using (5.46) and (5.47), the Melnikov integral for the present case becomes

$$M(t_0) = \int_{-\infty}^{\infty} y_h(\tau)[f\cos\omega(T + t_0) - \alpha y_h(\tau)]d\tau, \qquad (5.50)$$

where the homoclinic orbit is given by

$$\vec{X}_h(t) = (x_h(t),\ y_h(t)) = (\sqrt{2}\operatorname{sech}(\tau/\sqrt{2}),\ -\operatorname{sech}(\tau/\sqrt{2})\tanh(\tau/\sqrt{2})),$$
$$\tau = (t - t_0). \qquad (5.51)$$

Evaluating the integral (5.50) by using (5.51), we obtain

$$M(t_0) = 4\pi\omega f\operatorname{sech}(\pi\omega/\sqrt{2})\sin\omega t_0 - 4\sqrt{2}\alpha/3. \qquad (5.52)$$

Then the Melnikov criterion for the occurrence of horse-shoe chaos is

$$\alpha \leq \alpha_c = \frac{3}{4} f\omega\operatorname{sech}(\frac{\pi\omega}{2}). \qquad (5.53)$$

In Fig. 5.4, the Melnikov threshold curve for the damping parameter $\alpha = 0.1$ in the $(f\text{–}\omega)$ plane is also shown. It is clear that in this case the chaotic regime is rather well within the Melnikov criterion region and the approximate theory discussed in the previous section gives a closer bound for the chaotic region.

5.4. Analytic Structure of the Duffing Oscillator

An altogether different approach to investigating the nonintegrability aspects of the Duffing oscillator and other dynamical systems is the study of analytic structure of singularities of the solutions in the complex time plane, popularly known as the Painlevé singularity structure analysis (Refs. [25, 27, 28]). This method was originally applied by S. Kovalevskaya to identify integrable cases of the dynamics of a rigid body in a gravitational field. Recent applications to find integrable cases include many nonlinear coupled oscillators and related systems realized after the advent of soliton theory (Ref. [25, 27, 28]). For nonintegrable cases, while the procedure itself is in an evolving stage, very interesting aspects have already been gleaned out of such analysis (Refs. [105–108]). In this section, we give a brief introduction to these aspects with Duffing oscillator as a typical example.

5.4.1. *Movable Singularities and Nonlinear Dynamical Systems*

Considering the solutions of ordinary differential equations in the complex time plane $(t \in C^1)$, one may identify the *singularities or singular points* as those points at which the solutions cease to be analytic. They include poles, branch points and essential singularities. These singularities can be either fixed at some point of the complex t-plane or *movable* (that is, they depend on the integration constants). Of all the singularities, branch points and essential singularities are classified as *critical points*.

One can easily check that the solutions of linear differential equations can admit only fixed singularities determined by the form of the differential equations. Examples: Singularities of the solutions of Legendre, Bessel, Hermite, etc. differential equations. On the other hand, solutions of nonlinear odes can also admit movable singularities (in addition to fixed singularities). Some simple examples are given in Table 5.1. Thus the distinguishing feature of nonlinear odes is that the movable singularities exhibited by their solutions can be placed anywhere in the complex t-plane since they are dependent on the initial conditions.

Table 5.1.

Equation	Solution	Type of movable singularity (t_0)
$\frac{dx}{dt} + x^2 = 0$	$x = (t - t_0)^{-1}$	pole
$\frac{dx}{dt} + x^3 = 0$	$x = [2(t - t_0)]^{-(1/2)}$	branch point (square root)
$\frac{dx}{dt} - e^{-x} = 0$	$x = \ln(t - t_0)$ (t_0: integration constant)	branch point (logarithmic)

In the mathematics literature, there has been considerable development in the classification of odes according to the nature of singularities admitted by them, that is, movable critical or not (Ref. [109]). Considering the first-order odes $\frac{dx}{dt} = F(x,t)$, where F is rational in x and analytic in t, the only equation which is free from movable critical points is the generalized Riccati equation. Similarly, for second-order odes of the form $\frac{d^2x}{dt^2} = F(x, \frac{dx}{dt}, t)$, where F is rational in ($\frac{dx}{dt}$), algebraic in x and analytic in t, Painlevé and coworkers have shown that there exist 50 canonical types which are free from movable critical points. Among these 50 equations, 44 are integrable in terms of elementary functions including elliptic functions and the remaining six require new transcendental functions, namely the Painlevé transcendents. Partial classifications for higher order, especially third-order odes, are also available in the literature (Ref. [25]). In recent times it has been shown that soliton possessing nonlinear partial differential equations on similarity reduction to odes lead to solutions free from movable critical points (Refs. [25, 110]). Such a procedure of isolating differential equations free from movable critical points (to within certain allowed transformations) has led to the identification of a large number of integrable nonlinear dynamical systems such as specific parametric choices of coupled nonlinear oscillators, Lorenz systems, and so on (Refs. [27, 28]).

Then the question naturally arises as to what will be the nature of singularities exhibited by nonintegrable and chaotic systems, particularly for the damped and forced nonlinear oscillators discussed in this book. In particular one finds that the Duffing oscillator, the damped and forced pendulum, Morse oscillator, and so on are nonintegrable variants of second order Painlevé transcendental equations. The solutions of these equations have been found to exhibit an immensely complicated, multi-armed, infinite-sheeted structure of singularities, in contrast to the regular and ordered pattern of singularities in the case of integrable and regular dynamical systems. We will briefly demonstrate these aspects for the Duffing oscillator equation, taking it as a typical example.

5.4.2. Singularity Clustering in the Duffing Oscillator

Considering the Duffing oscillator equation

$$\frac{d^2x}{dt^2} + \alpha\frac{dx}{dt} + \omega_0^2 x + \beta x^3 = f\cos\omega t, \qquad (4.1)$$

we investigate the nature of the singularities by making a local Laurent expansion in the neighbourhood of a movable singular point t_0 (which is arbitrary) in the form

$$x(t) = \sum_{j=0}^{\infty} a_j \tau^{j+q}, \quad \tau = t - t_0 \to 0 \qquad (5.54)$$

so that the leading order is

$$x(t) \approx a_0 \tau^q . \qquad (5.55)$$

Using (5.45) in Eq. (4.1) and equating the dominant terms we find that

$$q = -1, \quad a_0^2 = -\frac{2}{\beta}, \quad a_0 = \pm i\sqrt{2/\beta}, \qquad (5.56)$$

and therefore t_0 represents a movable pole.

Proceeding with the full series (5.54), along with (5.56), substituting it in (4.1), and equating the dominant terms we further obtain

$$a_1 = -\frac{1}{6}\alpha a_0, \quad a_2 = \frac{1}{36}(6\omega_0^2 - \alpha^2)a_0 , \qquad (5.57a)$$

$$a_3 = -\frac{1}{108}(2\alpha^2 - 9\omega_0^2)\alpha a_0 - \frac{1}{4}f\cos\omega t_0, \tag{5.57b}$$

$$0.a_4 - \frac{1}{27}(2\alpha^2 - 9\omega_0^2)\alpha^2 a_0 - f(\alpha\cos\omega t_0 - \omega\sin\omega t_0) = 0, \tag{5.57c}$$

and so on.

In order that the Laurent expansion locally represents the general solution of the Duffing equation (4.1) (which is of second order in nature), the series (5.54) should admit two arbitrary constants and it should converge in a suitable sense. The series already has an arbitrary constant t_0, which defines the location of the singular point itself. The second arbitrary constant can only be a_4 as may be seen from Eq. (5.57c). However, this is possible only for the following choices of parameters: (i) $f = 0$ and $\alpha = 0$, (ii) $f = 0, \alpha = \pm(3/\sqrt{2})\omega_0$. The first choice corresponds to a free, undamped nonlinear oscillator whose solution can be given in terms of elliptic functions (as discussed in Chapter 2, Sec. 2.2). The second choice corresponds again to the integrable cases of the free damped oscillator discussed in Sec. 2.2. For all other choices of the parameters, Eq. (5.57c) is not compatible and the Laurent series (5.54) with (5.56) corresponding to a movable pole does not represent the correct form of the solution, which can now admit a movable logarithmic branch point at t_0.

To examine this possibility for the general case of the Duffing equation (4.1), following the method of Fournier, Levine and Tabor (Ref. [105]), we introduce the logarithmic psi series

$$x(t) = \sum_{j=0}^{\infty}\sum_{k=0}^{\infty} a_{jk}\tau^{j-1}(\tau^4 \ln\tau)^k, \quad \tau = (t - t_0) \to 0 \tag{5.58}$$

into the equation of motion. Equating the coefficients of powers of $\tau^{j+4k-3}(\ln\tau)^k$, we obtain the recursion relation for the coefficients a_{jk}'s as

$$(j + 4k - 1)(j + 4k - 2)a_{jk} + (k + 1)(2j + 8k - 3)a_{j-4,k+1}$$
$$+(k + 1)(k + 2)a_{j-8,k+2} + \alpha(j + 4k - 2)a_{j-1,k}$$
$$+\alpha(k + 1)a_{j-5,k+1} + \omega_0^2 a_{j-1,k} + \beta \sum_{\substack{p,q \\ r,s}} a_{j-r,k-s}a_{r-p,s-q}a_{pq}$$

$$= fT_{j-3}\delta_{0k}, \quad 0 \le p \le r \le j, \quad 0 \le q \le s \le k, \tag{5.59a}$$

where

$$T_n = \frac{1}{n!}\frac{\partial^n}{\partial t^n}(\cos\omega t)|_{t=t_0}. \tag{5.59b}$$

One can easily check by analysing the recursion relation (5.59) systematically that the coefficient a_{40} is arbitrary, while the other coefficients are given

uniquely. Thus the psi series is indeed a suitable local representation of the
solution in the neighbourhood of the now movable critical point t_0.

In order to show that the analytic structure of singularities in the nonintegrable case of Eq. (4.1) is indeed complicated, one has to analyse the nature of
the coefficients a_{jk} as given by the recursion relation (5.59) and the full series
(5.58). However, useful informations can be extracted by considering a subset
of the recursion relation corresponding to the coefficients a_{0k} (i.e. $j = r = 0$
in Eq. (5.59)). It reads as

$$(4k - 1)(4k - 2)a_{0k} + \beta \sum_{s,q} a_{0,k-s}a_{0,s-q}a_{0q} = 0.\tag{5.60}$$

Introducing now the generating function

$$\Theta(z) = \sum_{k=0}^{\infty} a_{0k}z^k, \ z = \tau^4 \ln \tau,\tag{5.61}$$

we can deduce the associated differential equation as

$$16z^2\frac{d^2\Theta}{dz^2} + 4z\frac{d\Theta}{dz} + 2\Theta + \beta\Theta^3 = 0.\tag{5.62}$$

One can indeed argue that Eq. (5.62) corresponds to the rescaled version of
Eq. (4.1) in the neighbourhood of the movable singularity t_0.

Remarkably Eq. (5.62) can be explicitly integrated. By making the trans-
formation

$$\Theta(z) = \xi g(\xi), \quad \xi = z^{1/4},\tag{5.63}$$

one finds that g satisfies the differential equation

$$\frac{d^2g}{d\xi^2} + \beta g^3 = 0.\tag{5.64}$$

On integrating Eq. (5.64) and identifying the integration constant appropriately, we find that

$$g(\xi) = \left[\frac{2i}{3^{3/4}\sqrt{\beta}}\right]\left\{(A - iB)^{1/4}\right\}ds\left[\frac{\sqrt{2}}{3^{3/4}}(A - iB)^{1/4}\xi\right],$$

where

$$A = 2\alpha^2(2\alpha^2 - 9\omega_0^2),\tag{5.65a}$$
$$B = 27\sqrt{2}f \sqrt{\beta}(\alpha \cos \omega t_0 - \omega \sin \omega t_0).\tag{5.65b}$$

The elliptic function $dsu = \frac{dnu}{snu}$ has standard poles with real and imaginary periods $4K(\frac{1}{2})$ and $4iK'(\frac{1}{2})$, where K and K' are complete elliptic integrals of the first kind and their associate respectively. The modulus square of the elliptic function is $1/2$ in the present case. Since $K = K'$, the poles lie on a square lattice in the u-plane with a spacing $2K(\frac{1}{2})$.

Using Eq. (5.55), we can obtain the pole positions of $g(\xi)$ in the ξ-plane as

$$\xi_{lm} = (108)^{1/4}K(\frac{1}{2})(A - iB)^{-1/4}(l + im), \quad l, m \in \mathbb{Z}, \qquad (5.66)$$

where l and m are lattice site integers. The corresponding singularity positions in the z-plane (see Fig. 5.5) are then

$$z_{lm} = 108[K(\frac{1}{2})]^4 \cdot \frac{(AC - BD) + i(AD + BC)}{A^2 + B^2}, \qquad (5.67)$$

where

$$C = (l^2 - m^2)^2 - 4l^2m^2, \quad D = 4lm(l^2 - m^2).$$

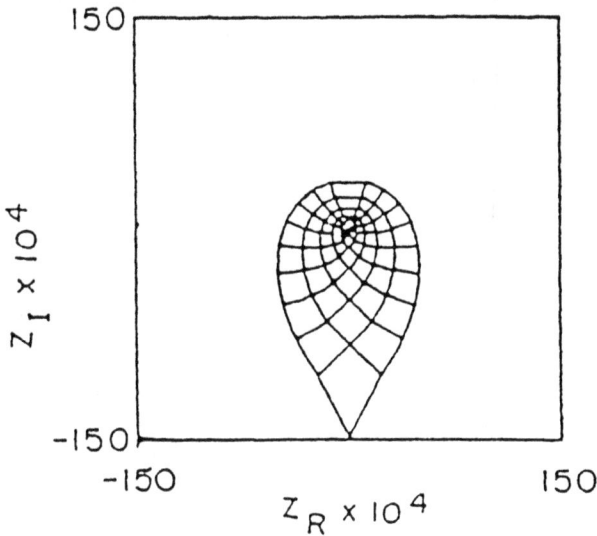

Fig. 5.5. Singularity structure of Eq. (4.1) in the complex z-plane (Eq. (5.67)) for $\omega_0^2 = 0.01, \alpha = 3\omega/\sqrt{2}, \beta = 10.0, f = 1.5, \omega = 1.0$, and $l, m = -10, -9, \ldots, -1, 0, 1, \ldots, 9, 10$.

Now to see how the singularity pattern, given by (5.67) in the z-plane, can be mapped back to the complex t-plane, we set the singularity position, for convenience, at $t_0 = 0$ in Eq. (5.58) and work with

$$z = t^4 \ln t. \tag{5.68}$$

By using polar coordinates in both the z- and t-planes,

$$z = \rho e^{i\phi}, \quad t = r e^{i\theta}, \tag{5.69}$$

we can easily check that

$$r = \exp[-(\theta + 2\pi n) \cot(4\theta - \phi)] \tag{5.70}$$

and so

$$\rho = -(\theta + 2\pi n) \exp[-4(\theta + 2\pi n) \cot(4\theta - \phi)].\mathrm{cosec}(4\theta - \phi), \tag{5.71}$$

where n is the Riemann sheet index in the t-plane. Equations (5.70) and (5.71) completely determine the mapping $z \to t$. Then a given pole in the ξ-plane (and so the z-plane) maps to four points in the t-plane for each sheet. The cumulative effect for large positive n (with the mirror image for negative n) is a four-arm 'star-like' structure as shown in Fig. 5.6. One notices that these arms have a slight twist, and the singularities become more densely packed along each arm as they approach the centre of the star as n increases.

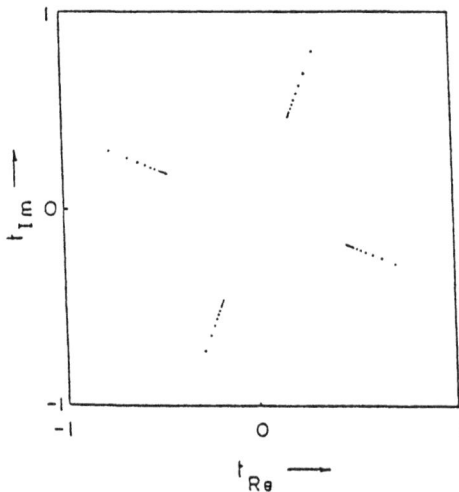

Fig. 5.6. Local singularity structure in the complex t-plane in the neighbourhood of a typical singularity, determined from the analytic mapping (5.70) and (5.71) for $\omega_0^2 = 0.01, \alpha = 3\omega_0/\sqrt{2}, \beta = 10.0, f = 1.5$ and $\omega = 1.0$.

The above local analysis can be confirmed further by carrying out a numerical analysis of the Duffing equation (4.1) in the complex t-plane (Ref. [105]) using the ATOMCC integrator or its refined version ATOMFT 2.51(1991) developed by Chang and Corliss (see for example (Ref. [111])). With this program it is possible to integrate an ordinary differential equation on any piecewise linear path in the complex t-plane. At each step the program reports the solution and the position and order of the nearest singularity. By a suitable choice of the integration paths one can build up a picture of the complex t-plane on any desired scale.

Figure 5.7 shows a portion of the complex t-plane for the completely integrable case ($f = 0, \alpha = 0$, undamped free oscillator) giving the expected square lattice of poles. Figure 5.8 depicts a typical t-domain picture for damped but free oscillator, ($\alpha \neq 0, f = 0$ in Eq. (4.1)). As damping α is increased the singularities are 'swept' away from the axis as shown in Fig. 5.8. To leading order the singularities are first-order poles but, as expected, demonstrably multivalued when repeated circuits are made about them. However, they do not exhibit any clustering.

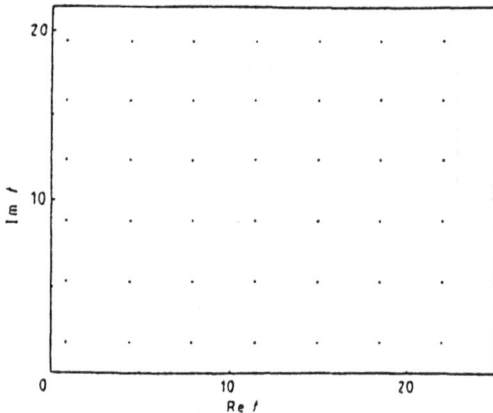

Fig. 5.7. Square lattice of poles found in typical solution of free undamped Duffing oscillator ($\alpha = 0, f = 0$).

Now as the driving term is added the regular lattice gradually gets distorted into a complicated structure of clustering, leading to the global complex t-domain structure as shown in Fig. 5.9. Further, the structure shows the presence of 'chimney-like' patterns observed originally by Bountis *et al.* (Ref. [106]), and the singularities shown in Fig. 5.9 forming the main 'tunnel' structures are all on the lowest Riemann sheet. The chimney-like pattern seems to be characteristic of non-integrable systems.

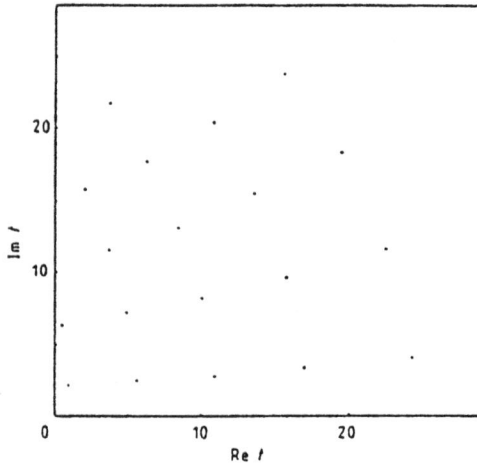

Fig. 5.8. Singularity distribution in the complex t-domain of a solution to the damped oscillator ($f = 0$ and $\alpha = 0.1$).

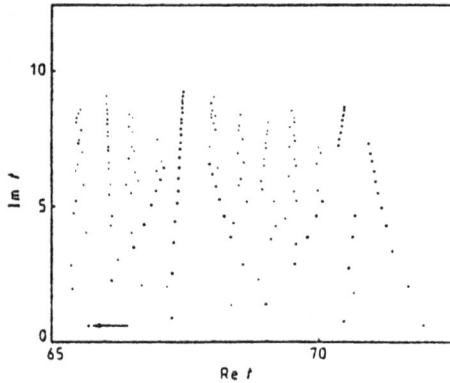

Fig. 5.9. Typical singularity distribution in complex t-domain for a solution of the undamped Duffing oscillator (4.1) with $\beta = 5.0$. Accentuated poles (heavy dots) indicate commonly observed 'tunnel' structures (adapted from Ref. [105]).

One can now look at the local structure in the neighbourhood of a given singularity in Fig. 5.9, by making repeated circuits of the 'central' singularity. One obtains in general a star-like structure, as was expected from the earlier analytic mapping (cf. Eqs. (5.69)–(5.71)). A typical local singularity structure in the neighbourhood of a given singularity from numerical analysis and analytical mapping is given in Fig. (5.10) (Ref. [105]).

(a)

(b)

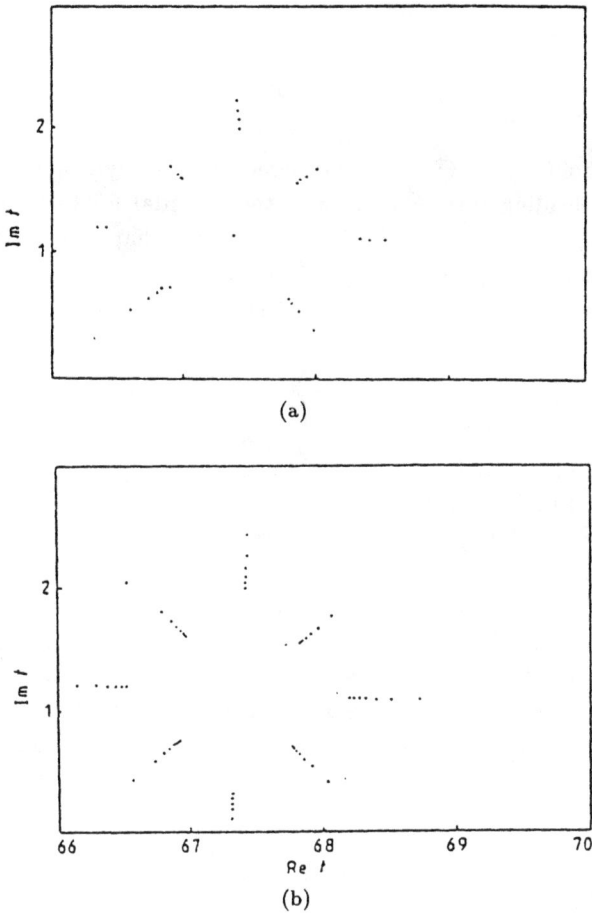

Fig. 5.10. (a) Typical local singularity structure, found using ATOMCC for damped, driven oscillator (4.1) with $\alpha = 0.1, \beta = 5.0$. (b) Corresponding structure determined by analytical mapping (Ref. [105]).

Finally it is of interest to note that the equation for the generating function $\Theta(z)$(Eq. (5.62)) can also be obtained by the substitution

$$x(t) = \frac{1}{\tau}\Theta_0(z), \quad z = \tau^4 \ln \tau, \quad \tau = t - t_0 \tag{5.72}$$

into Eq. (4.1). In the limit $t \to t_0$, it can be easily seen (Ref. [105]) that $\Theta_0(z)$ again satisfies equation (5.62), provided that there is an ordering in which $|t - t_0| \ll |z|$. Then (5.72) can be seen to be just the first term in an asymptotic expansion of the form

$$x(t) = \sum_{j=0}^{\infty} \Theta_j(z)\tau^{j-1} , \tag{5.73}$$

where the whole set of Θ_j's can be determined analytically in a recursive fashion. The leading term Θ_0 captures the essential nonlinearities, and the subsequent Θ_j's ($j \geq 1$) satisfy linear equations that can be integrated in terms of Lame functions (Ref. [105]). If the above expansion is compared with the original psi series (5.58), we can see that each Θ_j is just

$$\Theta_j(z) = \sum_{k=0}^{\infty} a_{jk} z^k , \tag{5.74}$$

which are the generating functions for the coefficient sets $a_{jk}(j, k = 0, 1, 2, .. \infty)$. Thus the series (5.73) constitutes a systematic resummation of the psi series (5.58) and the singularity structure appears to be the same in all orders of this expansion.

The above analysis indicates then that there is a close connection between nonintegrability, chaos and clustering of singularities in the Duffing oscillator equation, even though no precise one-to-one correspondence with real-time dynamics has been established. We also note that the exact convergence properties of the various psi-series expansions discussed above are still unknown. In spite of these, it appears that much relevant and useful informations can be extracted regarding nonintegrable and chaotic nonlinear systems using singularity structure analysis. Further interesting aspects will be discussed in the next chapter (Sec. 6.3) in connection with the analytic structure of the Duffing–van der Pol oscillator.

CHAPTER 6

BIFURCATION, CHAOS AND PHASE-LOCKING IN BVP AND DVP OSCILLATORS

In the earlier chapters, 2–5, we have made a detailed investigation of the bifurcation and chaos aspects of the Duffing oscillator from numerical, circuit theoretical and analytical investigations. In this chapter, we wish to extend our considerations to other physically and biologically interesting oscillator systems. In particular, we introduce in Sec. 6.1 the Bonhoeffer–van der Pol (BVP) oscillator (Ref. [112]), which can be considered as a generalization of both the Duffing oscillator and the well-known van der Pol oscillator (Ref. [113]) (see Appendix B). It is pointed out that apart from the familiar period-doubling bifurcations leading to chaotic motions, the system also exhibits resonance or *phase-locking* phenomena when external constant and periodic forces are applied (Refs. [17, 114]). An interesting feature associated with phase-locking is that a plot of the winding number versus the driving frequency consists of flat steps, the so-called *devil's staircase*, corresponding to different winding numbers. Then, we show that the BVP oscillator can under simple transformations be considered as equivalent to a Duffing–van der Pol (DVP) oscillator and discuss its dynamics (Sec. 6.2) in some detail. Finally, in Sec. 6.3, we give the salient features of the singularity structure analysis of the DVP oscillator, and show how it exhibits globally an infinite-sheeted structure of singularities in spite of the fact that its solution has locally square-root type movable branch point singularity.

6.1. The Bonhoeffer–van der Pol Oscillator

The nerve system in living organisms is an information processing system with external and internal inputs and outputs. External information is received from the environment through sense organs as well as from other parts of an organism. The nerve system consists of a huge number of constituents called nerve cells or neurons. Nerve cells are compartments filled with electrolyte surrounded by a membrane and embedded in extracellular fluid. In a neural axon there are circulating currents that cross the membrane and flow lengthwise inside and outside the axon and the membrane current and potential vary with distance as well as with time. The circulating currents provide the mechanism of conduction.

One can set up a model of the axon by combining the equations for an excitable membrane with the differential equations for an electrical core conductor cable, assuming the axon to be an infinitely long cylinder. A well-known approximation suggested by FitzHugh and Nagumo *et al.* (for details, see Refs. [112, 115]) to describe the propagation of voltage pulse $V(x,t)$ along the membranes of nerve cells is the set of coupled partial differential equations

$$V_{xx} - V_t = F(V) + R - I, \qquad (6.1a)$$
$$R_t = c(V + a - bR), \qquad (6.1b)$$

where $R(x,t)$ is the recovery variable, I is the membrane current, and a, b, c are the membrane radius, specific resistivity of the fluid inside the membrane and temperature factor, respectively. Here the suffices stand for partial differentiation.

When the spatial variation of V, namely $V_{xx'}$, is negligible, Eq. (6.1) reduces to a set of two coupled first-order ordinary differential equations known as the Bonhoeffer–van der Pol oscillator (Ref. [112]) system,

$$\dot{V} = V - V^3/3 - R + I, \qquad (6.2a)$$
$$\dot{R} = c(V + a - bR), \quad \left(\cdot = \frac{d}{dt} \right) \qquad (6.2b)$$

with $F(V) = -V + V^3/3$. Normally the constants in Eq. (6.2) satisfy the inequalities $b < 1$ and $3a + 2b \geq 3$, though from a purely mathematical point of view this may not be insisted upon. Various theoretical and numerical analyses show that the cubic form for $F(V)$ is reasonably adequate from a physiological point of view. Further, as seen from actual experimental considerations, it is of great importance to investigate the dynamics of the BVP oscillator under the influence of both constant and periodic external stimuli. In the following we give a brief account of the salient features based on Ref. [17].

6.1.1. *Bifurcations, Chaos and Phase-locking in the BVP Oscillator*

With a periodic membrane (ac) current $A_1 \cos \omega t$ along with a dc-bias A_0, the BVP oscillator (6.2) can be written as

$$\dot{V} = V - V^3/3 - R + A_0 + A_1 \cos \omega t, \tag{6.3a}$$
$$\dot{R} = c(V + a - bR). \tag{6.3b}$$

To start with, we study the dynamics of the BVP oscillator (6.3) in the absence of external force and then investigate the effect of a periodic current in conjunction with the bias A_0. As noted earlier, a two-dimensional autonomous system does not show chaotic behaviour. Consequently, the only attractors that the BVP oscillator, in the absence of periodic force, can admit are point attractors and periodic attractors. Thus the qualitative behaviour of the solution of (6.3) with $A_1 = 0$ can be examined by a linear stability analysis in a small neighbourhood of the stationary points.

A. *Linear stability analysis of a force-free system*: The fixed points of (6.3) in the absence of external stimuli ($A_0 = A_1 = 0$) are determined by equating $\dot{V} = \dot{R} = 0$, which then gives

$$V^3 - qV - p = 0, \tag{6.4}$$
$$V + a - bR = 0, \tag{6.5}$$

where $p = -3a/b$ and $q = 3(b-1)/b$. Roots of Eqs. (6.4)–(6.5) are the fixed points of (6.3). The cubic Eq. (6.4) has three real roots for $27p^2 < 4q^3$ and one real and two complex roots for $27p^2 > 4q^3$. Let (V_0, R_0) be a fixed point of (6.4)–(6.5). To examine the linear stability of this fixed point, we assume as usual that

$$V = V_0 + \varepsilon \xi(t), \tag{6.6a}$$
$$R = R_0 + \varepsilon \eta(t), \qquad \varepsilon \ll 1. \tag{6.6b}$$

Substituting (6.6) in (6.3) and to order ε we have

$$\begin{pmatrix} \dot{\xi} \\ \dot{\eta} \end{pmatrix} = \begin{pmatrix} 1 - V_0^2 & -1 \\ c & -bc \end{pmatrix} \begin{pmatrix} \xi \\ \eta \end{pmatrix} = J \begin{pmatrix} \xi \\ \eta \end{pmatrix}, \qquad J = \begin{pmatrix} 1 - V_0^2 & -1 \\ c & -bc \end{pmatrix}. \tag{6.7}$$

Fixed points are obviously classified into distinct types, depending on the eigenvalues λ_1 and λ_2 of the Jacobian matrix, satisfying the characteristic equation

$$\det(J - \lambda I) = 0, \tag{6.8}$$

where I is the unit matrix. From (6.7) we obtain the quadratic equation

$$\lambda^2 + a_1\lambda + a_2 = 0, \tag{6.9}$$

where $a_1 = V_0^2 + bc - 1$ and $a_2 = c(1 + bV_0^2 - b)$. Thus the eigenvalues are given as

$$\lambda_{1,2} = [-a_1 \pm (a_1^2 - 4a_2)^{1/2}]/2. \tag{6.10}$$

For a fixed point to be a stable node both the eigenvalues must be negative; for a stable focus the eigenvalues must be complex conjugate of each other with negative real part. Hence a fixed point is stable only if all the eigenvalues have a negative real part. Thus for the fixed point (V_0, R_0) to be stable, the necessary and sufficient condition, from (6.10), is $a_1 > 0$ and $a_2 > 0$, that is,

$$(V_0^2 + bc - 1) > 0 \quad \text{and} \quad c(1 + bV_0^2 - b) > 0. \tag{6.11}$$

In other words, for a given set of parameters the fixed points of (6.3) can be obtained from (6.4)–(6.5), while their stability can be identified from (6.11).

Now let us consider the stability of the fixed points for specific choices of the parametric values. For $a = 0.7, b = 0.8$ and $c = 0.1$, which are the values used by FitzHugh, the system can have only one real fixed point $(V_0, R_0) = (-1.1995, -0.6244)$. From (6.10) the eigenvalues of this fixed point are found to be $(-0.2594 \pm i0.26603)$. So the fixed point is a stable focus. It is interesting to note that for some parametric choices the BVP equations (6.3) can admit a saddle fixed point whose presence is found to be crucial in the mechanism of onset of chaos in weakly perturbed dynamical systems. One such parametric choice is $a = 0, b = 1.5, c = 0.2$, for which there are three fixed points

$$(V_0, R_0) = (0,0), (\pm 1, \pm 0.667). \tag{6.12}$$

For the fixed point $(0,0)$, from (6.10), we find $\lambda_1 = -0.1217$ and $\lambda_2 = 0.8217$ and so it is a saddle, while the remaining two fixed points are stable spiral points because the corresponding eigenvalues are $-0.015 + i0.4213$ and $-0.15 - i0.4213$, respectively. In Fig. 6.1 we have plotted the numerically computed trajectories for various initial conditions, where one clearly sees the attraction of trajectories to one of the stable spiral points (A or B) and the divergence of orbits, with an initial condition taken in the neighbourhood of the saddle S, from the saddle point.

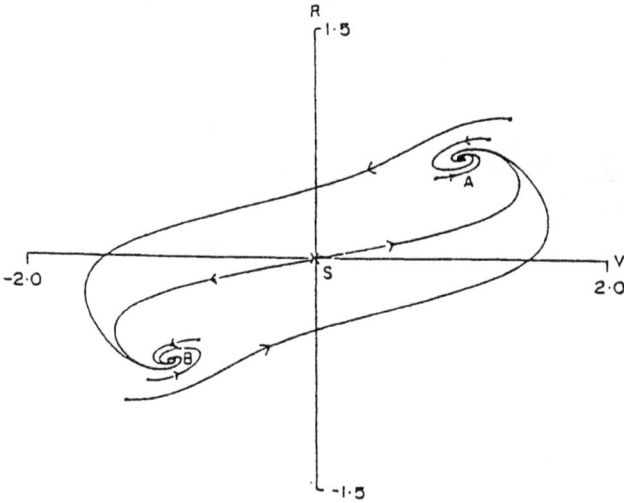

Fig. 6.1. The phase plane of the BVP equation for $a = 0, b = 1.5, c = 0.2$. A and B: Stable spiral points. S: Saddle fixed point.

B. *Linear stability analysis under influence of bias*: In this subsection we extend the fixed point analysis to the BVP oscillator with a constant membrane current A_0. Proceeding as in the force-free case with $A_1 = 0$ and $A_0 > 0$, the fixed points (V_0, R_0) are now roots of the equations

$$V^3 - qV - p = 0, \qquad (6.13a)$$
$$V + a - bR = 0, \qquad (6.13b)$$

where $q = 3(b - 1)/b$ and $p = 3(A_0 - a/b)$. Since the bias term A_0 does not explicitly occur in the linearized equation, the stability-determining eigenvalue equation turns out to be exactly the same as Eq. (6.7) obtained in the absence of A_0. Further, Eq. (6.11) is again the necessary and sufficient condition for stability of the fixed point of the dc-driven BVP oscillator with V_0 being the solutions of (6.13). Hence using (6.10) one can study the behaviour of the system in the neighbourhood of the fixed points obtained from (6.13).

Moreover, in order to understand the influence of A_0 on the dynamics of the BVP oscillator, we now look for changes in the stability property of the fixed points into limit-cycle motion through a Hopf bifurcation. Naturally for a Hopf bifurcation to occur in Eq. (6.3) both the eigenvalues must cross the imaginary axis, so that we have the condition

$$P(\lambda) = (\lambda + i\beta)(\lambda - i\beta)$$
$$= \lambda^2 + \beta^2 = 0, \tag{6.14}$$

where β is the imaginary part of the eigenvalues λ_1 and λ_2. Comparing (6.14) with (6.9) we find that

$$a_1 = V_0^2 + bc - 1 = 0 \tag{6.15a}$$

and

$$\beta^2 = a_2 = c(1 + bV_0^2 - b) > 0. \tag{6.15b}$$

From (6.15a) we have

$$V_{0\pm} = \pm(1 - bc)^{1/2}, \quad \beta^2 = c(1 - b^2c) > 0. \tag{6.16}$$

Using the above values of $V_{0\pm}$ in (6.13a), we obtain

$$A_{0\pm} = V_{0\pm}^3/3 - (b-1)V_{0\pm}/b + a/b. \tag{6.17}$$

Thus there are two values of A_0, A_{0-} and A_{0+}, with $A_{0-} < A_{0+}$, at which the eigenvalues are pure imaginary. Further analysis shows that for $A_0 < A_{0-}$ or $A_0 > A_{0+}$ the eigenvalues have negative real parts. Thus in these ranges the fixed point is asymptotically stable, being either a node or a spiral point. For $A_0 \in (A_{0-}, A_{0+})$ the eigenvalues have positive real parts so the fixed point is unstable and one may expect a stable limit cycle in this range.

We now demonstrate the existence of a limit cycle for a specific choice of the parameters, namely $a = 0.7, b = 0.8, c = 0.1$. For $A_0 > 0$ the system has only one real fixed point (V_0, R_0) determined by (6.16). For $0 < A_0 < 0.341$, the eigenvalues associated with this fixed point are complex with negative real part and the fixed point corresponds to a stable spiral (Fig. 6.2(a)). At $A_0 = 0.341$, the real part of the eigenvalues vanishes and the system undergoes a Hopf bifurcation. For $0.341 < A_0 < 1.397$, the real part is positive and the fixed point is an unstable spiral. However, as shown in Figs. 6.2(b) and 6.2(c) a stable limit cycle exists in this range. At $A_0 = 1.397$ the real part of the eigenvalues again vanishes and a second Hopf bifurcation occurs. For $A_0 > 1.397$ the real part is negative and the fixed point becomes a stable spiral. Thus the BVP oscillator, in the absence of the periodic membrane current, can exhibit stationary solution or limit cycle motion depending on the magnitude of the dc current, when the other parameters are held fixed.

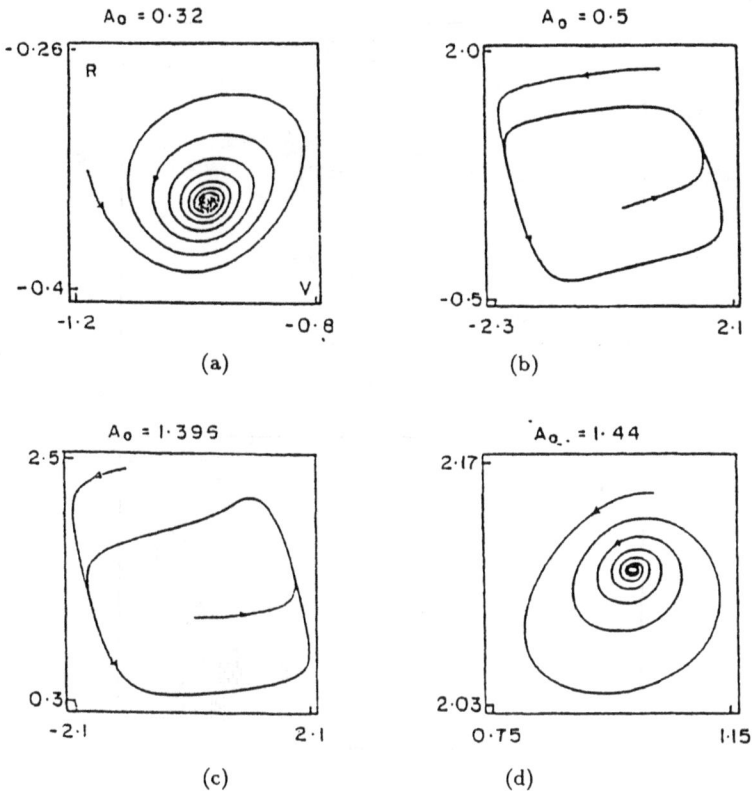

Fig. 6.2. Phase-portraits of BVP Eq. (6.3) for various A_0 values. The other parameters are fixed at $a = 0.7, b = 0.8, c = 0.1$, and $A_1 = 0$.

C. *Responses to periodic input current: bifurcations and chaos*: We now investigate numerically the effect of periodic current $A_1 \cos \omega t$ while keeping the bias $A_0 = 0$. For our study we again fix the parameters at $a = 0.7, b = 0.8, c = 0.1$, $\omega = 1$, and vary A_1 from 0 upwards. As before, numerical integration of Eq. (6.3) has been performed using a fourth-order Runge-Kutta method with a time step of $(\frac{2\pi}{\omega})/200$. Figures 6.3 show the bifurcation phenomenon in the interval $A_1 \in (0, 1.8)$, $A_1 \in (0.6, 0.8)$ and $A_1 \in (1.0, 1.3)$ separately. From Fig. 6.3(b) it is clear that for $0 < A_1 < 0.648$ there is a limit-cycle attractor of period $T = (\frac{2\pi}{\omega})$. At $A_1 = 0.648$, period-doubling bifurcation occurs and a period $2T$ limit cycle develops. Further period doublings occur at $A_1 = 0.7, 0.714$ and 0.717 giving rise to $4T, 8T$ and $16T$ period limit

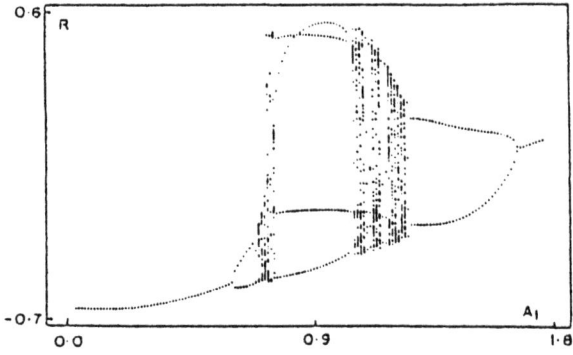

(a)

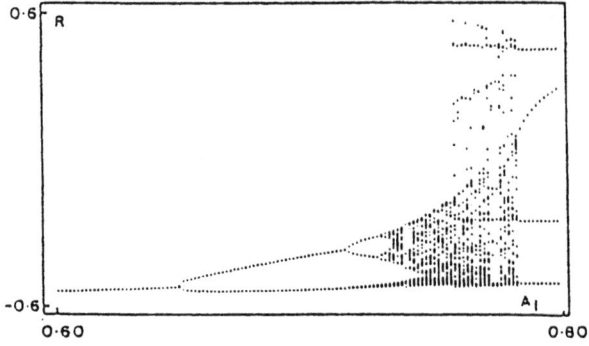

(b)

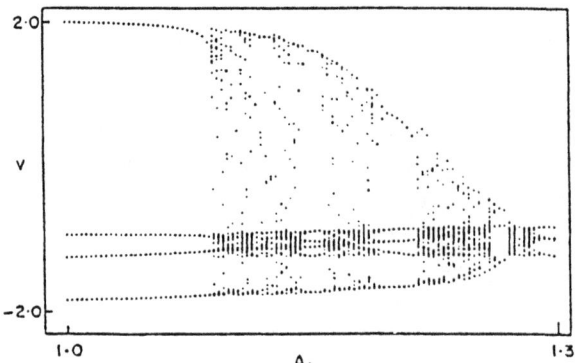

(c)

Fig. 6.3. Bifurcation diagram of the BVP Eq. (6.3) in the ranges (a) $A_1 \in (0, 1.8)$, (b) $A_1 \in (0.6, 0.8)$, and (c) $A_1 \in (1.0, 1.3)$.

(b)

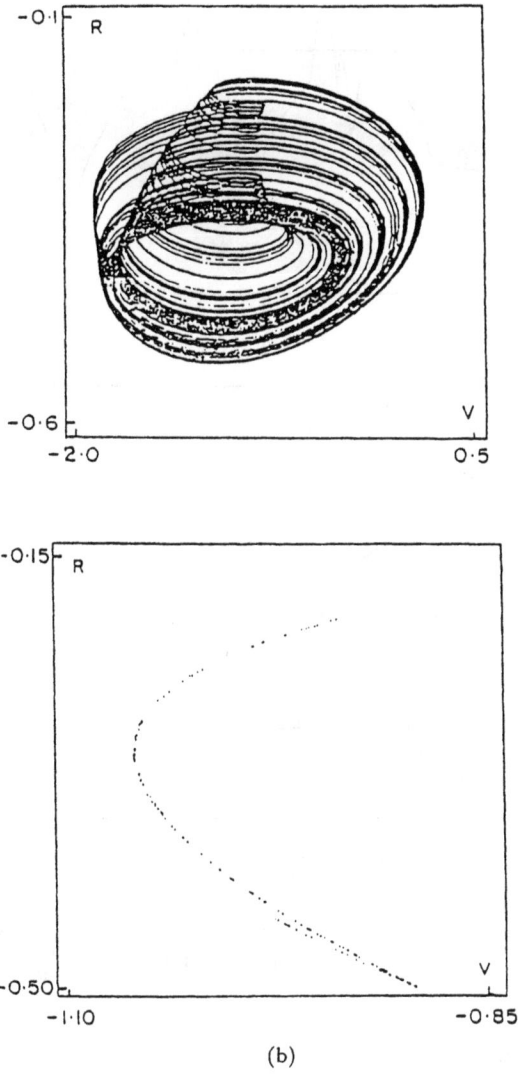

Fig. 6.4. (a) Chaotic orbit in (V,R) plane for $A_1 = 0.74$, $A_0 = 0.0$ (b) Poincaré map of (a).

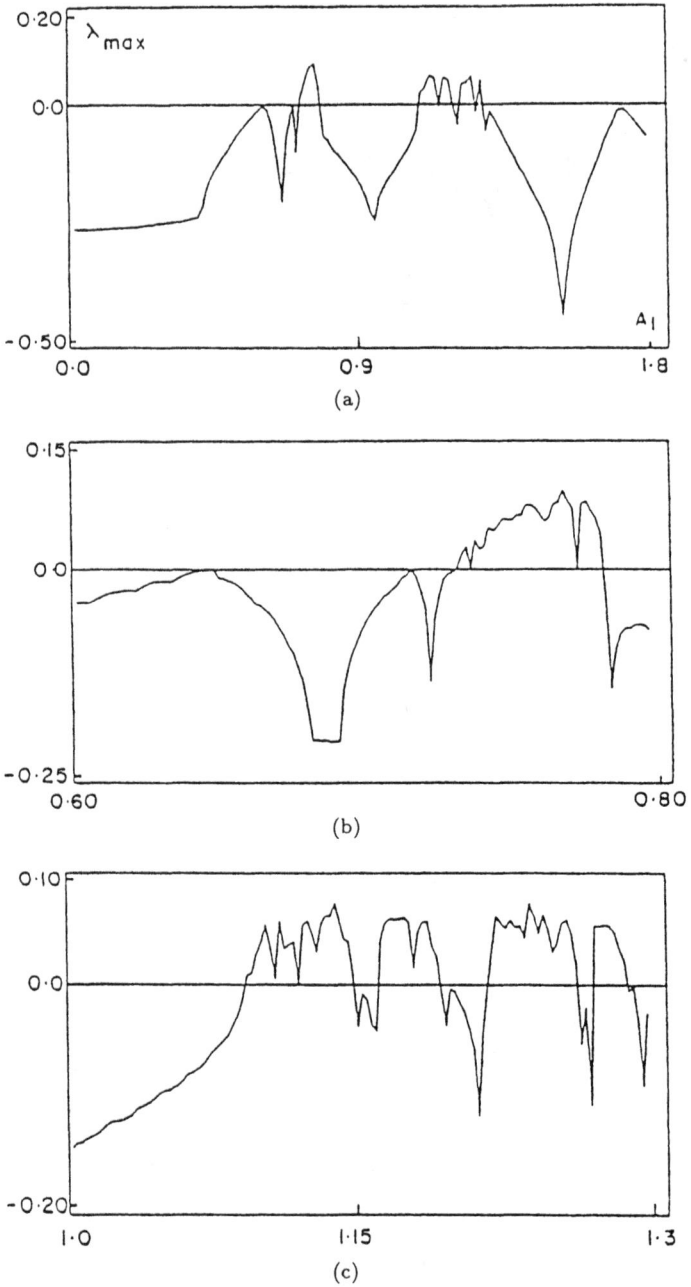

Fig. 6.5. The largest Lyapunov exponent against A_1 corresponding to Figs. 6.3.

cycles, respectively. This sequence is found to accumulate at $A_{1c} \approx 0.7182$ leading to a fairly satisfactory value for the Feigenbaum ratio $\delta = 4.667$. Figure 6.4(a) shows the chaotic orbit in the (V, R) plane for $A_1 = 0.74$ and Fig. 6.4(b) shows the corresponding Poincaré map, while Figs. 6.5 depicts the associated largest Lyapunov exponent against A_1. Further in the chaotic regime, there are a number of intermittency regions of type I as noted in Ref. [116], especially when crisis-type transition occurs. Some of the possible ranges of intermittent routes to chaos identified are $A_0 = 0$, $\omega = 1$, $A_1 \in$ (0.771375, 0.7717), (0.780, 0.78144), (1.09184, 1.094), (1.12198, 1.1225) and so on.

6.1.2. *Mode-locking and Complete Devil's Staircase Structures*

Having studied the effects of bias and periodic membrane current separately, we now discuss the mode-locking phenomenon under the simultaneous action of both the currents.

A. *Frequency or mode-locking*: In general, one would expect a forced nonlinear oscillator to exhibit oscillations which occur at frequencies corresponding to natural frequencies of the oscillator (ν_n) and the frequency of the driving force (ν_d). The resulting motion of the oscillator will then have the appearance of beating. As the frequency of the driving force approaches the natural frequency of the oscillator, beating disappears and there remains only the frequency ν_d, that is, the frequency ν_n is locked or entrained by the driving frequency. This is the phenomenon of *frequency-locking*, or *phase-locking*, or *mode-locking* (Refs. [36, 113]), and it plays an important role in synchronization of clocks, electric motors and so on.

B. *Winding number and devil's staircase*: In typical damped and driven nonlinear oscillator systems, it has been found (Ref. [117]) that if one plots the frequency of the applied force against the frequency of the oscillator the resulting curve may consist of an infinity of steps, the devil's staircase, where the ratio of their frequencies known as the winding number

$$W = \nu_d/\nu_n \,, \tag{6.18}$$

attains rational values. This interesting phenomenon has been observed in many dynamical systems (Ref. [117]) such as the one-dimensional circle map, Josephson junction simulators, Belousov–Zhabotinsky reaction, and in the driven electrical conductivity of barium sodium niobate crystals and so on.

C. *Complete devil's staircase in the BVP oscillator*: In the following we discuss the observed mode-locking in the BVP oscillator. With only A_0 applied (and $A_1 = 0$) to Eq. (6.3) the system has a frequency ν_0 and the corresponding driving frequency is $\nu_d = A_0\nu_0$. When A_1 is also switched on, the frequency of oscillation changes to a different value ν_0'. The winding number is the ratio of ν_d and ν_0'. We have studied the frequency-locking in Eq. (6.3) with $A_1 = 0.74$ by varying the driving frequency. Since the system has a stable focus for $0 < A_0 < 0.341$, as noted earlier the winding number for this interval is 0. For $A_1 > 0.341$, by numerical integration of (6.3) at different values of A_0 we have calculated both ν_0 and ν_d from the relation $\nu_d = A_0\nu_0$ and ν_0'. Using (6.19) the corresponding winding numbers can be calculated. We have found evidence for frequency-locking of (6.3) at several rational values of W. In Fig. 6.6 we have plotted the winding number as a function of A_0, the so-called devil's staircase. From Fig. 6.6 we find that the frequency locked intervals with different rational values fill up the whole A_0 axis.

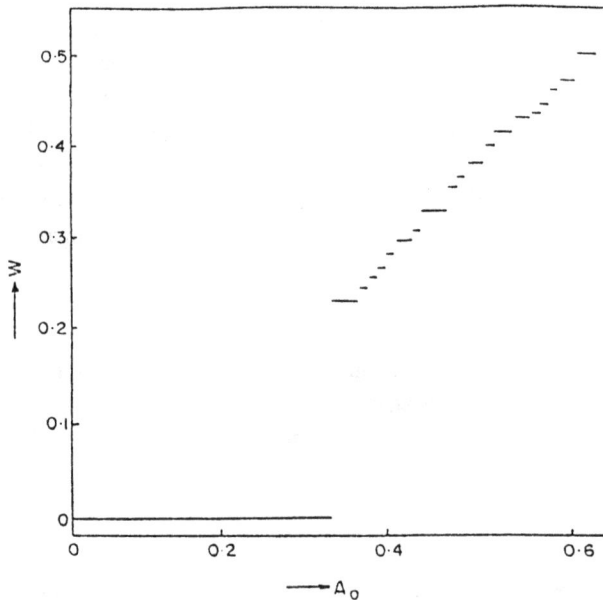

Fig. 6.6. The mode-locking structure at $A_1 = 0.74$ for the BVP oscillator (6.3). The devil's staircase is complete.

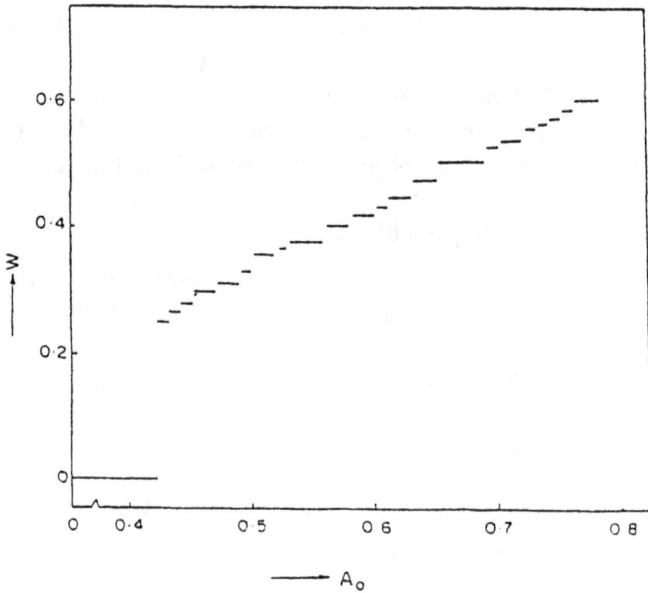

Fig. 6.7. The complete devil's staircase for Eq. (6.3) at $A_1 = 0.88$ with $a = 0.8, b = 0.7, c = 0.09$, and $\omega = 1.0$.

Also we have found that for certain intervals of A_0 the winding number assumes a constant rational value; that is, the system exhibits frequency-locking phenomenon. We have observed the following winding numbers within an error of 4%:

$$4/7, \ 1/4, \ 4/15, \ 3/11, \ 2/7, \ 3/10, \ 5/16, \ 1/3,$$
$$5/14, \ 4/11, \ 3/8, \ 2/5, \ 5/12, \ 3/7, \ 7/16, \ 4/9, \qquad (6.19a)$$
$$6/13, \ 8/17 \text{ and } 1/2.$$

Figure 6.7 shows the devil's staircase for $A_1 = 0.88$, $a = 0.8$, $b = 0.7$ and $c = 0.09$. The winding numbers observed in this case are

$$1/4, \ 4/15, \ 2/7, \ 3/10, \ 5/16, \ 1/3, \ 5/14, \ 4/11, \ 3/8,$$
$$2/5, \ 5/12, \ 3/7, \ 4/9, \ 8/17, \ 1/2, \ 9/17, \ 8/15, \ 5/9, \qquad (6.19b)$$
$$9/16, \ 4/7, \ 7/12 \text{ and } 3/5.$$

The fractal dimension D, is a useful way of characterizing the staircase. $D = 1$ represents the complete devil's staircase and $0 < D < 1$ implies the self-similarity of the devil's staircase. This has been supported by the numerical calculation of Jensen *et al.* in a circle map (Ref. [117]). From Figs. 6.6 and 6.7 we find that the total length of A_0 interval, that is the space between the staircase, is of zero-measure. So the dimension of the devil's staircases shown in Figs. 6.6 and 6.7 is 1, which implies that the devil's staircase is complete.

Finally, we note that further detailed phase-lockings and staircases have been observed for other regions of parametric space in Ref. [116].

6.2. Duffing–van der Pol Oscillator

It is of interest to note that the BVP oscillator discussed in the previous section is essentially equivalent to a combination of the ubiquitous Duffing and van der Pol oscillators. (A brief discussion of the properties of the van der Pol oscillator is given in Appendix B.) To see this clearly, we differentiate Eq. (6.3) with respect to time and replace R and $\dot{R}$ from the right-hand sides of Eqs. (6.3a) and (6.3b), respectively, and obtain

$$\ddot{V} - (1 - bc)\{1 - (V^2/(1 - bc))\}\dot{V} - c(b - 1)V + (bc/3)V^3$$
$$= c(A_0 b - a) + A_1 \cos(\omega t + \phi), \quad (\cdot = d/dt) \tag{6.20}$$

where $\phi = \tan^{-1}(\omega/bc)$. Using the transformations $x = V/\sqrt{1 - bc}$ and $t \to t' = t + (\phi/\omega)$, Eq. (6.20) can be rewritten (after dropping the prime) as

$$\ddot{x} + p(x^2 - 1)\dot{x} + \omega_0^2 x + \beta x^3 = f_0 + f_1 \cos \omega t, \tag{6.21}$$

where

$$p = (1 - bc), \quad \omega_0^2 = c(1 - b), \quad \beta = bc(1 - bc)/3, \tag{6.21a}$$
$$f_0 = c(A_0 b - a)/\sqrt{1 - bc}, \quad f_1 = A_1/\sqrt{1 - bc}. \tag{6.21b}$$

Equation (6.21) is easily seen to be a special case of the combined Duffing–van der Pol (DVP) oscillator equation

$$\ddot{x} + p(kx^2 + g)\dot{x} + \omega_0^2 x + \beta x^3 = f_0 + f_1 \cos \omega t, \tag{6.22}$$

where k and g are parameters. Equation (6.22) naturally reduces to the Duffing oscillator case and the van der Pol oscillator case in the $k = 0$ and $\beta = 0$ limits, respectively (with $f_0 = 0$). Obviously the properties of the BVP oscillator holds good for the DVP oscillator (6.22) also for appropriate choices of the parameters as given in Eq. (6.21). Equation (6.22) is often used to model optical bistability in a dispersive medium, in which the refractive index is dependent on the optical intensity (Ref. [54]). Other applications include certain flow-induced structural vibration problems (Ref. [118]) under the influence of periodic forcing and so on.

Considering the DVP oscillator, Eq. (6.22), one can again consider (as in the case of the Duffing oscillator, Chapter 4) at least three physically interesting situations, wherein the 'potential' $V = (\omega_0^2/2)x^2 + (\beta/4)x^4$ is a
(i) double-well ($\omega_0^2 < 0$, $\beta > 0$),
(ii) single-well ($\omega_0^2 > 0$, $\beta > 0$), or a
(iii) double-hump ($\omega_0^2 > 0$, $\beta < 0$).

We will discuss the nature of each one of these briefly.

A. *DVP oscillator with double-well potential:* The DVP oscillator equation with a double-well potential of the form (with the choice $k = 1$, $g = -1$ in Eq. (6.22))

$$\ddot{x} + p(x^2 - 1)\dot{x} - |\omega_0^2|x + \beta x^3 = f_0 + f_1 \cos \omega t, \quad \beta > 0, \qquad (6.23)$$

can be analyzed as in Sec. 6.1 and one may obtain the bifurcation routes, chaotic dynamics and phase-locking phenomenon as described earlier. One more interesting aspect (Ref. [119]) of this oscillator which we wish to point out here is the transverse intersection of the homoclinic orbits corresponding to Smale-horeshoe chaos.

In order to obtain the Smale-horseshoe chaos threshold, following the discussions in Chapter 5, Sec. 5.3, we evaluate the Melnikov integral for Eq. (6.23) by rewriting it as

$$\dot{x} = y, \qquad (6.24a)$$
$$\dot{y} = |\omega_0^2|x - \beta x^3 + \varepsilon[-p(x^2 - 1)y + f_0 + f_1 \cos \omega t]. \quad (\varepsilon \ll 1) \quad (6.24b)$$

Here ε is a small parameter so that the damping and forcing terms may be treated as perturbations on a free double-well anharmonic oscillator. The homoclinic orbit associated with Eq. (6.24) for $\varepsilon = 0$ is

$$(x_h(t), y_h(t)) = \left(\left(\sqrt{2|\omega_0^2|/\beta}\right) \mathrm{sech}\sqrt{|\omega_0^2|}\tau, \right.$$
$$\left. -|\omega_0^2|\sqrt{2/\beta}\,\mathrm{sech}\left(\sqrt{|\omega_0^2|}\right)\tau \tanh\left(\sqrt{|\omega_0^2|}\right)\tau \right), \quad \tau = t - t_0 \quad (6.25)$$

and so the Melnikov integral (cf. Eq. (5.47))

$$M(t_0) = \int_{-\infty}^{\infty} y_h[-p(x_h^2 - 1)y_h + f_0 + f_1 \cos \omega(\tau + t_0)].d\tau \qquad (6.26)$$

turns out to be

$$M(t_0) = \sqrt{2/\beta} f_1 \omega \pi \operatorname{sech}\left(\frac{\pi\omega}{2\sqrt{|\omega_0^2|}}\right) \sin \omega t_0$$
$$+ \left(4\left(p\frac{|\omega_0^2|^{3/2}}{3\beta}\right)\left(1 - \frac{4|\omega_0^2|}{5\beta}\right)\right). \qquad (6.27a)$$

Thus the threshold curve for Smale-horseshoe chaos is given by

$$f_1 \geq f_{1M} = ((2/3)\sqrt{2/\beta})\left(p\frac{(|\omega_0^2|)^{3/2}}{\pi\omega}\right)$$
$$\times \left(\left(\frac{4|\omega_0^2|}{5\beta}\right) - 1\right)\cosh\left(\frac{\pi\omega}{2|\omega_0^2|}\right). \qquad (6.27b)$$

The equality sign obviously corresponds to the tangential intersection of homoclinic orbits. Figure 6.8 depicts the threshold curves f_{1M} as a function of the frequency ω for $f_0 = 0, |\omega_0^2| = 1, \beta = 5$, and for various values of p. In the parameter regions above the threshold curves, transverse intersections of the stable and unstable manifolds of the saddle occur. From Fig. 6.8 it is evident that for a fixed value of p as ω increases the f_{1M} value decreases and approaches a limiting value.

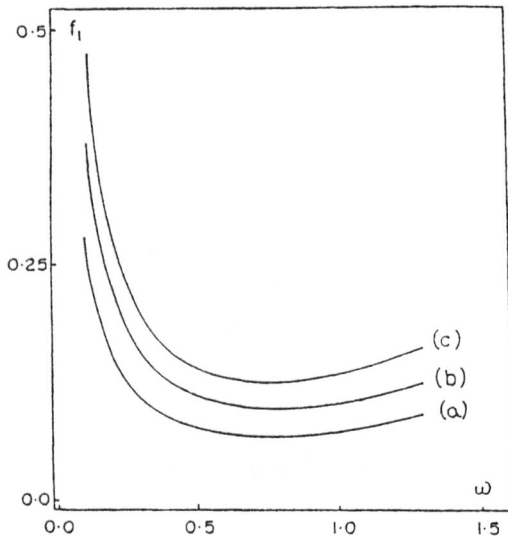

Fig. 6.8. Threshold curves for Smale-horseshoe chaos in the $(f_1 - \omega)$ plane for $|\omega_0^2| = 1.0, \beta = 5.0, f_0 = 0$, and (a) $p = 0.3$, (b) $p = 0.4$, and (c) $p = 0.5$.

The above analytical prediction f_{1M}, given by Eq. (6.27), can be compared with the numerical threshold f_{1N}, which is obtained through a numerical integration of Eq. (6.24), when the stable and unstable orbits become tangential to each other for various values of p. For $p = 0.4$, the Melnikov threshold value f_{1M} is (cf. Eq. (6.27) and Fig. 6.8) 0.1132. The corresponding f_{1N} value is 0.125 which is greater than the f_{1M} value, as expected. The numerical threshold was found by plotting the stable and unstable manifolds of the saddle in the Poincaré map for various values of f_1. To find both the stable and unstable manifolds of the perturbed system we have used a total of 400 initial conditions along the eigenvector associated with the stable and unstable manifolds of the saddle point. The unstable manifold is computed by integrating the system in the forward time direction. The stable manifold is obtained by backward time integration. For $p = 0.4$, Fig. 6.9 shows part of such orbits for three chosen values of f_1. For $f_1 < 0.125$, the stable and unstable orbits which are joined smoothly in the absence of perturbation are now separated, for which the Melnikov distance is always positive. This is shown in Fig. 6.9 (a) for $f_1 = 0.09$. In this parameter regime one may expect regular behaviour. When $f_1 \approx 0.125$ (Fig. 6.9(b)) the two orbits touch each other exactly and the Melnikov distance becomes zero at some $t = t_0$. Further, for $f_1 > 0.125$ we found transverse intersections of the two orbits where the Melnikov distance oscillates between positive and negative values. For example, Fig. 6.9(c) shows two such transverse intersections for $f_1 = 0.2$. Thus for $f_1 \geq 0.125$ it is possible to have either asymptotic chaos or transient chaos followed by asymptotically periodic motion.

In order to know the nature of attractors of the system near the horseshoe threshold value, Eq. (6.24) has been further numerically investigated for the onset of chaos therein. Figure 6.10 shows the nature of bifurcation and chaos exhibited by the system (6.24) for the range $0 < f_1 < 0.15$. From the bifurcation diagram (c.f. Fig. 6.10), the onset of chaos is seen to occur at $f_{1NC} = 0.125$. Further we note that at the onset of chaos a sudden change in the size of the chaotic attractor occurs. This bifurcation is nothing but a crisis. In the DVP oscillator, this sudden jump in the size of the attractor is associated with a global event, namely, the transverse intersection of the stable and unstable manifolds of the perturbed saddle. For clarity, the chaotic orbit in the x–$\dot{x}$ plane and the strange attractor in the Poincaré map are given in Fig. 6.11. The correlation dimension of the strange attractor is estimated as 1.06. Similar results for five different choices of p values are summarized in Table 6.1, from which it is clear that for all the p values chosen, the onset of chaos (f_{1NC}) is observed at the same numerical threshold (f_{IN}).

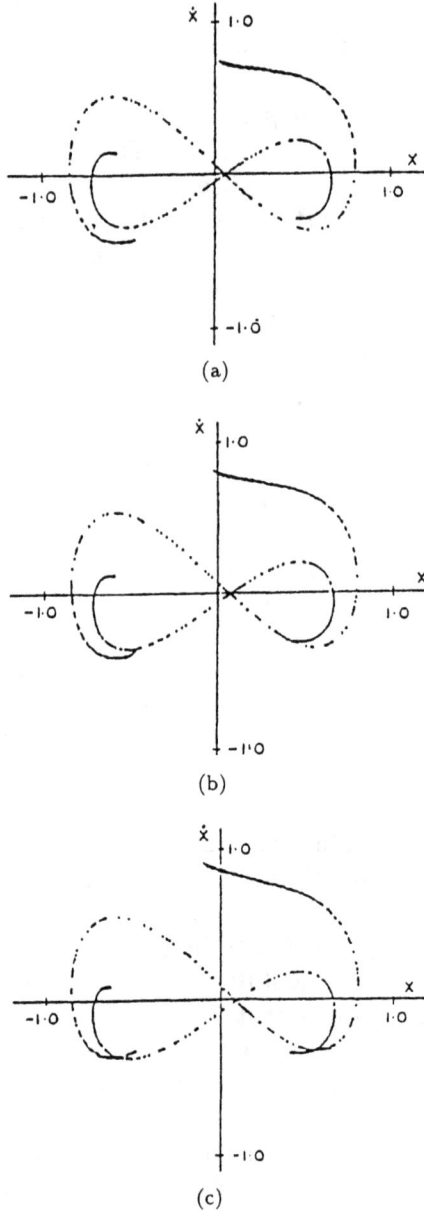

Fig. 6.9. Numerically computed stable and unstable manifolds for the DVP oscillator (6.24) in the $(x$–$\dot{x})$ plane for $p = 0.4, |\omega_0^2| = 1.0, \beta = 5.0, \omega = 1.0, f_0 = 0$ and (a) $f_1 = 0.09$, (b) $f_1 = 0.125$ and (c) $f_1 = 0.2$.

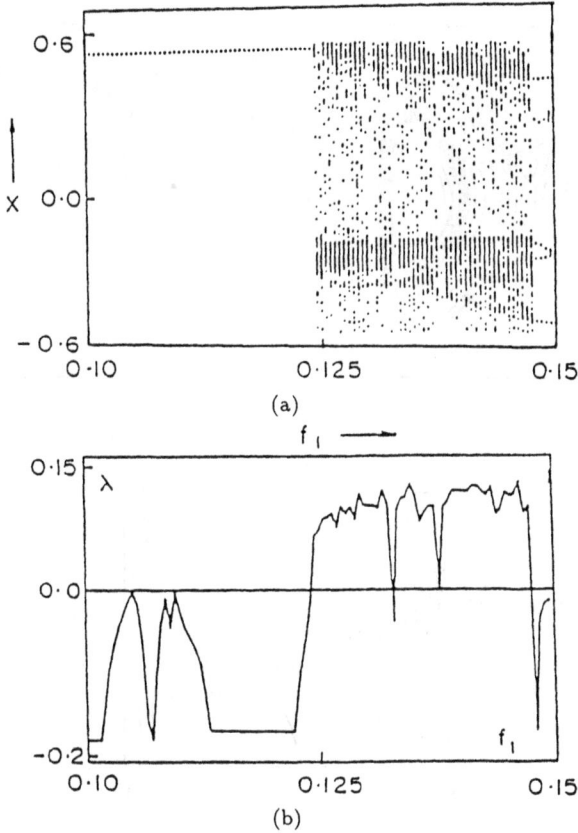

Fig. 6.10. (a) Bifurcation diagram of Eq. (6.24). (b) Maximal Lyapunov exponent $\lambda_{\max}$ versus amplitude f_1 of the external force corresponding to (a).

B. *DVP oscillator with single-well potential*: Kapitaniak and Steeb (Ref. [120]) have investigated the single-well DVP oscillator

$$\ddot{x} + p(x^2 - 1)\dot{x} + \omega_0^2 x + \beta x^3 = f \sin \omega t, \quad \omega_0^2 > 0, \quad \beta > 0, \qquad (6.28)$$

and discussed the chaotic behaviour numerically for the case $p = f = 5, \omega_0^2 = 1$ and $\omega = 2.466$. The nature of the attractors is briefly sketched in Fig. 6.12 for increasing values of β. In Ref. [120], it has been noted that periodic (symmetric as well as unsymmetric) and chaotic motions (again symmetric and unsymmetric) are observed in different parameter ranges and they are

(a)

(b)

(c)

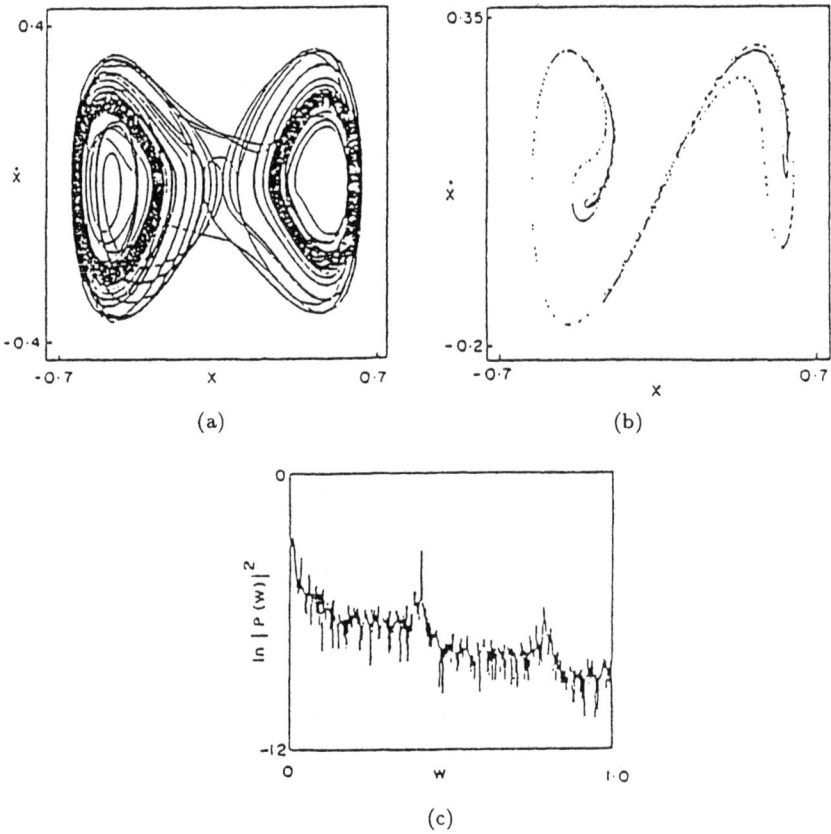

Fig. 6.11. (a) Phase-portrait, (b) Poincaré map of (a), and (c) Power spectrum of the chaotic attractor of the DVP oscillator(6.24) for $f_1 = 0.125$ (other parameters are fixed as in Fig. 6.10).

Table 6.1. Critical values of f_{1M}, f_{1N} and f_{1NC} for the DVP oscillator for various p values.

p	f_{1M}	f_{1N}	f_{1NC}
0.30	0.0849	0.119	0.119
0.35	0.0990	0.121	0.121
0.40	0.1132	0.125	0.125
0.45	0.1273	0.128	0.128
0.50	0.1415	0.136	0.136
0.55	0.1556	0.149	0.149

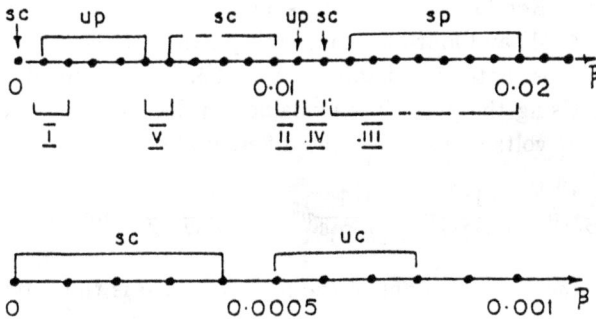

Fig. 6.12. Chaotic and periodic attractors on the parameter β line: *, chaotic; •, periodic; uc, unsymmetrical chaotic; sc, symmetrical chaotic; up, unsymmetrical periodic; sp, symmetrical periodic; I–V, zones where transitions from chaotic to periodic behaviour (or vice versa) take place (Ref. [120]).

indicated in Fig. 6.12 as sp, up, sc, uc, respectively. Typical forms of them are also shown in Figs. 6.13. For $\beta = 0$, that is for a purely forced van der Pol oscillator, one has a chaotic symmetrical attractor.

Two different routes to chaos have been reported in Ref. [120]. In the first one, starting from an unsymmetric periodic attractor (Figs. 6.13(a) and 6.14(a)) through period-doubling bifurcations (Figs. 6.13(b) and 6.14(b)) chaotic unsymmetrical (Fig. 6.13(c)) or symmetrical (Fig. 6.14(c)) attractors are obtained. For an unsymmetrical chaotic attractor ($\beta = 0.005$, Fig. 6.13(c)), the attractor evolves into a symmetrical one after a small decrease in β. This type of route to chaos takes place in the regions I and III in Fig. 6.12.

The second route starts from a symmetrical periodic attractor shown in Fig. 6.15(a) and it occurs in the region III of Fig. 6.12. With decreasing β one observes first that the structure of the small loop is changing (Fig. 6.15(b)) and then the system suddenly undergoes a crisis-type transition to the chaotic attractor shown in Fig. 6.15(c).

Since for $\beta = 0$, the attractor is chaotic, one would expect that a small additional nonlinearity $\beta x^3, \beta \ll 1$, will enhance the chaotic behaviour. But from the above analysis one finds that for as small a value for β as $\beta = 0.013$ the chaotic behaviour disappears. Even for smaller values of β there exist windows with periodic attractors.

C. *DVP oscillator with double-hump potential*: Next, we consider briefly the dynamics of the DVP oscillator with a double-hump potential of the form

$$\ddot{x} + p(x^2 - 1)\dot{x} + \omega_0^2 x - |\beta|x^3 = f \sin \omega t, \quad (\omega_0^2 > 0). \qquad (6.29)$$

Kao and Wang (Ref. [54]) have analog simulated Eq. (6.29) using conventional amplifiers (LM741) and four-quadrant multipliers (AD534) as described earlier for the Duffing oscillator in Chapter 4, Sec. 4.5. The circuit diagram is given in Fig. 6.16. Using the procedure explained in Sec. 4.5, it can easily be seen that the output voltage v satisfies the differential equation

$$\ddot{v} = \frac{0.1}{RC}\dot{v} - \frac{0.1v^2}{(RC)^5}\dot{v} + \frac{0.1}{(RC)^6}v^3 - \frac{1}{(RC)^2}v + f\sin(2\pi\nu t), \qquad (6.30)$$

where RC is the time constant of the integrators, ν is the driving frequency. By setting

$$x = \frac{v}{RC^2}, \quad \tau = \frac{t}{RC}, \quad \omega = \frac{\nu}{2\pi RC}$$

and redefining $\tau \to t$, one can easily check that Eq. (6.30) is nothing but the DVP oscillator equation (6.29) with a double-hump for the specific choice of the parameters $p = 0.1, \omega_0^2 = 1, |\beta| = 0.1$. The dynamic phenomena include mode-locking, multiple hysteresis, period-doubling route to chaos, intermittent hopping and crisis. The phase diagram deduced by Kao and Wang (Ref. [54]) is given in Fig. 6.17 and it includes all the above type of attractors.

New interesting features among the various types of attractors include

(i) the existence of a Farey sequence with winding numbers $W = \frac{1}{2}, \frac{2}{3}, \frac{3}{4}, \frac{1}{1}, \frac{4}{3}, \frac{3}{2}$, and $\frac{2}{1}$ for $f = 0.25$ (obtained during frequency scanning) as shown in Fig. 6.18(a) and $W = \frac{3}{2}, \frac{10}{7}, \frac{7}{5}$, and $\frac{4}{3}$ for $\omega = 0.554$ as given in Fig. 6.18(b) (obtained during amplitude scanning),

(ii) double hysteresis loops obtained during amplitude scanning, given in Fig. 6.19, and

(iii) swallow-tailed bifurcation structure of subharmonic locking.

More details are given in Ref. [54].

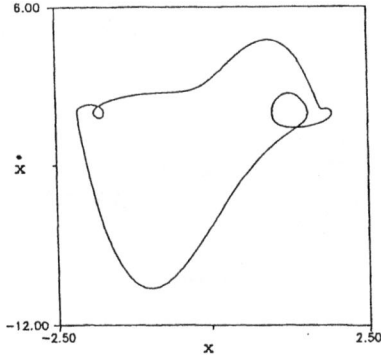

(a)

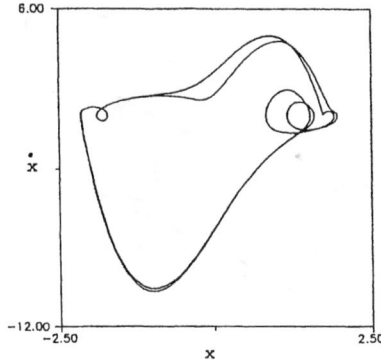

(b)

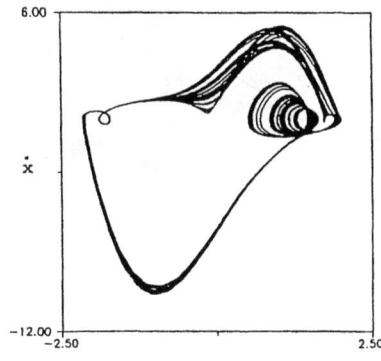

(c)

Fig. 6.13. Transition to chaos in region I. Phase-portraits: (a) $\beta = 0.002$, unsymmetric periodic orbit (*up*); (b) $\beta = 0.001$, unsymmetric periodic orbit (*up*); (c) $\beta = 0.0005$, unsymmetric chaotic orbit (*uc*).

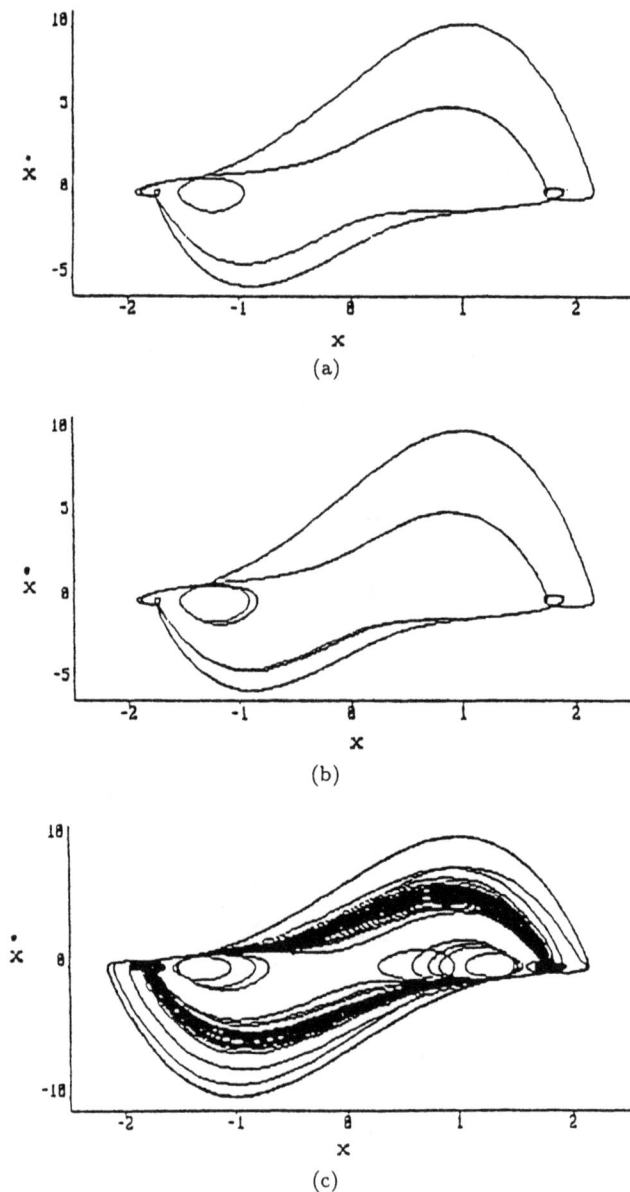

Fig. 6.14. Transition to chaos in region II. Phase-portraits: (a) $\beta = 0.011$, unsymmetric periodic orbit (*up*); (b) $\beta = 0.0109$, unsymmetric periodic orbit (*up*); (c) $\beta = 0.0108$, symmetric chaotic orbit (*sc*).

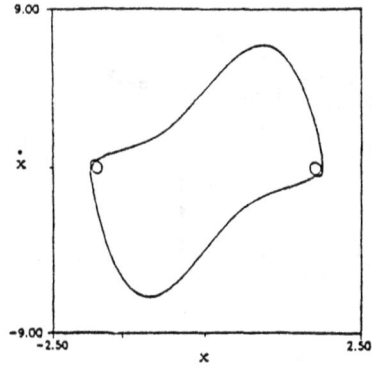

(a)

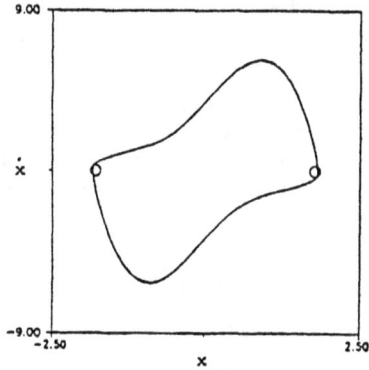

(b)

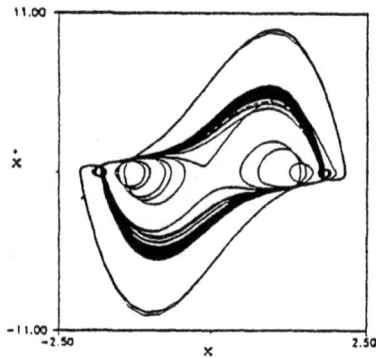

(c)

Fig. 6.15. Transition to chaos in region III. Phase-portraits: (a) $\beta = 0.02$, symmetric periodic orbit (sp); (b) $\beta = 0.0128$, symmetric periodic orbit (sp); (c) $\beta = 0.012$, symmetric chaotic orbit (sc).

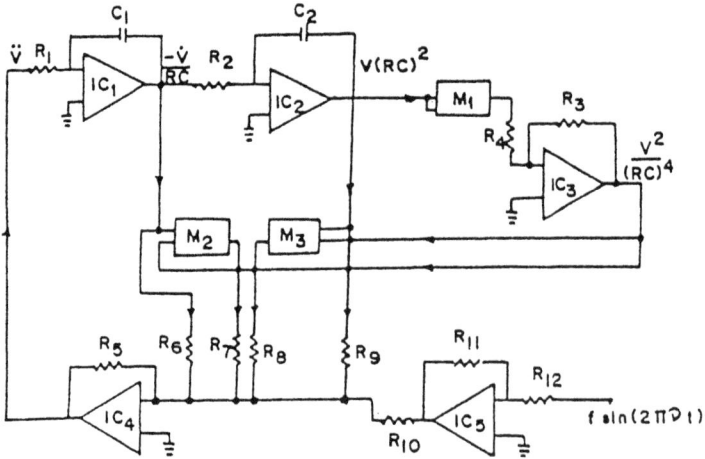

Fig. 6.16. The circuit diagram for Eq. (6.29). IC_1 and IC_2 (LM741) are operational amplifier integrators. The resistors and capacitors are set as $R = R_1 = R_2 = 10k\Omega$ and $C = C_1 = C_2 = 0.0982\mu F$. The resistors $R_3 = R_6 = 100k\Omega$ and the others are equal to $10k\Omega$. M_1, M_2, and M_3 are multipliers(AD534) with output $-xy/10$ for input x and y. (see Ref. [54]).

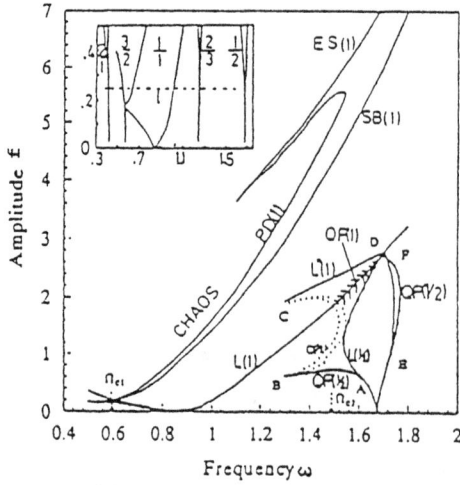

Fig. 6.17. The state diagram with parameters $p = 0.1, \omega_0^2 = 1.0, |\beta| = 0.1$, and $0.4 < \omega < 1.8$. Curves $L(\lambda), QP(\lambda), SB(\lambda)$, and $C(\lambda)$ are the thresholds of mode locking, quasiperiodic, symmetry breaking, and crisis, respectively, with winding number 1. Curve $PD(\lambda)$ is the threshold of the first period doubling. Curve $ES(\lambda)$ is the threshold for solutions escaping from the potential well. (Inset): Several small locking islands with ultrasubharmonic resonance (Ref. [54]).

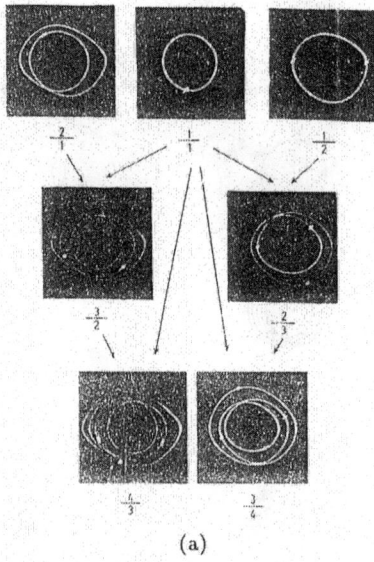

(a)

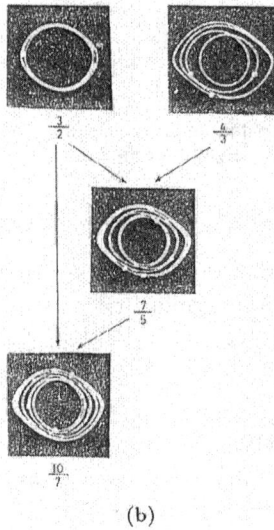

(b)

Fig. 6.18. Farey sequence for (a) $\lambda = 1/2, 2/3, 3/4, 1/1, 4/3, 3/2$, and $2/1$ with $f = 0.25$ and (b) $\lambda = 3/2, 10/7, 7/5$, and $4/3$ with $\omega = 0.554$. The number of dot points accounts for the driving periods (Ref. [54]).

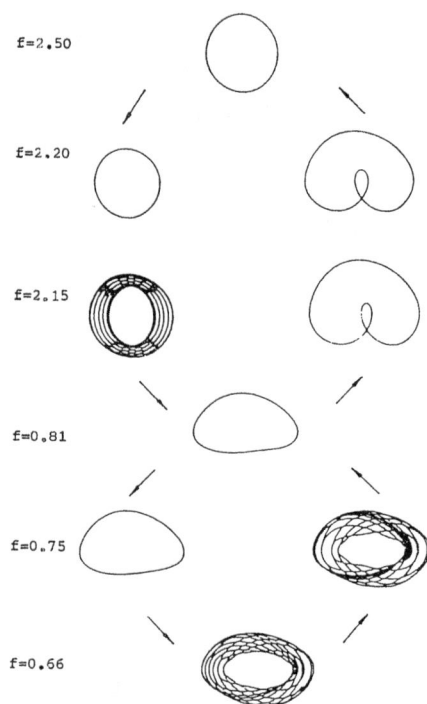

Fig. 6.19. The phase-portraits of double hysteresis loop with $\omega = 1.57$ in the subharmonic region. The arrows stands for the direction of amplitude scanning (Ref. [54]).

6.3. Analytic Structure of the DVP Oscillator (Ref. [121])

In Sec. 5.4, it was pointed out that the analytic structure of the solution of the Duffing oscillator is quite complicated. It locally admits a logarithmic movable branch point around which an immensely complicated clustering of singularities occurs on multi-arms in an infinitely-sheeted Riemann structure in the complex t-plane. The question arises as to what the corresponding structure in the BVP and DVP oscillators is. For this purpose we take up the DVP oscillator (6.22) and investigate the analytic structure properties (Ref. [121]) of the equation

$$\ddot{x} + p(x^2 - 1)\dot{x} + x + \beta x^3 = f \cos \omega t \qquad (6.31)$$

in the complex t-plane. One finds here that the solutions of (6.31) have no worse-than-algebraic branch point singularities at $t = t_0$ with only $(t - t_0)^{1/2}$

terms present in their local series expansion (Ref. [121]), unlike the case of the Duffing equation where $\ln(t - t_0)$ terms occur. Still, when integrating (6.31) around efficiently long contours, a remarkably intricate pattern of square-root type singularities emerges on different sheets, which appears to prevent solutions from ever returning to the original sheet. Such evidence of infinitely-sheeted solutions, termed the ISS property, has also been observed in certain Hamiltonian systems (Ref. [122]).

6.3.1. Singularity Analysis of the DVP Equation

As in the case of the Duffing equation, we look for a local series expansion of the form (5.54) around a movable singular point and deduce the leading orders. For Eq. (6.31) this turns out to be

$$x(t) \sim \left(\frac{3}{2p}\right)^{1/2} \tau^{-1/2}, \quad \tau = t - t_0 \to 0, \tag{6.32}$$

for $p \neq 0$. Proceeding to the determination of higher-order terms in the series expansion

$$x(t) = \left(\frac{3}{2p}\right)^{1/2} \tau^{-1/2} + a_1 + a_2\tau^{1/2} + a_3\tau + a_4\tau^{3/2} + \dots, \tag{6.33}$$

one can easily find, upon equating in (6.31) terms of order $\tau^{-2}, \tau^{-3/2}, \tau^{-1}$, separately,

$$a_1 = 0, \quad a_2 = \left(\frac{3}{2p}\right)^{1/2} \left[\frac{3\beta}{2p} + \frac{p}{2}\right], \quad a_3 = \text{arbitrary}. \tag{6.34}$$

Thus the series expansion (6.33) is compatible with the solution of Eq. (6.31) for all β and f. No logarithmic or any other movable critical singular points arise, except for the square-root type algebraic branch point, and the general solution of (6.31) near $t = t_0$ can be written as

$$x(t) = \tau^{-1/2} \sum_{n=0}^{\infty} a_n\tau^{n/2}, \quad \tau = (t - t_0) \to 0. \tag{6.35}$$

Clearly the periodic forcing term in (6.31) gives

$$f \cos\omega(\tau + t_0) = f \cos\omega t_0 \left(1 - \frac{\omega^2\tau^2}{2} + \dots\right)$$

$$- f \sin\omega t_0 \left(\frac{\omega\tau - \omega^3\tau^3}{6} + \dots\right)$$

and will start contributing in Eq. (6.31) at order τ^0, adding $f\cos\omega t_0$ to the equation from which the coefficient a_5 is calculated.

Furthermore, one can prove that the series (6.35) converges in the punctured disc

$$D = \{\tau \neq 0, \text{ with } |\tau| < R < \infty\}$$

within a finite radius of convergence, at least numerically, which can extend to the singularity nearest to t_0 in the complex t-plane. So the solutions of (6.31) are locally (that is near every t_0) finitely branched and finitely sheeted. Does this mean that they are also globally finitely sheeted?

To answer the above question, one can integrate Eq. (6.31) numerically in the complex t-plane along simple rectangular contours enclosing more than one singularity of the primary Riemann sheet. As mentioned in Chapter 5, Sec. 6.3, this can be done using the ATOMFT 2.51 (1991) algorithm, which is a more recent version of the original ATOMCC program of Chang and Corliss.

6.3.2. *Infinite-sheeted Solution (ISS) Structure*

First let us consider the free van der Pol oscillator equation ($\beta = f = 0$)

$$\ddot{x} + p(x^2 - 1)\dot{x} + x = 0, \ p > 0. \tag{6.36}$$

Equation (6.36) is itself nonintegrable and is of non-Painlevé type. Choosing an initial value to the limit cycle solution of (6.36),

$$x(0) = 2.0, \quad \dot{x}(0) = 0,$$

and taking $p = 1$, when the period of the limit cycle is nearly 2π, we see that the singularities of the solution, on the primary sheet, form a row of conjugate pairs placed along the t-axis, with nearly distance π between each pair, as shown in Fig. 6.20. Now let us enclose the pair of singularities in a rectangular contour C (Fig. 6.20), which intersects the Re t axis at $Q = -1.8$ whose vertices on the Im axis are at P and $-$P.

One can now make several turns around C and compute the values of $x(t)$ and $\dot{x}(t)$ at P after the Nth turn: $x_P(N)$ and $\dot{x}_P(N)$ respectively. Then we look for the quantities

$$\Delta x_P(N) = |x_P(N) - x_P(0)|, \quad \Delta\dot{x}_P(N) = |\dot{x}_P(N) - \dot{x}_P(0)|. \tag{6.37}$$

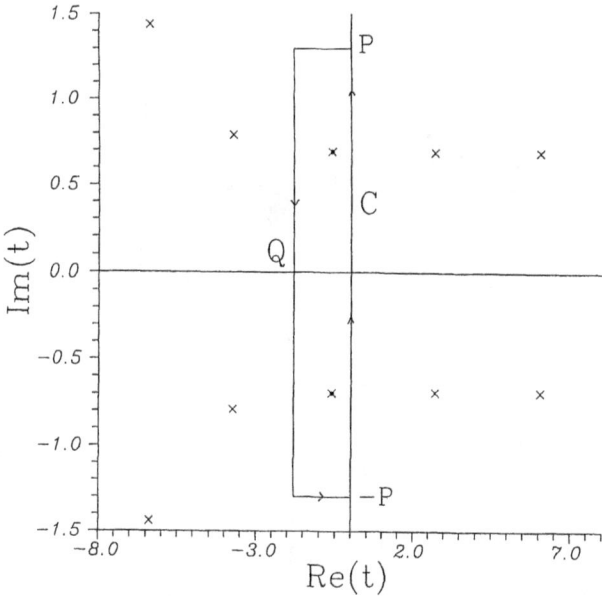

Fig. 6.20. Arrangement of singularities on the primary Riemann sheet of the solutions of (6.36) with $p = 1$ and initial conditions $\dot{x}(0) = 2.0, \ddot{x}(0) = 0$.

Table 6.2.

P	N	Number of singularities in Contour C	$\Delta x_P(N)$	$\Delta \dot{x}_P(N)$
$1.09i$	7	6	7.8×10^{-14}	3.2×10^{-14}
$1.18i$	7	6	1.0×10^{-12}	5.7×10^{-13}
$1.24i$	7	6	7.0×10^{-13}	3.6×10^{-13}
$1.25i$	7	6	1.6×10^{-11}	1.6×10^{-11}
$1.255i$	9	8	1.9×10^{-8}	1.7×10^{-8}
$1.27i$	9	8	8.5×10^{-10}	6.5×10^{-10}
$1.30i$	9	8	2.7×10^{-9}	2.9×10^{-9}

These are listed in Table 6.2 for several values of P between $1.1i$ and $1.3i$. For these contours there is clear evidence of *finitely-sheeted solutions* (FSS), which return to their starting values after a number N of turns which is equal to the number of singularities enclosed in C plus one. In Fig. 6.21(a), for $P = 1.3i$, these singularities belong to different sheets (the singularities of the primary sheet are marked x in Fig. 6.21(a)). This situation continues to be qualitatively analogous, until a dramatic change occurs when $P \geq 1.33194i$

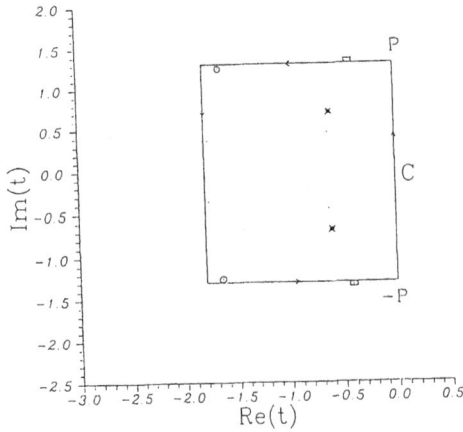

(a)

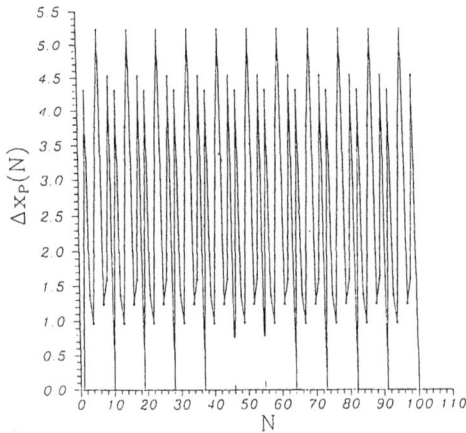

(b)

Fig. 6.21. Complex time integration of (6.36), with $p = 1$ and initial conditions $x(0) = 2, \dot{x}(0) = 0$ around two of the primary sheet singularities of Fig. 6.20 (marked by $\times$ here). (a) The upper right corner of the contour is at $P = 1.3i$ and no new singularities, other than the ones shown here, appear, even after 200 turns around C. (b) The solution differences $\Delta x_p(N)$, cf. (6.37), showing clear evidence of FSS as they return to 0, within 10^{-9}, after every nine turns.

(Fig. 6.22). A cloud of new singularities has now emerged, whose branching does not allow the solution to return to its original value even after more than 200 turns. One might say that the nonintegrability nature of (6.36) becomes apparent via the ISS property.

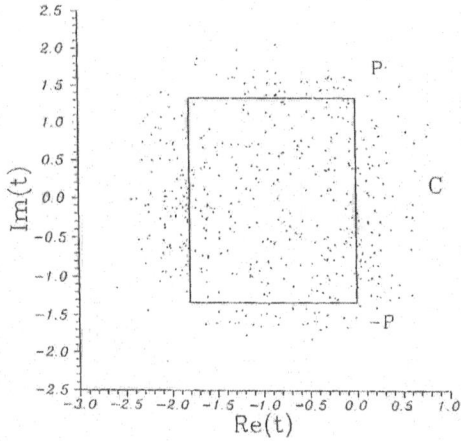

(a)

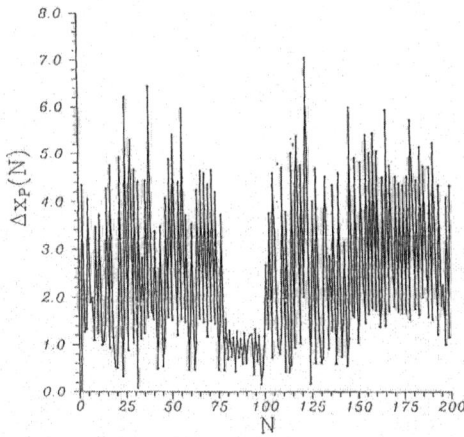

(b)

Fig. 6.22. (a) Same as Figs. 6.20(a), with $P = 1.33197i$, that is, just after the singularity has been included in C. An 'explosion' of singularities has occurred and the figures are now completely different. (b) New singularities now prevent the solution from recovering its initial value, even after $N = 200$ turns.

Now let us consider the effect of adding βx^3 to Eq. (6.36):

$$\ddot{x} + p(x^2 - 1)\dot{x} + x + \beta x^3 = 0. \qquad (6.38)$$

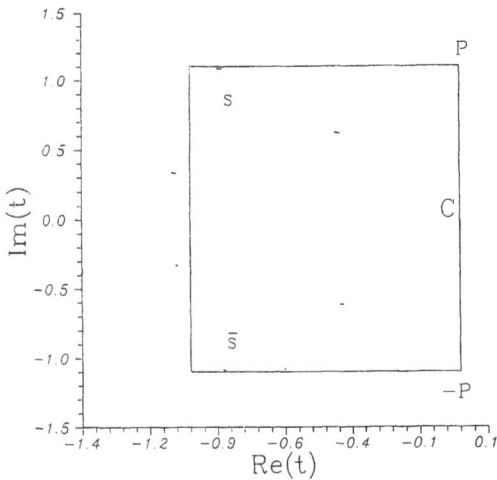

(a)

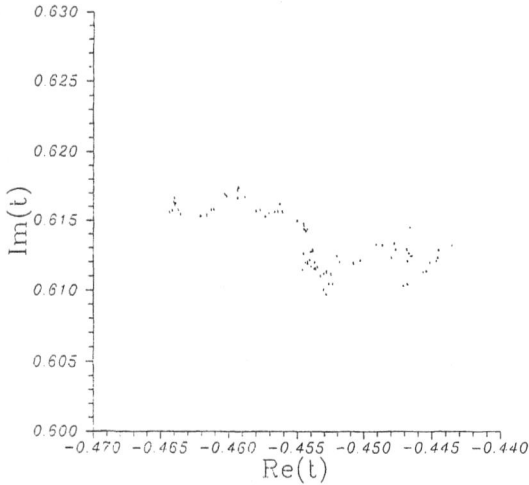

(b)

Fig. 6.23. (a) A singularity pattern of 'clusters' formed by the projections of singularities lying on different sheets of (6.38), with $p = 1$ and $\beta = 0.25$. (b) A magnification of one of these 'clusters'. (c) Solution differences grow almost linearly with N, showing clear evidence of ISS, in this case.

The results of integration in the complex t-plane are shown in Fig. 6.23. When the integration is carried around a contour C containing two singularities marked S and $\bar{S}$ in Fig. 6.23(a), new singularities appear on different

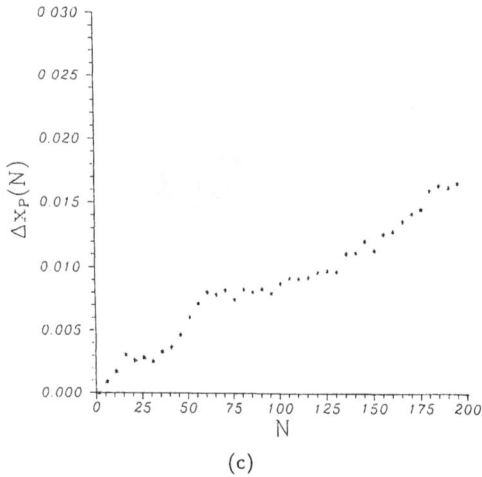

(c)

Fig. 6.23. (*Continued*)

sheets, which are grouped together in the form of clusters. In Fig. 6.23(b), a magnification of one of these clusters is exhibited and it shows a complicated (possibly) fractal structure. The result of the accumulation of these singularities is shown in Fig. 6.23(c), showing clear evidence of the ISS property. Again, as before, by increasing the vertical length of the contour a sudden transition to a 'scatter' of singularities in a cloud-like pattern is observed.

Finally, when the forcing $f \cos \omega t$ is also included as in Eq. (6.31), apart from the above, one observes a different type of singularity clustering, even though the solution near a singularity in the present case is of a square-root type branch point. This is essentially because of the presence of the $\cos \omega t$ term in Eq. (6.31) (Ref. [121]), and one expects an accumulation of singularities along the Im(t) direction. Figure 6.24(a) portrays such accumulation observed near primary sheet singularities of the DVP oscillator (6.31) for $p = 0.1, \beta = 0.01$ and $f = 0.001$ and $\omega = 1.0$ near the Im(t) axis (and sufficiently far away from the t-axis). The path followed to observe these singularities is shown by the dashed line in the figure. Further, one notes that on increasing the strength of f ten times (Fig. 6.24(b)) the relative distances between singularities of Fig. 6.24(a) decreases and the pattern becomes denser. In this sense one may say that the real-time motion of the dynamical system (6.31) is reflected in some way by the analytic properties of the solutions in the complex time.

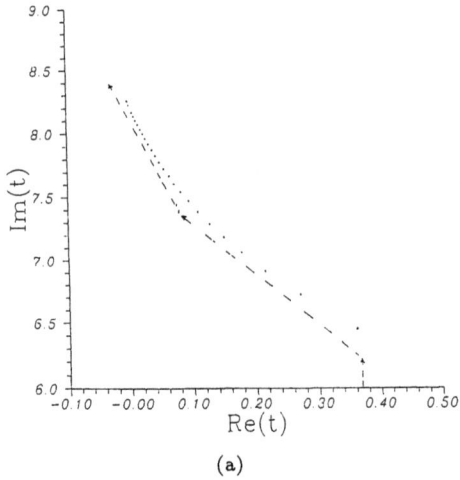

(a)

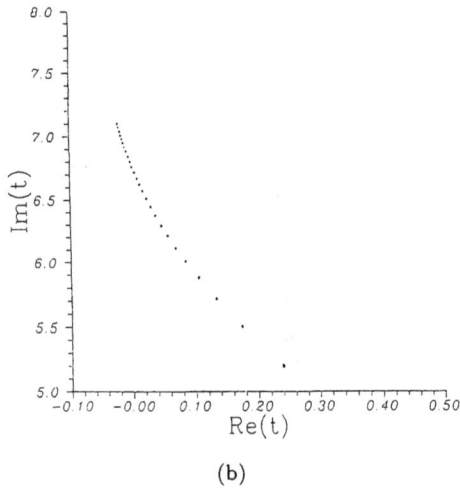

(b)

Fig. 6.24. (a) Singularity accumulation in the Im $t > 0$ direction for (6.31) with $p = 0.1$, $\beta = 0.01$ and $f = 0.001$. (b) Note how the pattern of (a) 'condenses' when the value of f is increased to $f = 0.01$. The dashed line indicates the integration path followed by ATOMFT to 'see' these singularities.

CHAPTER 7

CHAOTIC OSCILLATORS WITH CHUA'S DIODE

We now turn our attention to the study of nonlinear circuits and the possibility of adding convenient nonlinear circuit elements to the linear circuits discussed in Chapter 3 and studying the associated circuit dynamics. As pointed out in Sec. 3.4, the most natural extension of linear circuit theory to the world of nonlinear circuits is through piecewise-linear circuit modelling. In this chapter we take up a detailed study of nonlinear oscillators and circuits by including the nonlinear resistor, namely Chua's diode (Refs. [19, 73–75, 123–129]), introduced briefly in Sec. 3.4 as a nonlinear element. We first consider the case in which the third-order autonomous circuit of Fig. 3.5 of Chapter 3 is generalized by including a Chua's diode, so that we have the famous Chua's autonomous circuit, and a Chua's autonomous oscillator (Refs. [19, 123, 125, 127]). Further we consider the non-autonomous version of Chua's circuit (Refs. [18, 130, 131]) and its chaotic dynamics. Finally, we discuss the chaotic dynamics of the simplest dissipative non-autonomous circuit, namely, the Murali–Lakshmanan–Chua (MLC) circuit (Refs. [77–79]). In addition, by considering a simple analytical version of Chua's circuit, namely, the autonomous van der Pol–Duffing (ADVP) oscillator circuit model, we will briefly discuss its chaotic dynamics.

7.1. Chua's Diode and Its Characteristics

Chua's diode, which was mentioned briefly in Chapter 3, is used to denote any two-terminal nonlinear resistor with a piecewise-continuous voltage–current $(V_R - i_R)$ characteristic synthesized by using standard circuit elements (Refs. [74, 75, 124]). This line of research was initiated in the mid-1960s, when Chua developed a theory of nonlinear circuit synthesis and realized that such a theory would be academic unless one were allowed to use practical two-terminal nonlinear resistors with a prescribed v–i characteristic. This in turn motivated the development of systematic synthesis procedures for two-terminal nonlinear resistors (Refs. [49, 56, 132]), ultimately leading to the realization of Chua's diode. The typical characteristic curve of such a nonlinear resistor consists of the odd-symmetric three-segment piecewise-linear form as shown in Fig. 7.1.

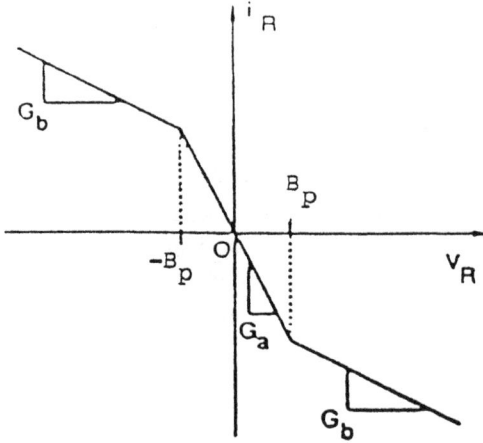

Fig. 7.1. Example of a piecewise-linear characteristic for Chua's diode.

Here G_a and G_b are the inner and outer slopes respectively, and breakpoints are located at $V_R = -B_P$ and $V_R = +B_p$ respectively. The mathematical representation of the characteristic curve of Chua's diode (Refs. [49, 74]) is given by

$$i_R = f(V_R) = G_b V_R + 0.5(G_a - G_b)[|V_R + B_P| - |V_R - B_P|], \qquad (7.1a)$$

or

$$f(V_R) = \begin{cases} G_b V_R + (G_b - G_a)B_p & \text{if } V_R < -B_p, \\ G_a V_R & \text{if } -B_p \leq V_R \leq B_p, \\ G_b V_R + (G_a - G_b)B_p & \text{if } V_R > B_p, \end{cases} \qquad (7.1b)$$

where $B_p > 0$.

Several implementations of a Chua's diode with three-segment-odd-symmetric piecewise-linear characteristics already exist in the literature; these use operational amplifiers (op amps) (Ref. [123]) or diodes (Refs. [134, 135]) or transistors (Ref. [136]). A systematic procedure for synthesizing precision piecewise-linear Chua's diodes with independently adjustable slopes and break-points is described by Kennedy in Ref. [74]. Recently, a single chip integrated circuit (IC) realization of Chua's diode using operational transconductance amplifiers has also been reported (Refs. [128, 129]).

7.2. Simple Practical Implementation of Chua's Diode

In this section, we describe a simple practical implementation of Chua's diode (N) using two op-amps and six resistors as reported by Kennedy (Refs. [74, 75, 124]). Figure 7.2 shows such an implementation. A complete list of the components required is given in Table 7.1. The Chua's diode circuit is then constructed using an Analog Devices AD712 dual BiFET op-amp, two 9V-batteries and six linear resistors. The op-amp circuit consisting of op-amps A_1, A_2 and linear resistors R_1–R_6 functions as a nonlinear resistor N with DP (driving point) characteristic as shown in Fig. 7.3. The use of two 9V-batteries to power the op-amps gives the voltages $V^+ = 9V$ and $V^- = -9V$. From measurements of the saturation levels of the AD712 outputs, $E_{\text{sat}} \approx 8.3V$, one obtains the breakpoint voltage as $B_p \approx 1V$. The slope of the resulting

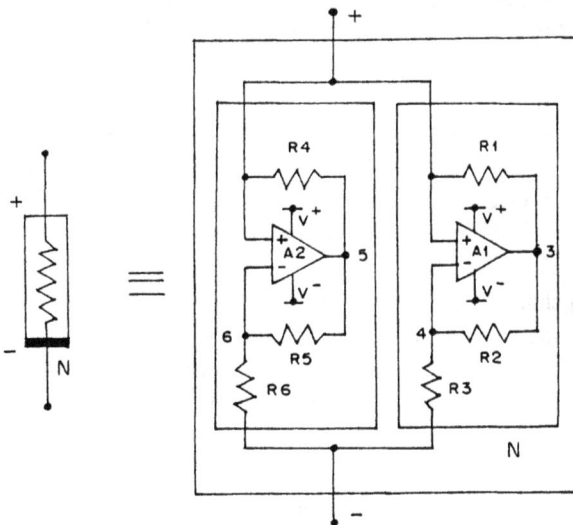

Fig. 7.2. A circuit realization of Chua's diode.

Table 7.1. Component list for the practical implementation of Chua's diode (Fig. 7.2) as used in Refs. [74, 75, 124].

Element	Description	Value	Tolerance
A_1 & A_2	Op-amp (1/2 AD712, TL082, or equivalent)		
R_1	1/4 W Resistor	220Ω	±5%
R_2	1/4 W Resistor	220Ω	±5%
R_3	1/4 W Resistor	2.2kΩ	±5%
R_4	1/4 W Resistor	22kΩ	±5%
R_5	1/4 W Resistor	22kΩ	±5%
R_6	1/4 W Resistor	3.3kΩ	±5%

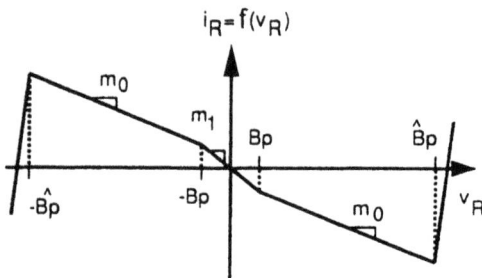

Fig. 7.3. Op amp realization consists of five piecewise-linear segments. If the dynamics is confined to the regions of negative slope, one need not consider the outer segments.

V_R-i_R characteristic curve is given by (see Table 7.1) $G_b = -(R_2/R_1 R_3) + (1/R_4) = -4.09$mA/V and $G_a = -(R_2/R_1 R_3) - (R_5/R_4 R_6) = -0.758$mA/V. The breakpoints are determined by the saturation voltages $E_{\rm sat}$ of the op-amps: $\hat{B}_p = (R_3/(R_2 + R_3))E_{\rm sat} \approx 7.61$V and $B_p = (R_6/(R_5 + R_6))E_{\rm sat} \approx 1.08$V (Refs. [74, 75, 124]). For an alternative realization using two diodes and an op-amp, see Sec. 7.4.2 below.

The V_R-i_R characteristic curve of the op-amp based Chua's diode (Fig. 7.3) differs from the desired piecewise-linear characteristic shown in Fig. 7.1 in that the former has five segments, the outer two of which have positive slopes $G_c = (1/R_2) = (1/220)$mA/V. Every physical resistor is eventually passive, meaning simply that for a large enough voltage across its terminals, the instantaneous power $P(t)(= V(t)i(t))$ consumed by a real resistor is positive. For large enough $|V|$ or $|i|$, therefore, the $V_R - i_R$ characteristic must lie only in the first and third quadrants of the V–i plane. Hence, the V_R-i_R characteristic of a real Chua's diode must include at least two outer segments with positive slopes in the first and third quadrants (Fig. 7.3). From a practical point of view, as long as the voltages and currents in a given nonlinear circuit with Chua's diode are restricted to the negative resistance region of the characteristic, the above-mentioned outer segments will not affect the circuit's dynamical behaviour. The measured V–i characteristic of Chua's diode is shown in Fig. 7.4 (see Refs. [74, 124] for measurement details).

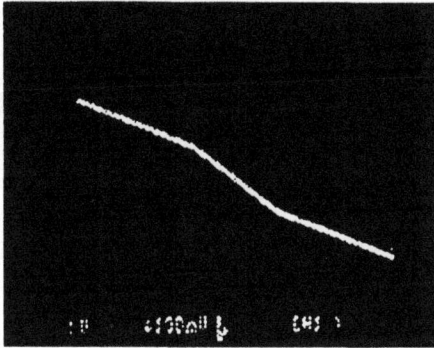

(a)

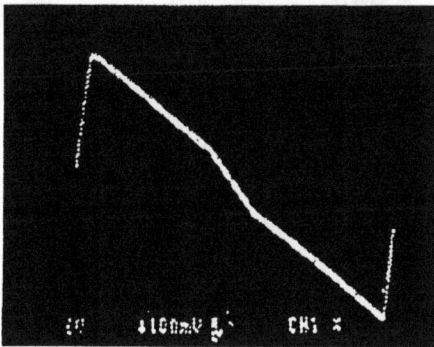

(b)

Fig. 7.4. Measured V–i characteristic of the Chua's diode of Fig. 7.2 (Refs. [74, 75, 124]). (a) Three-segment characteristic. (b) Five-segment characteristic.

7.3. Autonomous Circuits with Chua's Diode

7.3.1. *A. Chua's Autonomous Circuit*

The circuit shown in Fig. 7.5 was synthesized to be the *simplest* autonomous (that is, no input signals) electronic circuit generator of chaotic signals. A

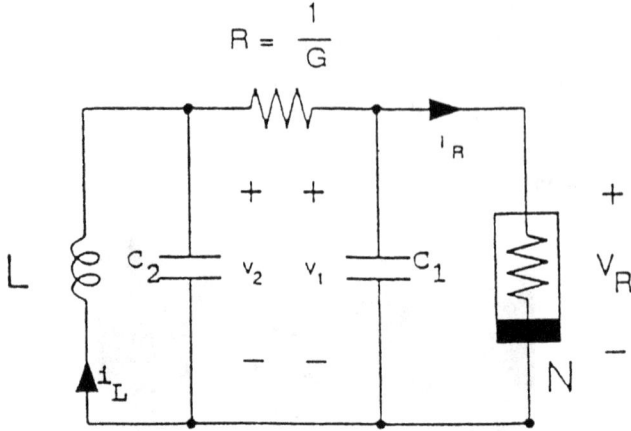

Fig. 7.5. Circuit model for Chua's autonomous circuit.

history of the conception of this circuit and its systematic synthesis procedure is well summarized in Refs. [19, 73]. The chaotic nature of this circuit
was first verified through computer simulation by Matsumoto, who named
it Chua's circuit (Ref. [134]), and confirmed experimentally by Zhong and
Ayrom (Ref. [133]). A comprehensive mathematical analysis of Chua's circuit
and the first rigorous proof of its chaotic property are given in Refs. [135,
137]. Because of its simplicity, and the rich dynamical behaviour, this circuit
has generated a very wide interest among electrical engineers, physicists and
mathematicians (Refs. [19, 73, 123–127]).

Chua's circuit, shown in Fig. 7.5, is a simple oscillator circuit which exhibits
a variety of bifurcations and chaos. It contains three linear energy storage elements (an inductor and two capacitors), a linear resistor, and a single nonlinear
resistor, namely the Chua's diode (N) discussed in the previous section. As
noted above, the operational region of the Chua's diode will be confined to a
three-segment piecewise-linear part of the V–i characteristic curve as shown
in Fig. 7.1.

By applying Kirchoff's laws to the various branches of the circuit of Fig. 7.5,
we obtain the state equations for the Chua's circuit as

$$C_1 \frac{dv_1}{dt} = (1/R)(v_2 - v_1) - f(v_1),$$

$$C_2 \frac{dv_2}{dt} = (1/R)(v_1 - v_2) + i_L,$$

$$(7.2)$$

$$L\frac{di_L}{dt} = -v_2\,,$$

where C_1, C_2, L, and R are all circuit parameters and v_1, v_2 and i_L are the voltage across capacitor C_1, the voltage across capacitor C_2, and the current through inductor L, respectively. In Eq. (7.2), $f(v_1)$ is the mathematical representation of the Chua's diode which is given in Eq. (7.1).

7.3.1. B. Bifurcations and Chaos: R Bifurcation Sequence

By fixing the circuit parameters at $C_1 = 10\text{nF}, C_2 = 100\text{nF}$ and $L = 18\text{mH}$ and by reducing the variable resistor (R) from 2000Ω towards zero, Chua's circuit is readily seen to exhibit a sequence of bifurcations, (which is called an R sequence) (Ref. [124]), from dc equilibrium through a Hopf bifurcation and period-doubling sequence to a spiral-Chua attractor and a double-scroll Chua attractor (see below), as illustrated in Fig. 7.6. In the experimental study, a two-dimensional projection of the attractor is obtained by connecting v_1 and v_2 to the X and Y channels, respectively, of an X–Y oscilloscope.

An alternative way to view the bifurcation sequence is by adjusting C_1. In this case, by fixing the value of R at 1800Ω and varying C_1 from 12.0nF to 6nF, the full range of bifurcations from equilibrium through Hopf bifurcation, period-doubling sequence to spiral-Chua attractor and double-scroll Chua attractor can be obtained (see Ref. [124] for complete results and discussions). Some of the important features associated with Chua's circuit are the following.

1. Chua's circuit exhibits a number of different scenarios in the appearance of chaos, namely, transition to chaos through period-doubling cascade (Refs. [124, 127]), through breakdown of invariant torus (Ref. [19]), etc. All of these bifurcation phenomena make the study of Chua's circuits a rather universal problem.

2. Chua's circuit exhibits a typical chaotic attractor called "double-scroll Chua's attractor" (Refs. [125, 127]). It appears at the conjunction of a pair of nonsymmetric spiral attractors. Three equilibrium states of saddle-focus type are visible in this attractor, which indicates that the double-scroll Chua's attractor is *multistructural*, which distinguishes it from other known attractors of three-dimensional systems.

3. As regards their mathematical nature, the attractors which occur in Chua's circuits are essentially more complicated objects than they had seemed before. This conclusion is based on new subtle results on systems with homoclinic tangencies and homoclinic loops of a saddle focus (Ref. [125]).

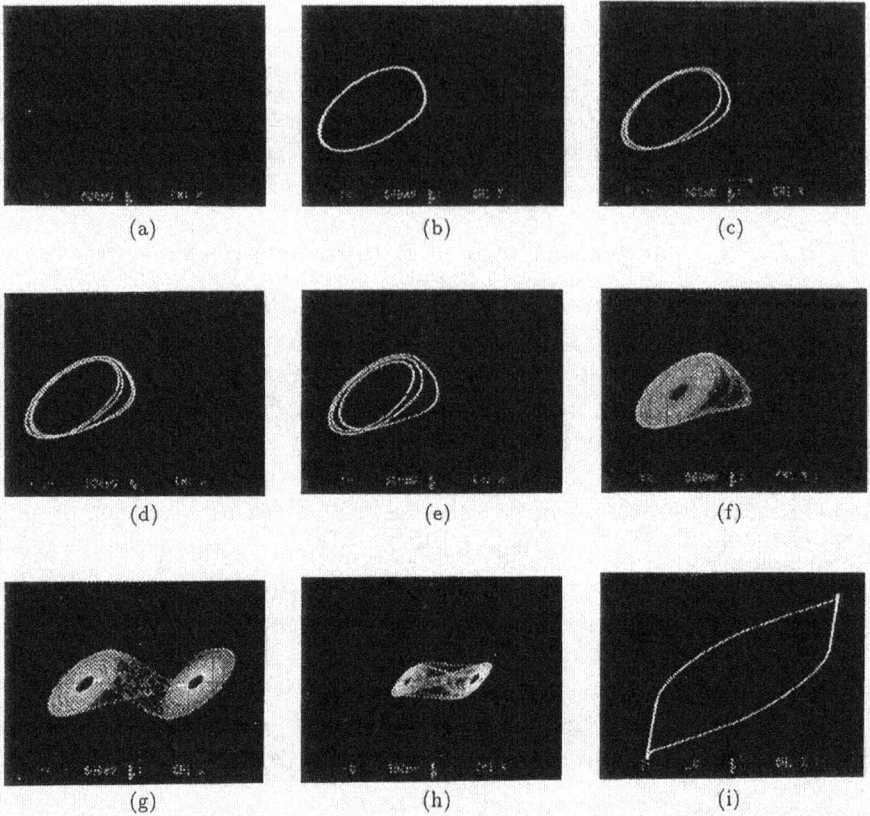

Fig. 7.6. Typical R bifurcation sequence in Chua's circuit (component values as in Table 7.1). (a) $R = 2.00\text{k}\Omega$, dc equilibrium; (b) $R = 1.88\text{k}\Omega$, period-1; (c) $R = 1.85\text{k}\Omega$, period-2; (d) $R = 1.84\text{k}\Omega$, period-4; (e) $R = 1.825\text{k}\Omega$, period-3 window; (f) $R = 1.79\text{k}\Omega$, Rössler-type chaotic attractor; (g) $R = 1.74\text{k}\Omega$, Double-scroll attractor; (h) $R = 1.49\text{k}\Omega$, Double-scroll attractor; (i) $R = 1.40\text{k}\Omega$, large limit cycle corresponding to outer segments of the $V–i$ characteristic.

4. The dimensionless form of Eq. (7.2) is

$$\dot{x} = \alpha(y - x - h(x)),$$
$$\dot{y} = x - y + z, \tag{7.3}$$
$$\dot{z} = -\beta y, \quad (\cdot = d/d\tau)$$

where $h(.)$ is the rescaled nonlinear function representation of $f(v_1)$ in Eq. (7.2), $h(x) = bx + 0.5(a - b)[|x + 1| - |x - 1|]$, and $v_1 = xB_p, v_2 = yB_p, i_L = (B_pG)z,$

$t = C_2\tau/G, G = 1/R, a = RG_a, b = RG_b, \alpha = (C_2/C_1), \beta = C_2R^2/L$. Equations (7.3) are rather close to the equations of a three-dimensional normal form for bifurcations (Ref. [32]) of an equilibrium state with three zero characteristic exponents (for the case with additional symmetry), and that for a periodic orbit with three multipliers equal to -1.

7.3.2. *Chua's Autonomous Oscillator*

By adding a linear resistor R_0 in series with the inductor in the Chua's circuit of Fig. 7.5, one can obtain an unfolded circuit called Chua's oscillator (Refs. [19, 74, 123–127]) as shown in Fig. 7.7(a). This circuit is *canonical* in the sense that every continuous three-dimensional odd-symmetric piecewise-linear vector field may be mapped onto it (Refs. [19, 123]). In other words, this circuit can exhibit every dynamical behaviour known to be possible in a third-order autonomous dynamical system described by a continuous odd-symmetric three-region piecewise-linear vector field (Refs. [19, 125, 126]).

The circuit equations of Fig. 7.7(a) are

$$C_1 \frac{dv_1}{dt} = (1/R)(v_2 - v_1) - f(v_1),$$

$$C_2 \frac{dv_2}{dt} = (1/R)(v_1 - v_2) + i_L, \qquad (7.4)$$

$$L \frac{di_L}{dt} = -(v_2 + R_0 i_L),$$

where $f(v_1)$ is again the mathematical representation of Chua's diode (see Fig. 7.7(b)).

Equation (7.4) is called a *global unfolding* of Chua's circuit (Ref. [125]). By an appropriate change of variables, one can transform Eq. (7.4) into the following dimensionless form:

$$\dot{x} = k\alpha(y - x - h(x)),$$

$$\dot{y} = k(x - y + z), \qquad (7.5)$$

$$\dot{z} = k(-\beta y - \gamma z),$$

where $\gamma = (RR_0C_2/L), \tau = t/|RC_2|$. Here x, y, z, G, a, b, α, and β are the same as in Eq. (7.3). The derivatives in Eq. (7.5) are with respect to t (by redefining τ as t).

In Eq. (7.5) k is defined as follows:

$$k = \begin{cases} 1 & \text{if } RC_2 > 0, \\ -1 & \text{if } RC_2 < 0. \end{cases}$$

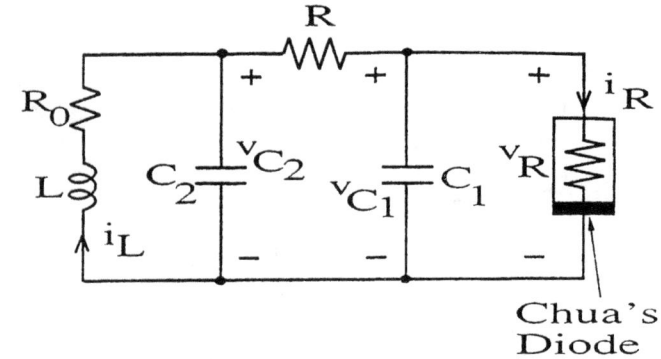

(a)

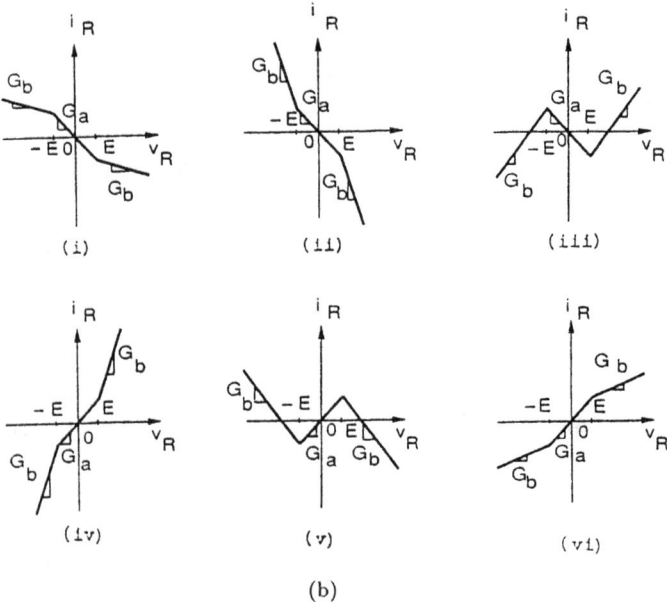

(b)

Fig. 7.7. (a) The Chua's oscillator. (b) Typical odd-symmetric v–i characteristic curves with three piecewise-linear segments having an inner slope G_a and an outer slope G_b: (i) $G_a < G_b \leq 0$, (ii) $G_b < G_a \leq 0$, (iii) $G_a \leq 0, G_b \geq 0$, (iv) $G_b > G_a \geq 0$, (v) $G_a > 0, G_b \leq 0$, (vi) $G_a > G_b \geq 0$. (Refs. [19, 125, 128]). (c) A gallery of some strange attractors from the Chua's oscillator (Refs. [19, 125, 128]).

To appreciate the global unfolding nature of Eq. (7.5) we may consider the characteristic exponents $\lambda = (\lambda_1, \lambda_2, \lambda_3)$ associated with the equilibrium state 0 at the origin and the set of characteristic exponents corresponding to $x > 1$

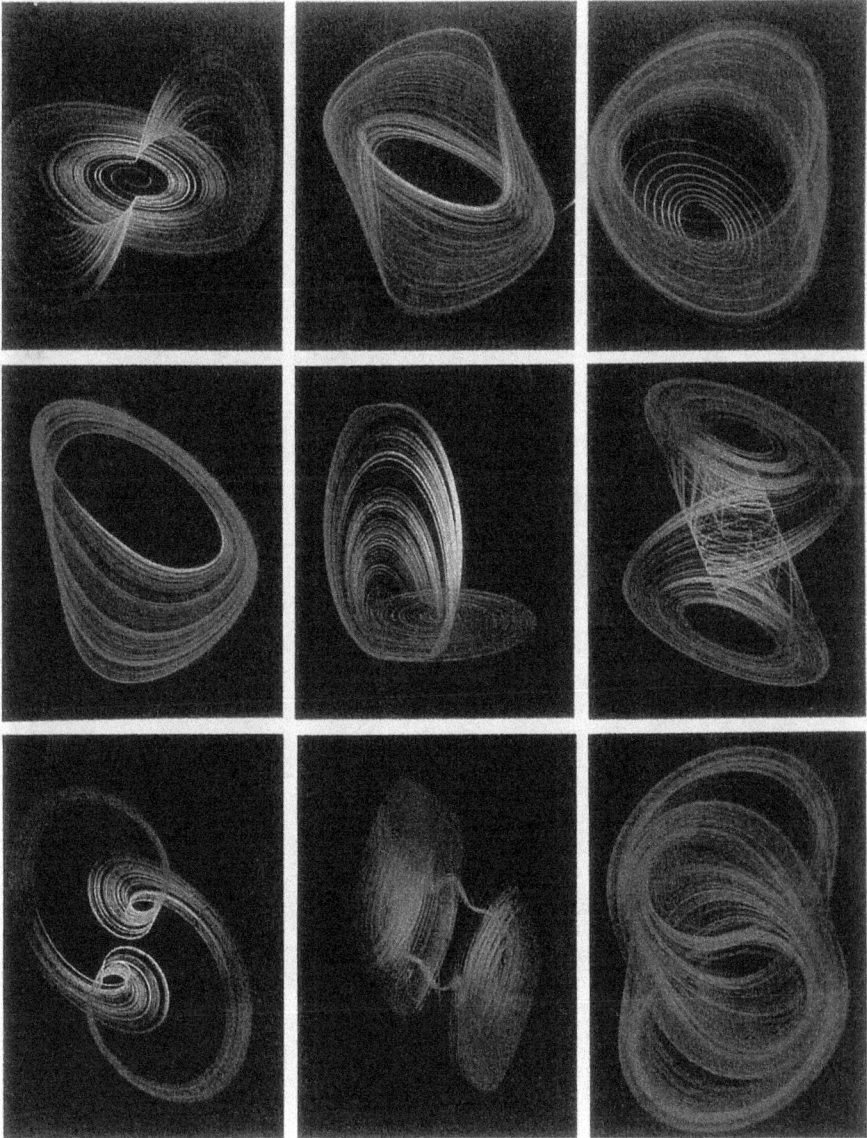

(c)

Fig. 7.7. (Continued)

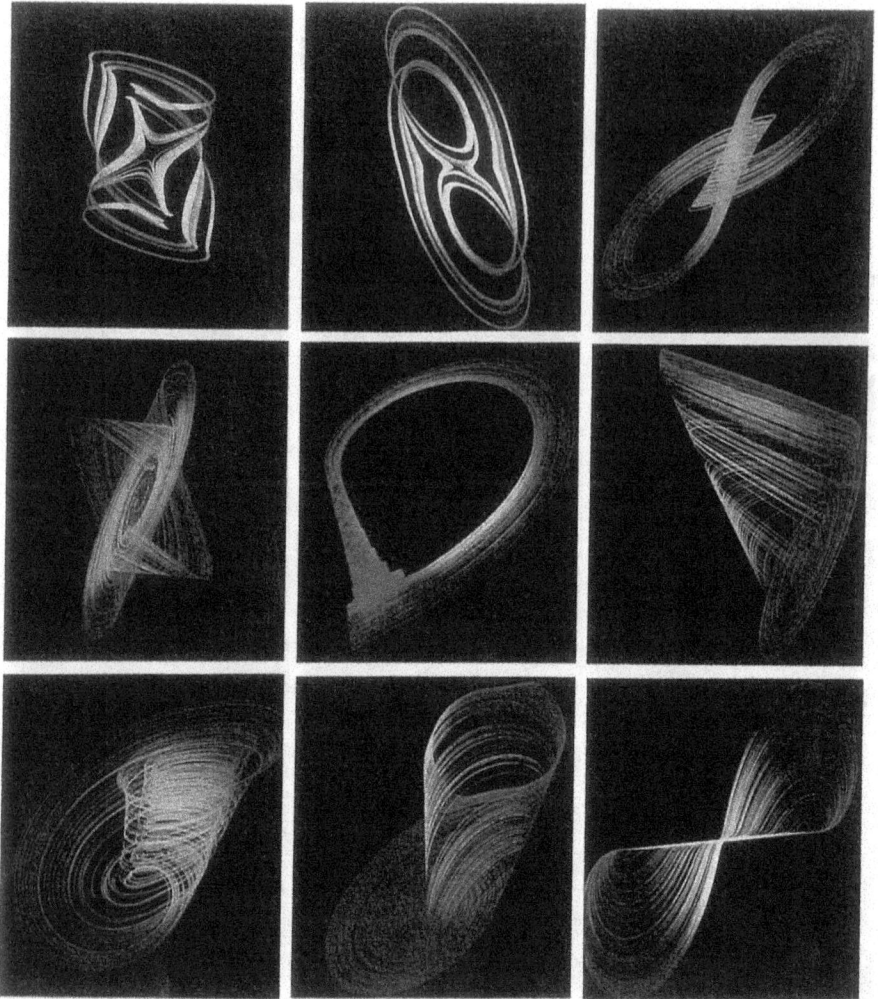

(c)

Fig. 7.7. (*Continued*)

(or, what is the same, to $x < 1$) as $\nu = (\nu_1, \nu_2, \nu_3)$. Then it is well known
(Refs. [19, 125]) that two Chua's circuits are equivalent if they have equal sets
of λ and ν. Therefore, it is natural to choose these sets as control parameters.
Moreover, all of the 12 piecewise-linear three-dimensional systems with three
regions of linearity (except for a set of zero measure in the parameter space)

are mapped into the Chua's oscillator, Eq. (7.4), in a natural way. A more detailed account of this aspect is presented in Refs. [19, 123], where a number of typical shapes of attractors existing at different values of λ and ν are also discussed in detail. A gallery of some strange attractors of Chua's oscillator is given in Fig. 7.7(c).

7.4. Non-autonomous Circuits with Chua's Diode [Refs. 130, 131]

It is of considerable interest to discuss the effect of additional periodic forcing on the Chua's autonomous circuit or oscillator. As an example, in this section, we study the dynamics of a nonautonomous Chua's circuit, first by adding a sinusoidal forcing in series with the inductor L of Fig. 7.5. Then we present a detailed investigation of the chaotic dynamics of the Chua's circuit with an additional inductor connected parallel to the forcing.

7.4.1. Simplified Driven Chua's Circuit

Apart from the interesting dynamics exhibited by the autonomous Chua's circuit of Fig. 7.5, the nonautonomous Chua's circuit, obtained by including an external periodic signal also turns out to be a veritable black box to study the various bifurcation sequences and chaotic structures. First let us consider a simplified driven Chua's circuit as given in Fig. 7.8. In this circuit $f(t)$ is an external periodic signal, $f(t) = F' \sin(\omega t)$, connected in series with the inductor of the autonomous Chua's circuit. By applying Kirchoff's laws to

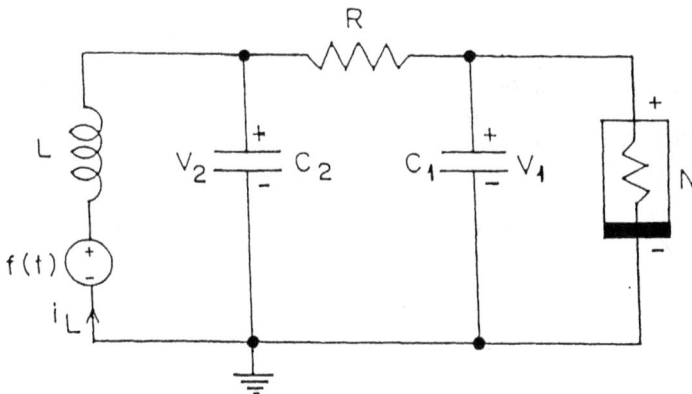

Fig. 7.8. Simplified driven Chua's circuit.

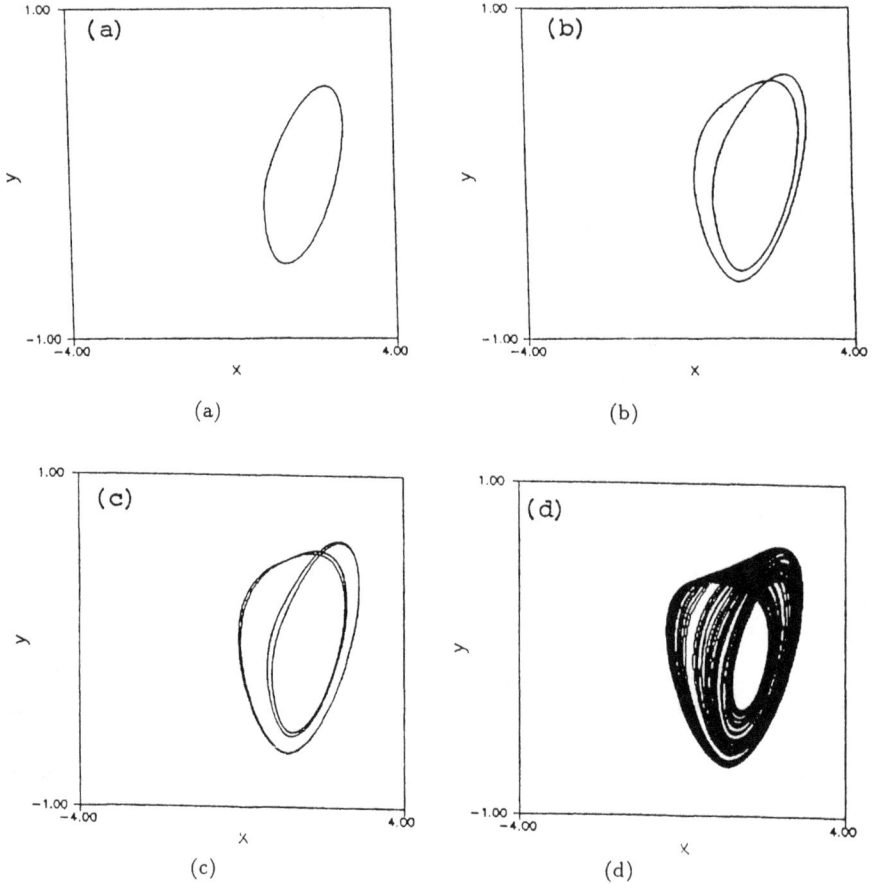

(a)

(b)

(c)

(d)

Fig. 7.9. Phase-portraits of Eq. (7.7). $\alpha = 7.0, \beta = 14.87, \Omega = 3.0..$ (a) $f = 0.0$ (Period-T limit cycle). (b) $f = 0.4$ (Period-$2T$ limit cycle). (c) $f = 0.478$ (Period-$4T$ limit cycle). (d) $f = 0.6$ (One-band chaos). (e) $f = 0.7$ (Double-band chaos). (f) $f = 1.25$ (Period-$3T$ window). (g) $f = 1.5$ (Double-band chaos).

various branches of this circuit (Fig. 7.8), the following circuit equations can be derived:

$$C_1 \frac{dv_1}{dt} = (1/R)(v_2 - v_1) - f(v_1),$$

$$C_2 \frac{dv_2}{dt} = (1/R)(v_1 - v_2) + i_L, \qquad (7.6)$$

$$L \frac{di_L}{dt} = -v_2 + F' \sin(\omega t),$$

(e)

(f)

(g)

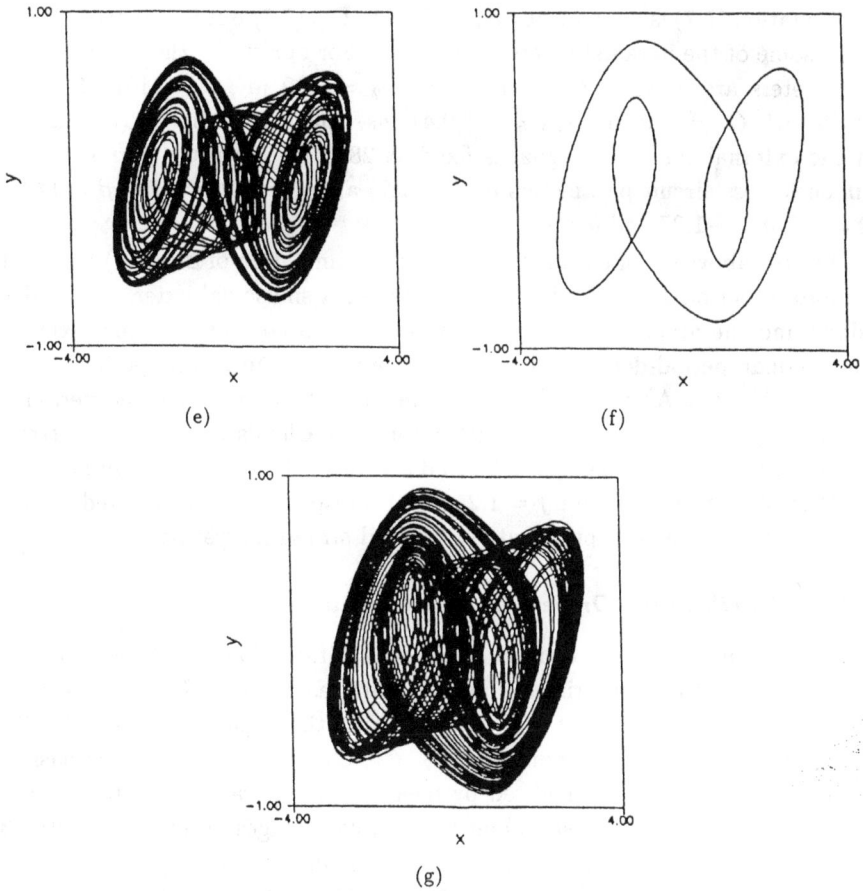

Fig. 7.9. (*Continued*)

where $f(v_1)$ is again the characteristic curve of the Chua's diode (Fig. 7.1) given by Eq. (7.1b)

Equation (7.6) can be recast into a dimensionless form, via rescaling as for Eq. (7.3) and with $f = F'\beta/B_p$, $\omega = G\Omega/C_2$,

$$\dot{x} = \alpha(y - x - h(x)),$$
$$\dot{y} = x - y + z, \qquad\qquad (7.7)$$
$$\dot{z} = -\beta y + f\sin(\Omega t),$$

where $h(x)$ is as defined in Eq. (7.3).

Equation (7.7) is dynamically equivalent to Eq. (7.6) but is more convenient since some of the parameters are normalized. For our study, the actual circuit parameters are fixed at $C_1 = 14.3$ nF, $C_2 = 100$ nF, $R = 1670 \ \Omega$, $L = 18.75$ mH, $G_a = -0.76$ ms, $G_b = -0.41$ ms, and the frequency ($\nu = \omega/2\pi$) of the external sinusoidal signal is fixed at 2860Hz. Then the corresponding dimensionless circuit parameters of Eq. (7.7) are given as $\alpha = 7.0$, $\beta = 14.87$, $\Omega = 3.0$, $a = -1.27$ and $b = -0.68$.

For the above set of parameters, numerical simulation of Eq. (7.7) for $f = 0$ exhibits a period-1 ($= \frac{2\pi}{\Omega}$) limit cycle. When a sinusoidal external signal is added and the amplitude f is varied from zero upwards, the system exhibits the familiar period-doubling bifurcation route to chaos, then periodic windows, and so on. As a typical example, in Fig. 7.9, we have plotted period-1, period-2, period-4, spiral Chua's attractor, and Chua's double-scroll attractor respectively at $f = 0.0, 0.4, 0.478, 0.6$ and 0.7. Also a period-3 window and another chaotic attractor for $f = 1.25$ and 1.5 respectively are depicted. These attractors can also be experimentally realized straightforwardly.

7.4.2. *Fourth-order Driven Chua's Circuit*

Further, in order to observe more exhaustive bifurcation structures, one can consider a fourth-order driven Chua's circuit (as given in Fig. 7.10(a)), by including the second inductor in parallel with the capacitor C_2 of Fig. 7.8 (Refs. [18, 130]). In this circuit, L_1, L_2, C_1, C_2 and R are all linear passive elements. Chua's diode is realized by means of an operational amplifier with a pair of diodes, seven resistors and DC supply voltages of $+9$V and -9V as shown in Fig. 7.10(b). This is an alternative way of realizing Chua's diode and was originally implemented by Matsumoto *et al.* (Refs. [135–137]) (see Sec. 7.2). The state equations are obviously

$$C_1 \frac{dv_1}{dt} = (1/R)(v_2 - v_1) - f(v_1), \qquad (7.8a)$$

$$C_2 \frac{dv_2}{dt} = (1/R)(v_1 - v_2) + i_{L2} - i_{L1}, \qquad (7.8b)$$

$$L_1 \frac{di_{L1}}{dt} = v_2 - F \sin(\omega t), \qquad (7.8c)$$

$$L_2 \frac{di_{L2}}{dt} = -v_2. \qquad (7.8d)$$

Here v_1, v_2, i_{L1} and i_{L2} are the voltage across C_1, the voltage across C_2, the current through L_1, and the current through L_2 respectively.

(a)

(b)

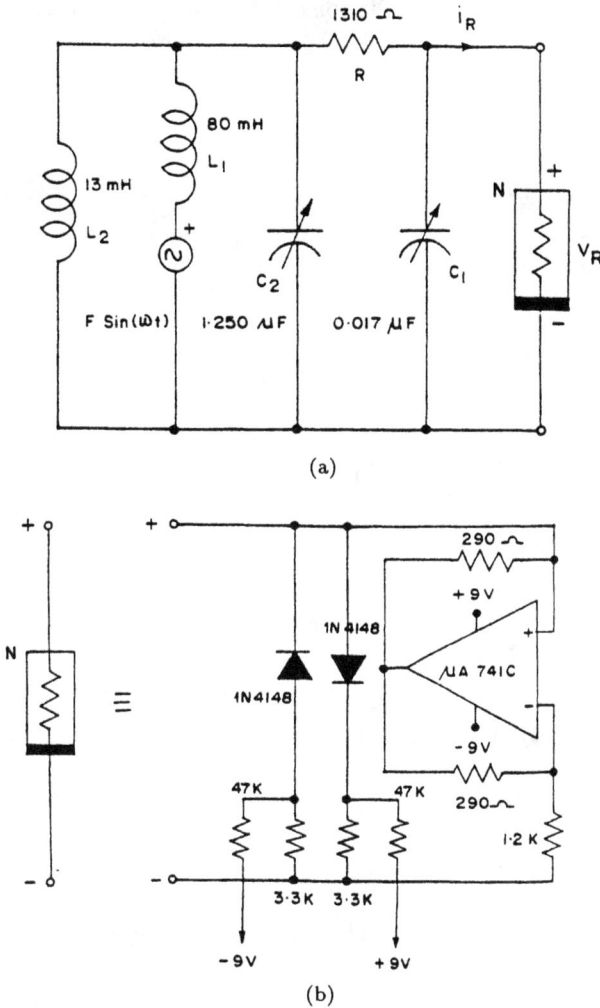

Fig. 7.10. (a) Circuit realization of the driven piecewise-linear circuit. (b) Circuit realization of the nonlinear element N of (a). This is an alternative to the realization of Fig. 7.2.

When the second inductor (L_2) is absent and the external force is zero in the circuit of Fig. 7.10(a), we have the standard Chua's autonomous circuit. Due to the inclusion of additional inductor L_2 and a periodic forcing, the behaviour of the circuit depends on seven controlling parameters, namely, C_1, C_2, L_1, L_2, R, the amplitude F, and the frequency $\nu(= \omega/2\pi)$ of the external force. It is also of interest to note that by appropriate transformations one can obtain the simplified driven Chua's circuit equations (7.6) from (7.8). To see this, we first multiply Eq. (7.8c) and Eq. (7.8d) by L_2 and L_1 respectively and obtain

$$L_1 L_2 \frac{di_{L1}}{dt} = L_2 v_2 - L_2 F \sin(\omega t), \qquad (7.8c')$$

$$L_1 L_2 \frac{di_{L2}}{dt} = -L_1 v_2. \qquad (7.8d')$$

By subtracting Eq. (7.8c′) from (7.8d′), we get

$$L_1 L_2 \frac{d}{dt}(i_{L2} - i_{L1}) = -(L_1 + L_2)v_2 + L_2 F \sin(\omega t)$$

or

$$\left[\frac{L_1 L_2}{(L_1 + L_2)}\right] \frac{d}{dt}(i_{L2} - i_{L1}) = -v_2 + \left[\frac{L_2}{(L_1 + L_2)}\right] F \sin(\omega t). \qquad (7.8e)$$

By replacing $\left[\frac{L_1 L_2}{(L_1 + L_2)}\right] = L$ and $(i_{L2} - i_{L1}) = i_L$ in Eq. (7.8e), we get

$$L \frac{di_L}{dt} = -v_2 + F' \sin(\omega t), \qquad (7.8f)$$

where $F' = \frac{L_2 F}{(L_1 + L_2)}$. Then the combined set of Eqs. (7.8a–d) can be represented as Eq. (7.6), while the equation for $I_L = L_2 i_{L2} + L_1 i_{L1}$ becomes decoupled into

$$\frac{dI_L}{dt} = -F \sin(\omega t), \qquad (7.8g)$$

so that

$$I_L(t) = (L_2 i_{L2} + L_1 i_{L1}) = -(F/\omega)[1 - \cos(\omega t)] + \text{constant}. \qquad (7.8h)$$

Thus knowing i_L, using (7.8h), i_{L1} and i_{L2} can be individually determined. Hence it is sufficient to consider Eqs. (7.6) alone for numerical studies. However for experimental studies, we find it more convenient to use the fourth-order system of Fig. 7.10(a) as such.

In our study, the circuit parameters are fixed at the values $C_1 = 0.017\mu F$, $C_2 = 1.250\mu F$, $L_1 = 80$ mH, $L_2 = 13$ mH and $R = 1310\Omega$. By varying the amplitude F from 0 to 800mV and the frequency $\nu(= \omega/2\pi)$ from 800 to 1500Hz, the dynamics of the circuit (Fig. 7.10) is studied. We note that for the present set of circuit parameters, the autonomous case ($F = 0$) corresponds to the usual fixed point attractor.

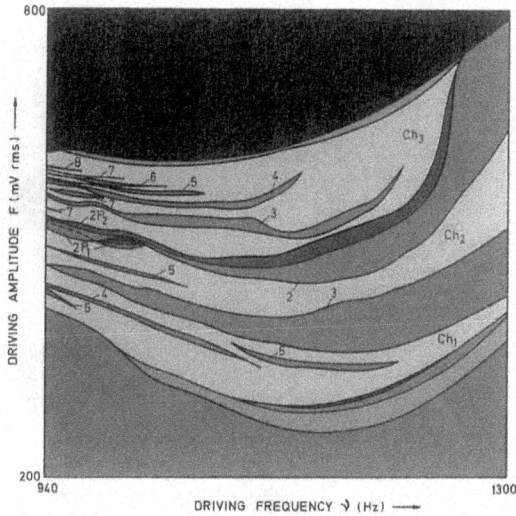

(a)

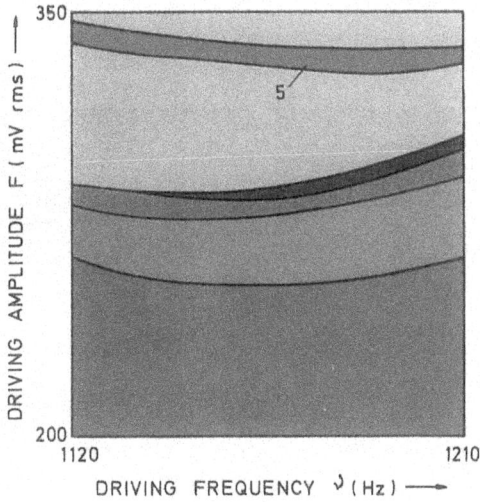

(b)

Fig. 7.11. (a) Global bifurcation diagram (driving amplitude-driving frequency). Numbers indicate the periods of window regions. Different attractors are colour-coded as follows: Period-1 =red; Period-2 =orange; Period-4 =green; Period-8 =brown; Chaos (Ch$_1$, Ch$_2$, and Ch$_3$) = yellow; Boundary crisis = blue; Periodic windows = pink; 2× period of windows = light blue. (b) Bifurcation diagram (blow-up) in the F–ν plane of the period-doubling region. (c) Bifurcation diagram (blow-up) in the F–ν plane of the period-adding region. (d) Bifurcation diagram (blow-up) in the F–ν plane of the equal-periodic bifurcation region.

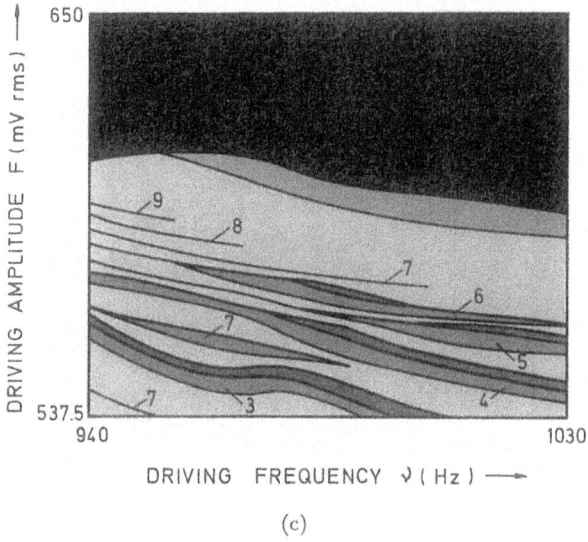

(c)

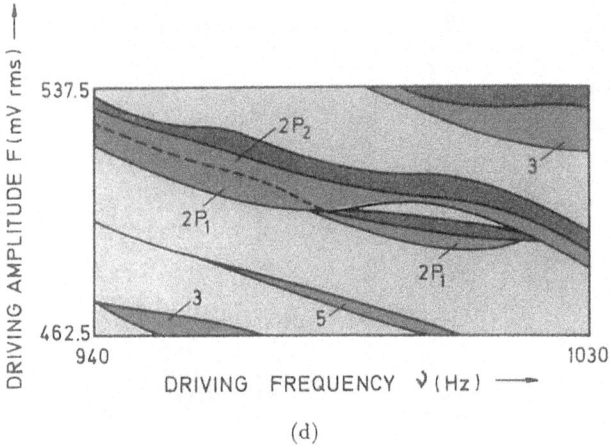

(d)

Fig. 7.11. (*Continued*)

A. *The bifurcation diagram*: Based on detailed measurements of voltage changes across the capacitors C_1 and C_2 and the associated transition to chaos with different forcing parameter values in the forcing frequency range 800Hz to 1500Hz, a profile of bifurcation diagram in the F–ν (drive amplitude–drive frequency) parameter plane has been constructed as shown in Fig. 7.11. This figure reveals the kind of oscillations which the circuit admits for each value of the driving amplitude F and the frequency ν. We have traced the trajectories of voltage v_1 across C_1 and voltage v_2 across C_2 and noted the changes in the attractor projected onto the v_1–v_2 plane directly on the oscilloscope screen. A live picture of the corresponding Poincaré map of the projected attractor has been produced using the Poincaré map circuit as shown in Fig. 7.12. The stroboscopic Poincaré maps are produced by triggering the beam of the oscilloscope at the driving frequency $\nu(= \omega/2\pi)$ (z-modulation) (Refs. [18, 130, 131]).

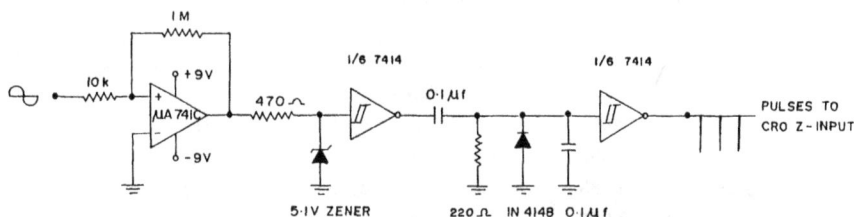

Fig. 7.12. Circuit realization of the Poincaré map circuit.

B. *Period-doubling scenario*: For the experimental circuit, when the driving amplitude $F = 0$ (corresponding to the autonomous Chua's circuit) only a fixed point attractor is observed. When F is slightly increased ($F > 0$), Hopf bifurcation occurs and only period-1 motion synchronized with the external signal is observed. As F is increased beyond 200mV the system undergoes a cascade of period-doubling bifurcations leading to chaos in the middle of the frequency region (see Fig. 7.11). These bifurcation sequences can be easily identified either by direct observation of the projected attractor or by observation of the Poincaré map displayed on the oscilloscope. As the order of the period in the period-doubling sequence increases, the difference between the drive amplitudes (F) corresponding to two successive bifurcations becomes smaller and smaller. The calculated convergence rate is close to the Feigenbaum's universal constant.

Beyond the period-doubling bifurcations, we observe three different chaotic regimes Ch_1, Ch_2, and Ch_3 interspersed with several periodic windows. These windows also undergo further period-doubling bifurcations, which are however too narrow to be depicted in Fig. 7.11, except for the windows lying in the region Ch_3. However, for higher drive amplitude values, boundary crisis (dark blue region) is usually observed. Table 7.2 along with Figs. 7.13(a) presents a brief summary of bifurcation phenomena as the amplitude F is varied with fixed frequency $\nu = 1150$ Hz. Also, we have depicted a typical chaotic attractor projected onto the (v_1-v_2) plane and its corresponding Poincaré map in the region Ch$_2$ in Fig. 7.13(b). Also, one can observe period-halving or reverse period-doubling bifurcations upon increasing or decreasing the forcing frequency values at lower drive amplitudes (see Fig. 7.11).

Table 7.2. Brief summary of bifurcation phenomena of the circuit of Fig. 7.10 (External forcing frequency $\nu = 1150Hz$).

Amplitude F V	Description of attractor	Fig. No. 7.13(a)
$0 < F < 0.01$	Stable fixed point	
0.01	Hopf bifurcation	
$0.01 < F \leq 0.265$	Period-1 limit cycle	(i)
$0.265 < F \leq 0.283$	Period-2 limit cycle	(ii)
$0.283 < F \leq 0.285$	Period-4 limit cycle	(iii)
$0.285 < F \leq 0.287$	Period-8 limit cycle	
$0.287 < F \leq 0.341$	One-band chaos	(iv)
$0.341 < F \leq 0.347$	Period-5 window	
$0.347 < F \leq 0.365$	Double-band chaos	(v)
$0.365 < F \leq 0.49$	Period-3 Window	(vi)
$0.49 < F \leq 0.515$	Chaos	
$0.515 < F \leq 0.579$	Period-2 window	
$0.579 < F \leq 0.59$	Period-4 window	
$0.59 < F \leq 0.605$	Chaos	
$0.605 < F \leq 0.62$	Period-3 window	
$0.62 < F \leq 0.64$	Chaos	(vii)
$0.64 < F$	Boundary crisis (Period-1)	(viii)

C. *Period-adding scenario and Farey sequence*: In our studies a complicated succession of periodic and chaotic oscillations has been observed experimentally in the circuit of Fig. 7.10. Beyond the period-doubling region described earlier, we have further noted periodic windows of all orders starting from period-2 to period-9 in the region Ch$_3$ at lower driving frequency values and higher driver amplitudes (see Fig. 7.11). If we look at Fig. 7.11 (at lower frequency region) a succession of large periodic windows, whose periods increase exactly by one, appears when the drive amplitude is increased. The waveforms v_1 of the periodic windows starting from period-4 in the period-adding sequence are depicted in Fig. 7.14. It is difficult to record further higher

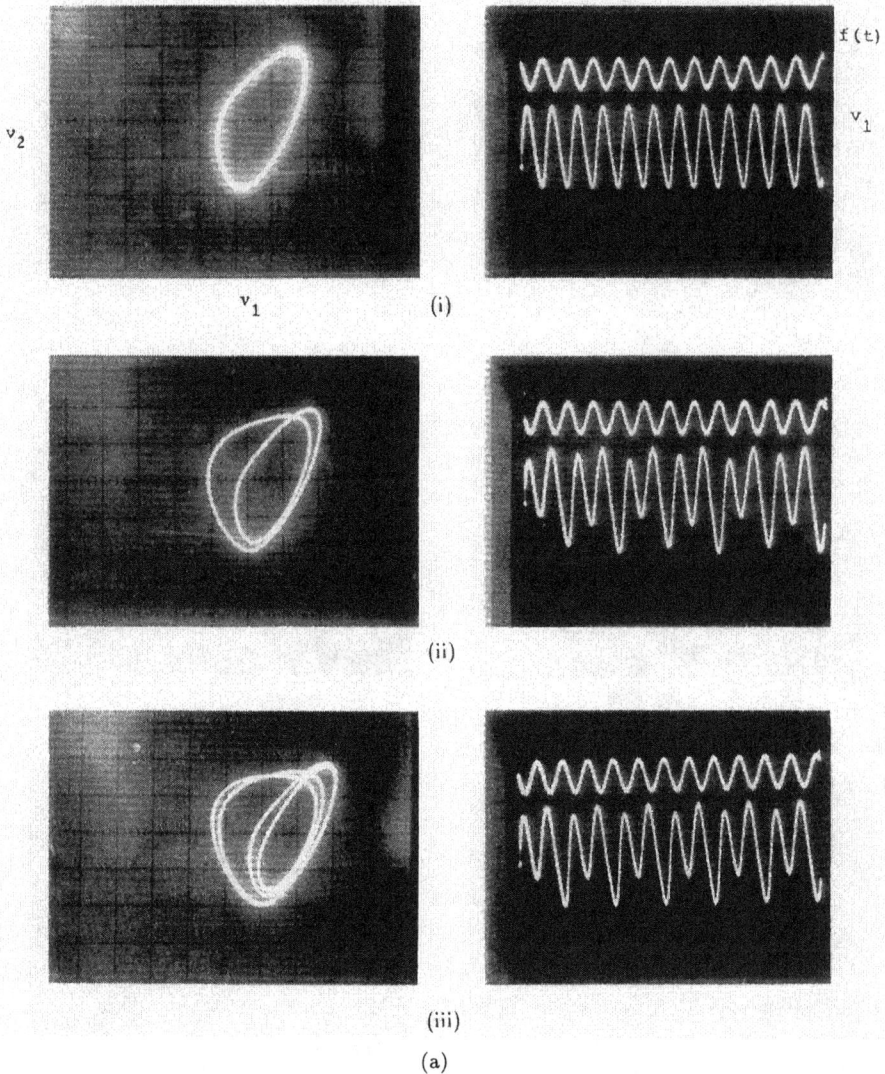

(a)

Fig. 7.13. (a): (i) Period-1 limit cycle; $F = 0.26V_{rms}$; (ii) Period-2 limit cycle; $F = 0.28V_{rms}$; (iii) Period-4 limit cycle; $F = 0.284V_{rms}$; (iv) One-band chaos; $F = 0.34V_{rms}$; (v) Double-band chaos; $F = 0.362V_{rms}$; (vi) Period-3 window; $F = 0.4V_{rms}$; (vii) Chaotic attractor; $F = 0.63V_{rms}$; (viii) Period-1 boundary; $F = 0.65V_{rms}$. (b): (i) Typical trajectory plot in the v_1–v_2 plane of a chaotic attractor ($F = 500mV$ and $\nu = 1200Hz$); (ii) Poincaré map of (i).

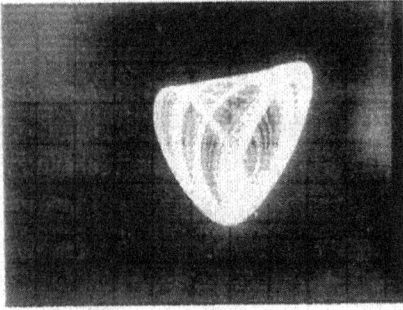 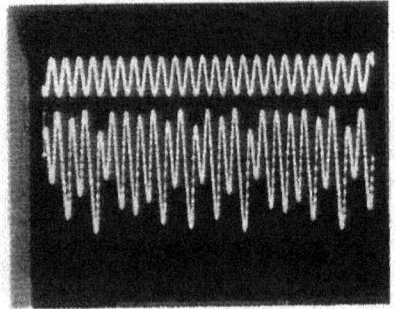

(iv)

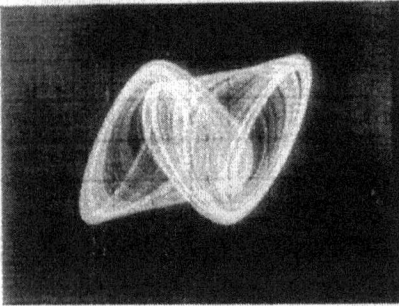 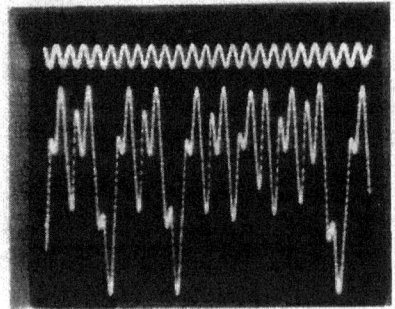

(v)

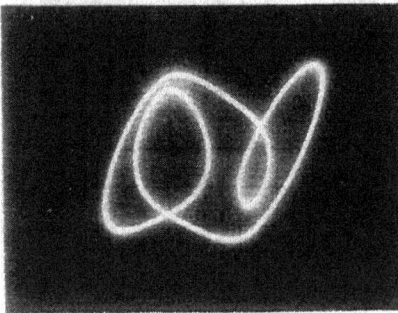 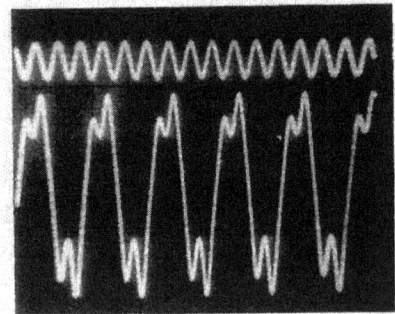

(vi)

(a)

Fig. 7.13. (*Continued*)

(vii) (viii)

(a)

(i) (ii)

(b)

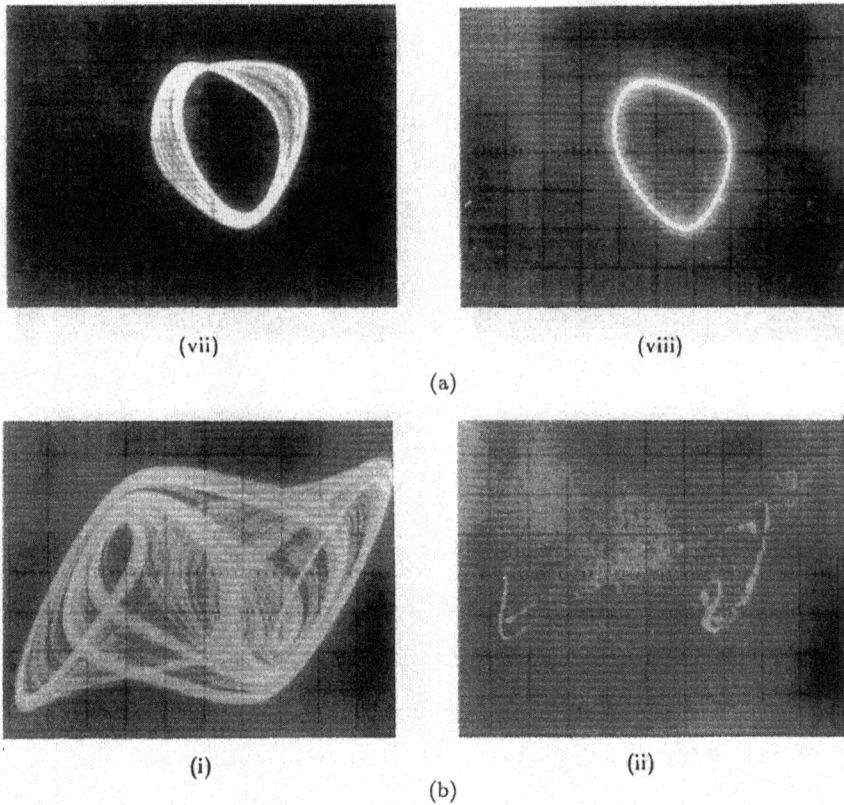

Fig. 7.13. (*Continued*)

period windows in this sequence since the separation between successive window regions decreases and the widths of the window regions diminish. The transition from one periodic window to another is initiated by a period-doubling bifurcation to chaos, followed by a recovery to the next periodic state in the period-adding sequence as shown in Fig. 7.11. The existence of period-adding sequences has also been reported in some driven negative resistance oscillators (Refs. [46, 69, 138]).

Also, we have found that the periods of some windows satisfy the familiar Farey sequence (Refs. [46, 69]) in the chaotic regions Ch_1, Ch_2 and Ch_3. Here we find that in between period-(n) window and period-$(n-1)$ window, there is a phase-locked window of period-$(2n-1)$. Specifically, we observed that for lower driving amplitude values, in the chaos region Ch_1 in Fig. 7.11, between period-3 and period-2 windows there exists a period-5$(= 3 + 2)$ window; between

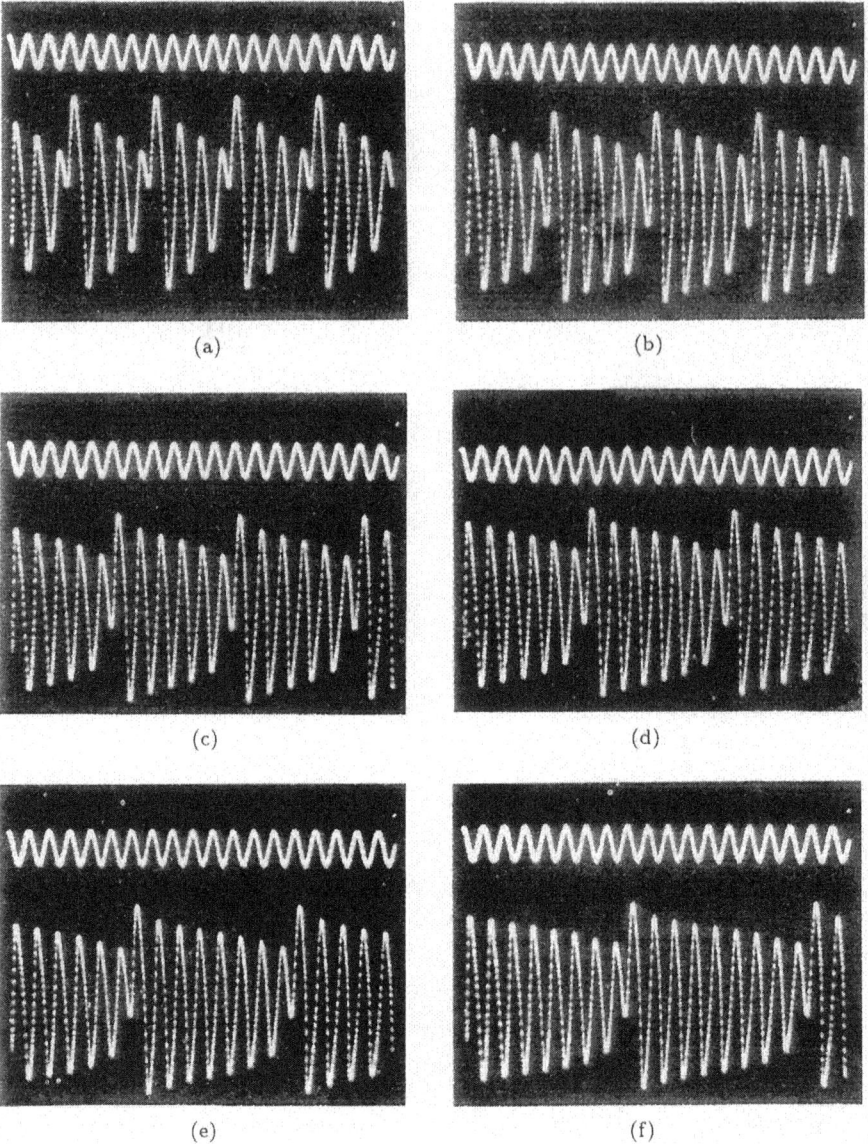

Fig. 7.14. The wave form v_1 (lower trace) along with the driving sine wave (upper trace) for the period-adding sequence for $\nu = 941$ Hz. Refer also to Figs. 7.11(c) and 7.15. (a) Period-4 ($F = 583.6$ mV), (b) Period-5 ($F = 593.6$ mV), (c) Period-6 ($F = 597.2$ mV), (d) Period-7 ($F = 602.6$ mV), (e) Period-8 ($F = 606.1$ mV), (f) Period-9 ($F = 609.4$ mV).

period-3 and period-1 windows there is a period-4 $(= 3 + 1)$ window, between period-4 and period-1 windows there is a period-5($=4+1$) window. In the chaos region Ch_2 there is a period-5($= 2 + 3$) window between period-2 and period-3 windows. In the chaos region Ch_3, a period-7($= 4 + 3$) window appears in between period-4 and period-3 windows.

The periodic windows are found to persist over a limited range of forcing amplitude (F) values, thereby creating a step-like bifurcation diagram when the forcing amplitude (F) versus the period of the observed window is plotted. In order to elucidate this structure the period diagram for $f = 941$ Hz is shown in Fig. 7.15.

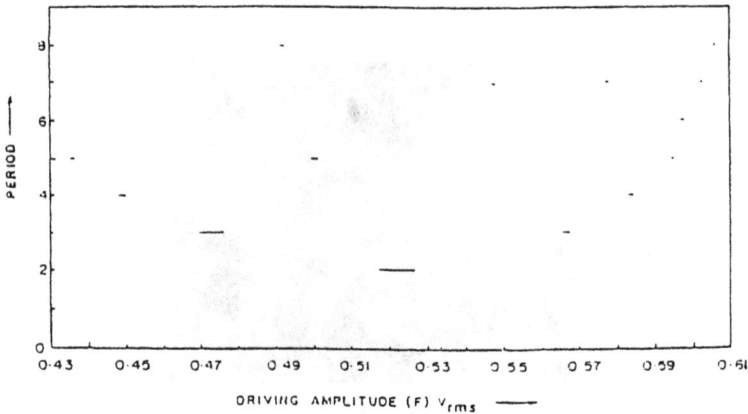

Fig. 7.15. Period diagram exhibiting the period-adding structure for $\nu = 941$ Hz.

D. *Intermittent behaviour*: Beyond certain critical forcing parameters, the system transits to intermittent behaviour. During this motion irregular bursts of chaos are interspersed with stable periodic window regions (laminar phases) lasting a variable length of time. Figure 7.16 depicts the waveform v_1, the trajectory plot in the (v_1-v_2) plane and its corresponding Poincaré map for an intermittency region near a period-2 window $(F = 589.7\text{mV}$ and $\nu = 840$ Hz). In the Poincaré map, the trace is continuous for long-time exposures, indicating chaotic behaviour. A portion of the time dependence of waveform v_1 is shown in Fig. 7.16(a), where the periodic oscillations are interrupted by intermittent voltage bursts. With further increase of amplitude F, the system gives birth to fully developed chaos. Another intermittency region near the period-2 window for $F = 465.9$ mV, $\nu = 1348$ Hz is shown in Fig. 7.17. In this case, as the frequency decreases further, the system transits to a complete chaotic

(a)   (b)

(c)

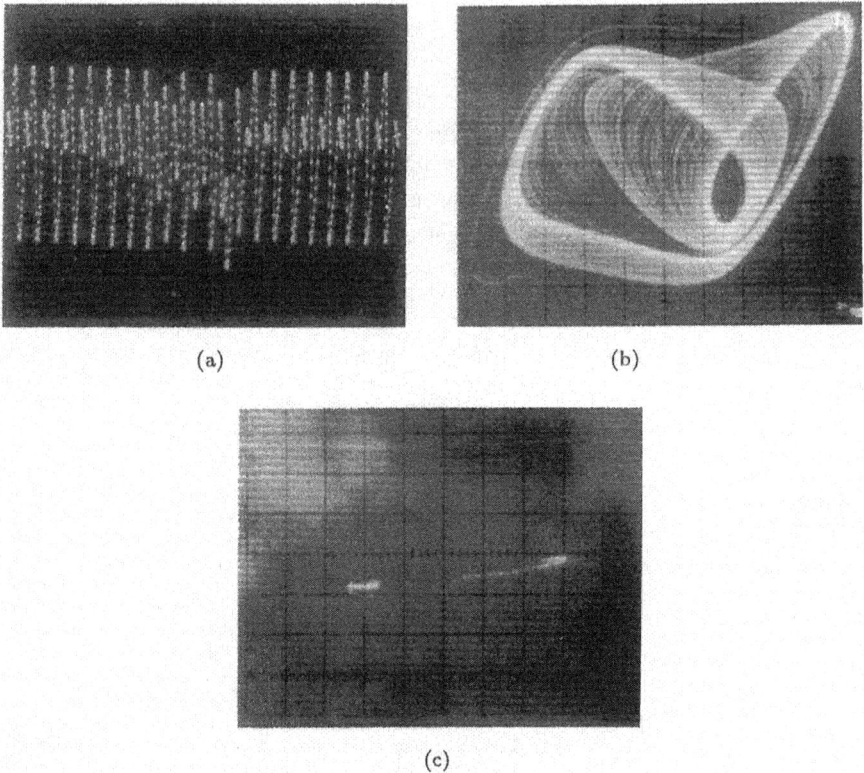

Fig. 7.16. (a) The wave form v_1. (b) the trajectory plot in the (v_1-v_2) plane, and (c) Poincaré map of an intermittency region for $F = 589.7$ mV and $\nu = 840$ Hz.

attractor. Also near the boundary crisis (Refs. [139, 140]) region beyond the period-adding sequence, the waveform or the attractors are dominated by intermittency. A typical waveform v_1, a trajectory plot in the (v_1-v_2) plane and its corresponding Poincaré map with intermittent transitions near the boundary crisis point is shown in Fig. 7.18 for $F = 620$ mV and $\nu = 941$ Hz. In these intermittent transition behaviours, the signal v_1 loses its regularity and chaotic burst appears at a particular sub-harmonic amplitude together with a decrease (or increase) of the fundamental amplitude. Immediately after this there is a reappearance of the regular (laminar) behaviour. This is the characteristic of type-III intermittency (Refs. [139, 140]).

E. *Equal-periodic bifurcation*: In a typical two-dimensional area-preserving map, if the control parameter is larger than some critical value, the original

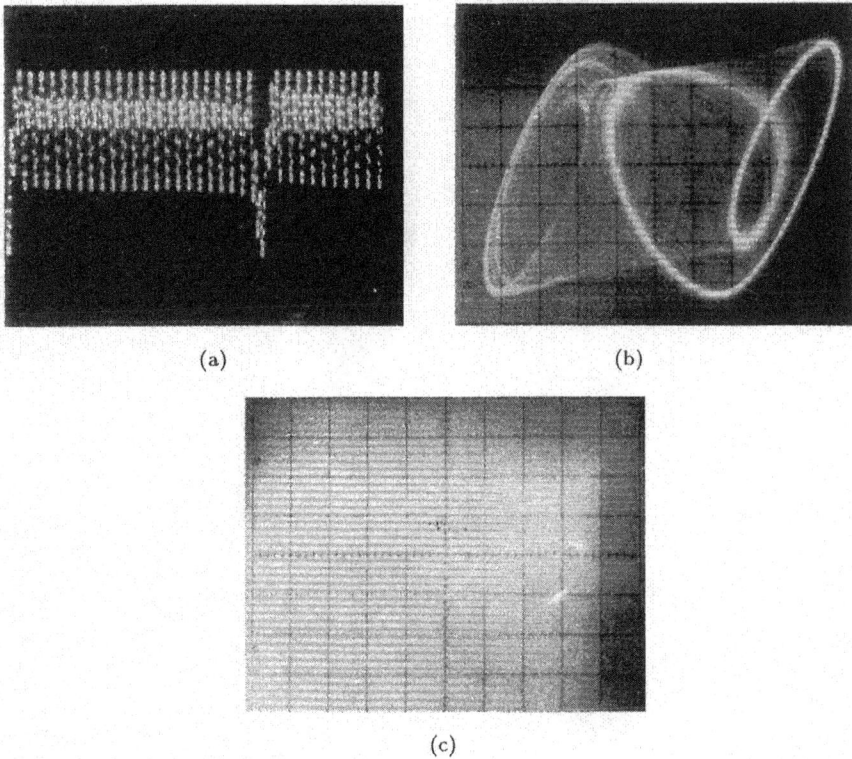

(a)

(b)

(c)

Fig. 7.17. (a) The wave form v_1. (b) the trajectory plot in the $(v_1$–$v_2)$ plane, and (c) Poincaré map of an intermittency region for $F = 465.9$ mV and $\nu = 1348$ Hz.

periodic motion is non-stationary and two new periodic motions appear together with the same period as the original one. This phenomenon is called equal-periodic bifurcation, and it is recently reported in some dissipative dynamical systems (Ref. [141]).

Now let us look at the period-2 window within the chaos regimes Ch$_2$ and Ch$_3$ in Fig. 7.11. There are two regions 2P$_1$ and 2P$_2$ divided by a dotted line for certain forcing parameter values, and a single 2P region for another set of parameters. The region below the dotted line is 2P$_1$ and that above the dotted line is 2P$_2$. For example, if we choose $\nu = 940$ Hz, the 2P$_1$ structure of Fig. 7.19(a) appears initially for $F = 518.6$ mV. This structure changes slightly as in Fig. 7.19(b) for $F = 523.3$ mV. Further increase of F gives birth to the 2P$_2$ structure as in Fig. 7.19(c) and it persists up to $F = 524.7$ mV. Here both the 2P$_1$ and 2P$_2$ structures have the same period 2 but their waveforms

(a)

(b)

(c)

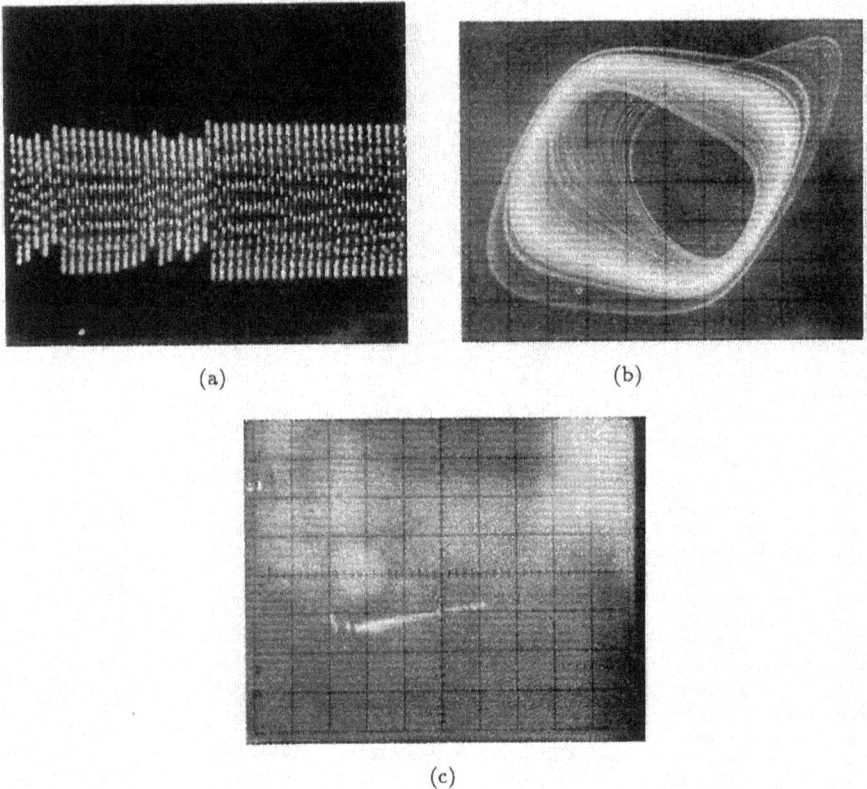

Fig. 7.18. (a) The waveform v_1. (b) the trajectory plot in the (v_1-v_2) plane, and (c) Poincaré map of an intermittency region for $F = 620$ mV and $\nu = 941$ Hz.

(v_1 or v_2) and their corresponding Poincaré maps in the (v_1-v_2) plane look different. An important point here is that there can be either $2P_1$ or $2P_2$ structure depending on the initial conditions in the area near the dotted line in Fig. 7.11. For example, if we initially cut off the external sinusoidal source and connect it instantly to the circuit, then either $2P_1$ or $2P_2$ oscillations appear randomly, but it is quite unpredictable. It is interesting to note that further change in the value of F after the equal-periodic bifurcation gives birth to period-doubling bifurcation in the period-$2P_2$ region. Further, the system also exhibits period-doubling bifurcations from $2P_1$ to $4P$, etc. for the set of parameters (F and ν) in the $2P_1$ region (see Fig. 7.11). Also a similar kind of equal periodic bifurcation has been observed in a small area of the period-3 window region in the chaos area Ch_3 for $F = 596.6$ mV, $\nu = 880$ Hz (Fig. 7.19(e)) and $F = 600.8$ mV, $\nu = 880$ Hz (Fig. 7.19(f)).

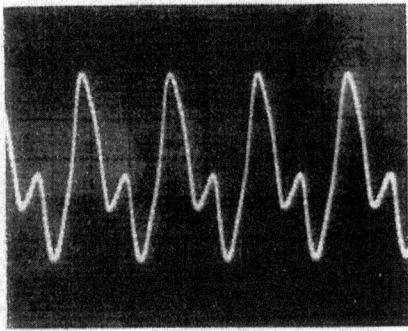

a(i)

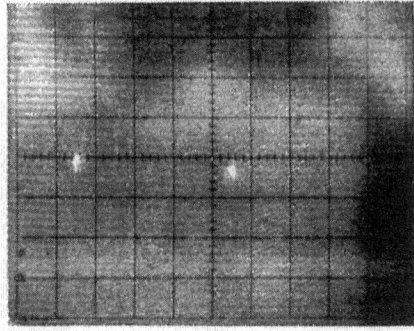

a(ii)

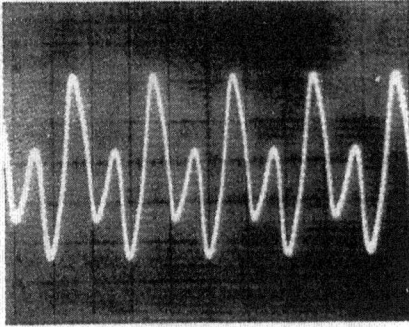

b(i)

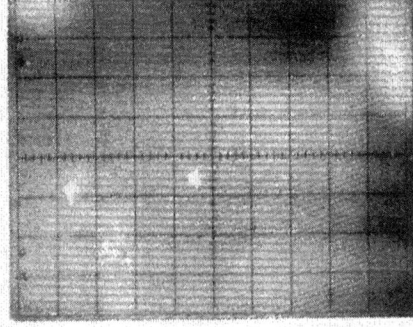

b(ii)

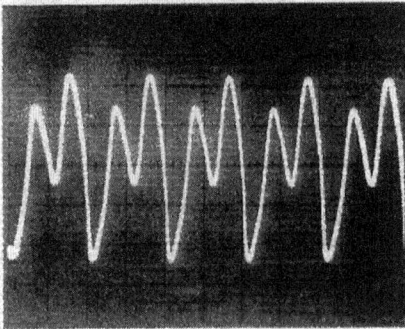

c(i)

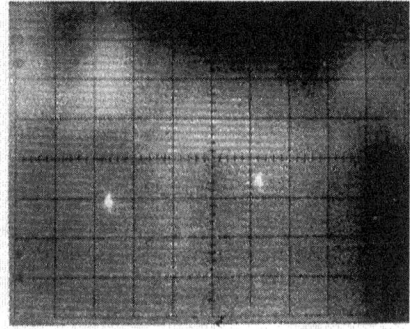

c(ii)

Fig. 7.19. (a): (i) The wave form v_1 and (ii) the strobed Poincaré map of the 2P$_1$ region ($F = 518.6$ mV, $\nu = 940$ Hz). (b) Same as (a) for the 2P$_1$ region ($F = 523.3$ mV, $\nu = 940$ Hz). (c) Same as (a) for the 2P$_2$ region ($F = 524.7$ mV, $\nu = 940$ Hz). (d) Same as (a) for the 2P$_2$ region ($F = 522.9$ mV, $\nu = 940$ Hz). (e) Same as (a) for the 3P$_1$ region ($F = 596.6$ mV, $\nu = 880$ Hz). (f) Same as (a) for the 3P$_2$ region ($F = 600.8$ mV, $\nu = 880$ Hz).

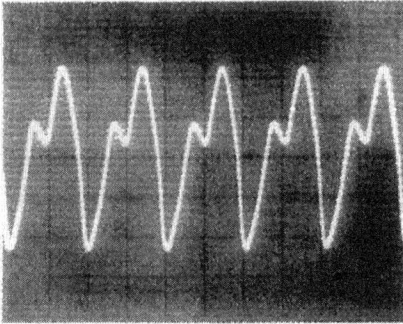

d(i)

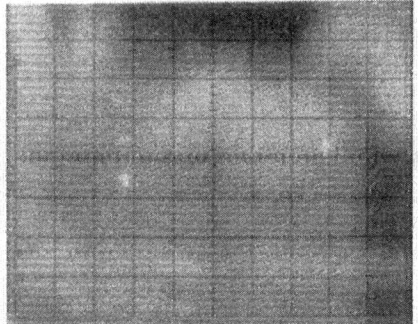

d(ii)

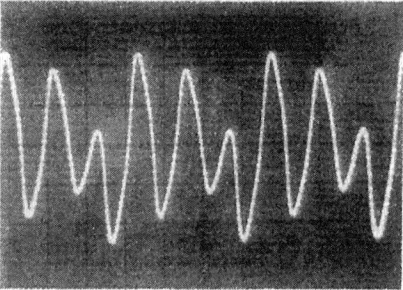

e(i)

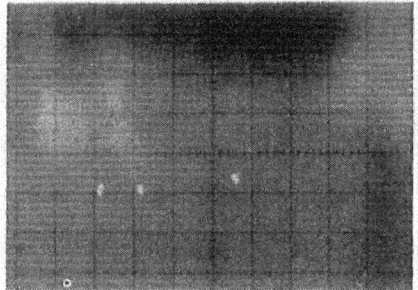

e(ii)

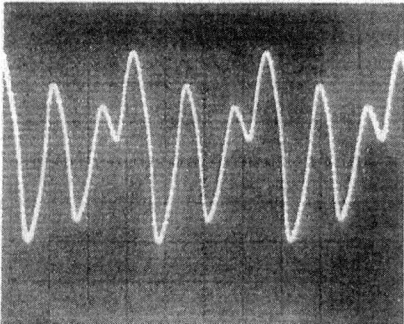

f(i)

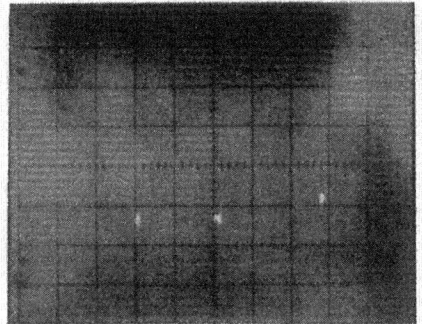

f(ii)

Fig. 7.19. (*Continued*)

F. *Hysteresis and coexistence of multiple attractors*: Interestingly hysteresis has also been observed in the driven Chua's circuit, but it does not seem to have any influence over the equal-periodic bifurcation. As discussed above, within the equal-periodic bifurcation region, the transition from $2P_1$ structure (Fig. 7.19(a)) to $2P_2$ structure (7.19(c)) is observed by increasing the amplitude F with fixed frequency $\nu = 940$Hz. However, the structure of Fig. 7.19(b) is not recovered after the equal-periodic bifurcation by resetting the value of F, that is the transition path is irreversible. Even if we decrease the F value from 524.7mV, the $2P_2$ structure of Fig. 7.19(c) persists initially and a slightly changed structure of Fig. 7.19(d) is observed for $F = 522.9$mV. With further decrease in the F value below 522.9mV, the $2P_1$ structure of Fig. 7.19(a) is observed instead of Fig. 7.19(b). Thus the system follows a different transition path while F is decreased rather than when F is increased, which is reminiscent of hysteresis (Refs. [141, 142]). The structure still possesses equal-periodic bifurcation even during this transition. This kind of hysteresis transition has also been observed in the period-3 window region of the Ch_3 area.

For nonlinear dissipative dynamical systems, two or more attractors may coexist; that is, more than one dynamical behaviour is possible (Ref. [143]), depending solely on the initial conditions. Coexistence of periodic and chaotic attractors has been observed in this circuit. Figure 7.20 depicts the coexistence of a period-4 attractor and a chaotic attractor for $F = 587.2$ mV, $\nu = 840$ Hz. These two attractors can be observed by altering the initial conditions (turning OFF and ON of the forcing source or power supply to the circuit). This interesting behaviour indicates that the system has initial value dependence and one can observe either chaotic or non-chaotic attractors in the $(F{-}\nu)$ parameter space depending upon the initial state.

Recently Lindberg (Ref. [144]) has carried out a detailed numerical analysis of the fourth-order driven Chua's circuit for the same circuit parameters used in the present section (Fig. 7.10(a)) and identified a variety of chaotic and limit cycle behaviours for different forcing parametric values. Lindberg demonstrates that it is indeed possible to obtain an insight into the physical behaviour of the driven Chua's circuit by means of simple computer-aided circuit analysis packages like ANP3 and NAP2 (Ref. [144]).

G. *Double-scroll Chua's attractor and Devil's staircase structures*: In a second set of experiments, the circuit parameters of Fig. 7.10(a) are fixed at the values $C_1 = 600$ pF, $C_2 = 0.005\mu$F, $L_1 = 2.8$mH, $L_2 = 10$mH, $R = 1430\Omega$. Then by varying the amplitude F from 0 to 0.2V and the frequency ν in the

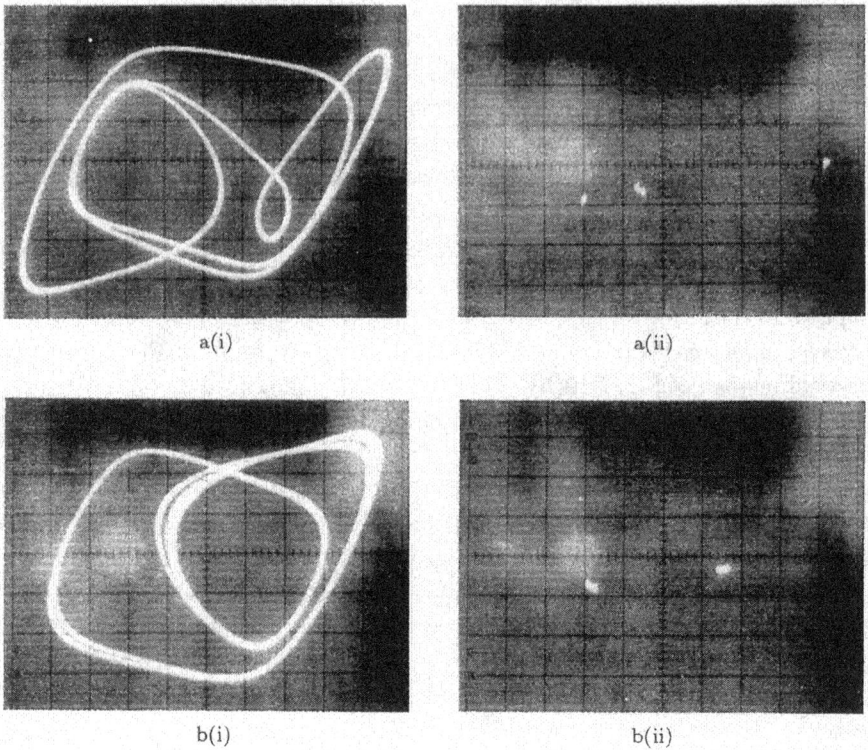

Fig. 7.20. Coexistence of multiple attractors for $F = 587.2$ mV and $\nu = 840$ Hz. a(i) Period-4 attractor; a(ii) The Poincaré map of a(i); b(i) Chaotic attractor; b(ii) The Poincaré map of b(i).

range 24KHz to 150KHz, the dynamics of the circuit is studied. When the amplitude F of the external sinusoidal signal is zero, the circuit of Fig. 7.10(a) becomes autonomous and the familiar double-scroll chaotic attractor has been noticed. Then by varying the amplitude F and frequency ν, reverse bifurcations, quasiperiodic attractors, phase-lockings, chaos, and devil's staircase structures have been observed for this choice of circuit parameters (Ref. [131]). Also, double-hook type chaotic attractors and period-adding sequences have been observed from this nonautonomous circuit. In Table 7.3, a brief summary of the bifurcation phenomena is depicted. For further details see Ref. [131].

Table 7.3. Summary of quasiperiodicity, phase-locking, devil's staircase structure and chaos of Fig. 7.10(a) with parameters (Ref. [131]) $C_1 = 600$pF, $C_2 = 0.005\mu$F, $R = 1430\Omega$, $L_1 = 2.8$ mH, $L_2 = 10$ mH and amplitude $F = 0.2V$.

Forcing frequency $\nu \times KHz$	Attractor description
126	Period-5 phase-locking
115	Quasiperiodicity
104	Period-4 phase-locking
91	Quasiperiodicity
81	Period-3 phase-locking
79.8–78.0	Period-doubling bifurcations of period-3 phase-locked state
68.2	Phase-locked state$(3/8)^+$
67.5	Phase-locked state$(5/13)^+$
66.0	Phase-locked state$(2/5)^+$
61.8	Phase-locked state$(3/7)^+$
61.5	Chaotic attractor
54.2	Period-2 phase-locking
36.8	Period-1 phase-locking

$^+$Winding number

7.5. Chaotic Dynamics of the Simplest Dissipative Non-autonomous Circuit: Murali–Lakshmanan–Chua Circuit

In this section we wish to point out that apart from higher order non-autonomous circuits, even a much simpler second order non-autonomous dissipative nonlinear circuit consisting of Chua's diode as the only nonlinear element, suggested by Murali, Lakshmanan and Chua (Refs. [77–79]), can exhibit a rich variety of bifurcation and chaos phenomena. In Ref. [58], Linsay had demonstrated chaos experimentally by driving a series circuit made up of a linear resistor, a linear inductor and a varactor diode with a sinusoidal voltage source. However, even though this physical circuit contains only three circuit elements, the circuit model used for computer simulation in Ref. [81] contains six circuit elements and the associated nonlinearities are exponential functions, thereby making any mathematical analysis intractable. However, the present circuit of Fig. 7.21(a) proposed by Murali, Lakshmanan and Chua (Refs. [77–79]) is much simpler from a circuit theoretic point of view due to the presence of only one nonlinear resistor, namely, Chua's diode, which is conceptually a much simpler circuit element than a nonlinear capacitor.

7.5.1. *Experimental Realization*

The circuit diagram shown in Fig. 7.21(a) is a classic configuration of a forced negative-resistance oscillator (Ref. [145]) where N denotes a voltage-controlled nonlinear resistor described by $i = h(v)$. A specific case where $h(v) = -av + bv^3$ has been shown to be *chaotic* in a seminal work by Ueda and Akamatsu (Ref. [146]). However, since it is difficult to realize this nonlinear function with the precise parameters "a" and "b", as well as a large enough dynamic range, the strange attractors presented in Ueda & Akamatsu (Ref. [146]) were derived from an analog computer. In contrast, the nonlinear function for the present circuit is characteristic of Chua's diode.

The actual circuit realization of the simple non-autonomous (MLC) circuit, shown in Fig. 7.21(a), is as follows. It contains a capacitor, an inductor, a linear resistor, an external periodic forcing, and only one nonlinear element, namely

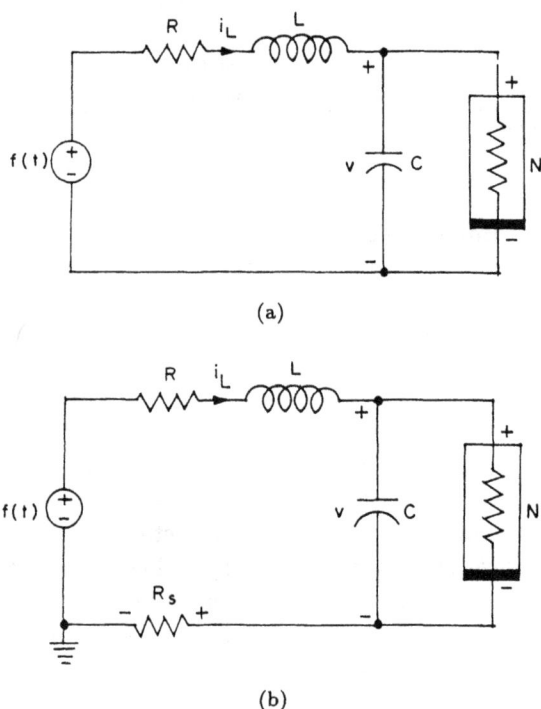

(a)

(b)

Fig. 7.21. (a) Circuit realization of the simple non-autonomous circuit. Here, N is the Chua's diode. (b) Experimental circuit model with the current sensing resistor R_s. $R = 1340\Omega, L = 18$ mH, $C = 10$ nF, $R_s = 20\Omega$.

Chua's diode (N). In order to measure the inductor current i_L in our experiments, we insert a small current sensing resistor R_s as shown in Fig. 7.21(b). In the corresponding computer simulations, this resistor is simply added to the resistor R. By applying Kirchoff's laws to this circuit, the governing equations for the voltage v across the capacitor C and the current i_L through the inductor L are represented by the following set of two first-order non-autonomous differential equations:

$$C\frac{dv}{dt} = i_L - f(v),$$
$$L\frac{di_L}{dt} = -Ri_L - R_s i_L - v + F\sin(\Omega t), \qquad (7.9)$$

where $f(.)$ is the piecewise-linear function defined by Eq. (7.1). In Eq. (7.9) F is the amplitude and Ω is the angular frequency of the external periodic force. The parameters of the circuit elements are fixed at $C = 10\text{nf}$, $L = 18\text{mH}$, $R = 1340\Omega$, $R_s = 20\Omega$, and the frequency $\nu(= \Omega/2\pi)$ of the external forcing source is 8890 Hz.

7.5.2. *Stability Analysis*

The actual values of the slopes and break-point voltage of the Chua's diode G_a, G_b and B_p, are fixed as -0.76ms, -0.41ms and 1.0V respectively [see Fig. (7.2)]. Rescaling Eq. (7.9) with $v = xB_p$, $i_L = GyB_p$, $G = 1/R$, $\omega = \Omega C/G$ and $t = \tau C/G$ and then redefining τ as t the following set of normalized equations are obtained:

$$\dot{x} = y - h(x), \quad \dot{y} = -\beta y - \nu\beta y - \beta x + f\sin(\omega t), \quad (. = d/dt) \qquad (7.10)$$

where $\beta = (C/LG^2), \nu = GR_s$, and $f = (F\beta/B_p)$. Obviously $h(x)$ is represented by Eq. (7.3) or

$$h(x) = \begin{cases} bx + a - b, & x \geq 1, \\ ax, & |x| \leq 1, \\ bx - a + b, & x \leq -1. \end{cases} \qquad (7.11)$$

Here $a = G_a/G$, $b = G_b/G$. Now the dynamics of Eq. (7.10) depends on the parameters ν, β, a, b, ω and f. The experimental circuit parameters used in the previous section are then rescaled to $\beta = 1$, $\nu = 0.015$, $a = -1.02$, $b = -0.55$ and $\omega = 0.75$.

One can easily establish that a unique equilibrium (x_0, y_0) for Eq. (7.10) exists in each of the following three subsets:

$$D_1 = \{(x,y)|x \geq 1\} : P^+ \doteq (-k_1, -k_2),$$

$$D_0 = \{(x,y)|\|x| \leq 1\} : 0 = (0,0), \tag{7.12}$$

$$D_{-1} = \{(x,y)|x \leq -1\} : P^- = (k_1, k_2),$$

where $k_1 = \frac{\sigma(a-b)}{(\beta+\sigma b)}$, $k_2 = \frac{\beta(b-a)}{(\beta+\alpha b)}$ and $\sigma = \beta(1+\nu)$.

In each of the regions D_1, D_0 and D_{-1}, Eq. (7.10) is linear when $f = 0$. It is then easy to see that the stability-determining eigenvalues for the equilibrium point $0 \in D_0$ are calculated from the matrix $A_0 = A(\beta, \sigma, a) = \begin{bmatrix} -a & 1 \\ -\beta & -\sigma \end{bmatrix}$ as $\lambda_1 = 0.1904$ and $\lambda_2 = -0.1854$, a hyperbolic fixed point (unstable). Similarly, the matrix $A_0 = A(\beta, \sigma, b) = \begin{bmatrix} -b & 1 \\ -\beta & -\sigma \end{bmatrix}$ associated with the regions D_1 and D_{-1} has a pair of complex-conjugate eigenvalues ($\lambda_1, \lambda_2 = \lambda_1^*$) with a negative real part, like $\lambda_1 = -0.2325 + i(0.623)$ and $\lambda_2 = -0.2325 - i(0.623)$, which indicates that P^+ and P^- are stable spiral fixed points (Ref. [20]). Naturally these fixed points can be observed depending upon the initial condition $x(0)$ and $y(0)$ of Eq. (7.10) when $f = 0$. As the forcing signal is included ($f > 0$) these fixed points give rise to limit cycles through Hopf bifurcation and as f is increased further the system exhibits period doubling bifurcations from the period-1 limit cycle to chaos as discussed below.

7.5.3. *Explicit Analytical Solutions*

Actually Eq. (7.10) can be explicitly integrated in terms of elementary functions in each of the three regions D_0, D_1 and D_{-1} and matched across the boundaries to obtain the full solution as shown below.

It is quite easy to see that in each one of the regions D_0, D_{+1}, D_{-1}, Eq. (7.10) can be represented as a single second-order inhomogeneous linear differential equation for the variable $y(t)$:

$$\ddot{y} + (\beta + \beta\nu + \mu)\dot{y} + (\beta + \mu\beta\nu + \beta\mu)y = \Delta + \mu f \sin\omega t + f\omega \cos\omega t, \tag{7.13}$$

where

$$\mu = a, \quad \Delta = 0 \text{ in region } D_0, \tag{7.14}$$

$$\mu = b, \quad \Delta = \pm\beta(a-b) \text{ in region } D_\pm. \tag{7.15}$$

The general solution of Eq. (7.13) can be written as

$$y(t) = C_{0,\pm}^1 e^{\alpha_1 t} + C_{0,\pm}^2 e^{\alpha_2 t} + E_1 + E_2 \sin \omega t + E_3 \cos \omega t, \qquad (7.16)$$

where $C_{0,\pm}^1$ and $C_{0,\pm}^2$ are integration constants in the appropriate regions $D_0, D_\pm$, and

$$\alpha_1 = \left(-A + \sqrt{A^2 - 4B}\right)/2, \quad \alpha_2 = \left(-A - \sqrt{A^2 - 4B}\right)/2,$$
$$E_1 = 0 \text{ in region } D_0 \text{ and } E_1 = \Delta/B \text{ in region } D_\pm,$$
$$E_2 = (f\omega^2(A - \mu) + \mu f B)/(A^2\omega^2 + (B - \omega^2)^2),$$
$$E_3 = f\omega(B - \omega^2 - \mu A)/(A^2\omega^2 + (B - \omega^2)^2),$$
$$A = \beta + \beta\nu + \mu \text{ and } B = \beta + \mu\beta\nu + \beta\mu. \qquad (7.17)$$

Knowing $y(t)$, $x(t)$ can be obtained from (7.10) as

$$\begin{aligned}
x(t) &= (1/\beta)\{-\dot{y} - \beta y(1 + \nu) + f \sin \omega t\} \\
&= (1/\beta)\{-\dot{y} - \sigma y + f \sin \omega t\}, \\
&= (1/\beta)\{-C_{0,\pm}^1 e^{\alpha_1 t}(\alpha_1 + \sigma) - C_{0,\pm}^2 e^{\alpha_2 t}(\alpha_2 + \sigma) - (\cos \omega t)(E_2\omega + E_3\sigma) \\
&\quad + (\sin \omega t)(f - E_2\sigma + E_3\omega) - E_1\sigma\}. \qquad (7.18)
\end{aligned}$$

Thus if we start with the initial condition in D_0 the arbitrary constants C_0^1 and C_0^2 in (7.16) get fixed. Then $x(t)$ evolves as given in (7.18) up to either $t = T_1$, when $x(T_1) = 1$ and $\dot{x}(T_1) > 0$, or $t = T_1'$, when $x(T_1') = -1$ and $\dot{x}(T_1) < 0$. Knowing whether $T_1 > T_1'$ or $T_1 < T_1'$ we can determine the next region of interest $(D_{\pm 1})$, and the arbitrary constants of the solutions of that region can be fixed by matching the solutions. This procedure can be continued for each successive crossing. In this way explicit solutions can be obtained in each of the regions $D_0, D_{\pm 1}$. However, it is clear that sensitive dependence on initial conditions is introduced in each of these crossings at appropriate parameter regimes during the inversion procedure of finding $T_1, T_1', T_2, T_2', \ldots$ etc. from the solutions, and thereby leading to chaos.

7.5.4. *Experimental and Numerical Studies*

In order to study the dynamics of this circuit, the amplitude F of the forcing signal is used as the bifurcation parameter. By increasing the amplitude F from zero upwards, the circuit of Fig. (7.21) is found to exhibit experimentally a sequence of bifurcations. Starting from a dc equilibrium, the solution bifurcates through a Hopf bifurcation to a limit cycle, and then by period-doubling

a(i)   a(ii)

b(i)   b(ii)

c(i)   c(ii)

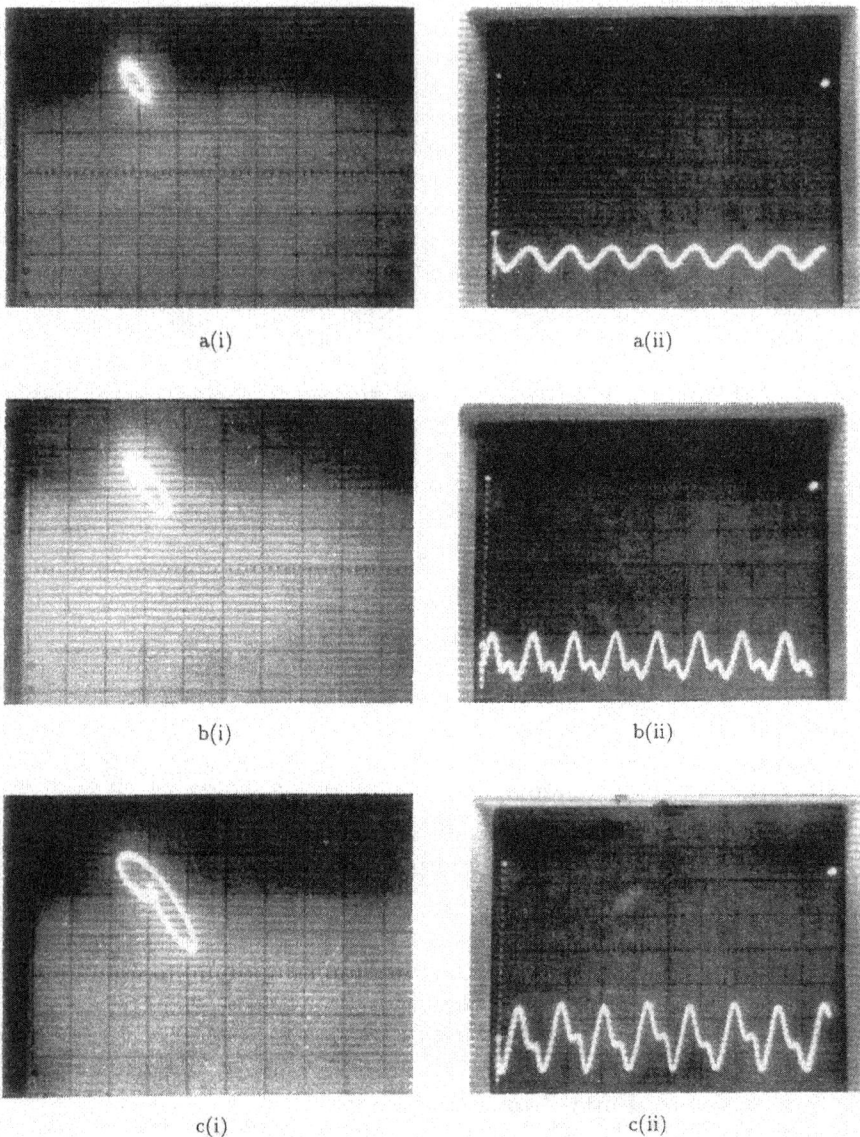

Fig. 7.22. Period-doubling bifurcation sequences. (i) Horizontal axis v; 0.5V/div, vertical axis $(v_s = R_s i_L)$ 5mV/div; (ii): Wave form $v(t)$. (a) $F = 0.0365\text{V}_{\text{rms}}$; Period-1. (b) $F = 0.0549\text{V}_{\text{rms}}$; Period-2 (c) $F = 0.064\text{V}_{\text{rms}}$; Period-4 (d) $F = 0.0723\text{V}_{\text{rms}}$; One-band chaos (e) $F = 0.107\text{V}_{\text{rms}}$; Double-band chaos; (i) Phase portrait v versus $v_s(= R_s i_L)$; (ii) Wave form $v(t)$; (iii) Poincaré map of (i). (f) $F = 0.145\text{V}_{\text{rms}}$; Period-3 window. (g) $F = 0.488\text{V}_{\text{rms}}$; Period-1 boundary.

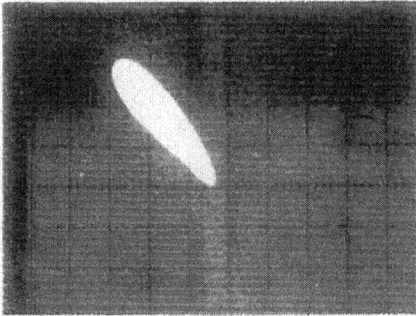

d(i)

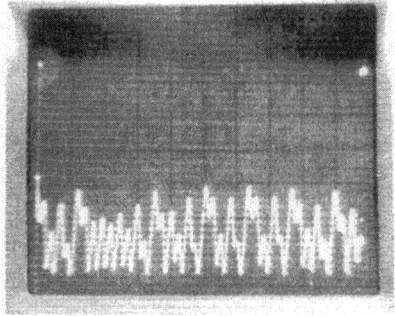

d(ii)

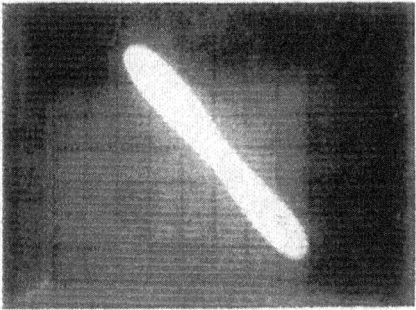

e(i)

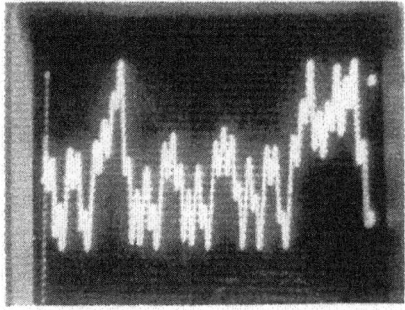

e(ii)

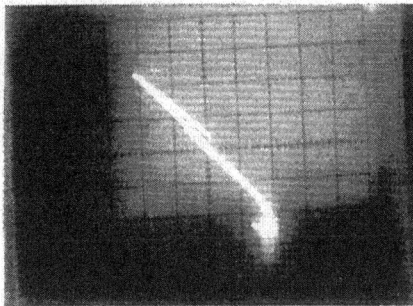

e(iii)

Fig. 7.22. (*Continued*)

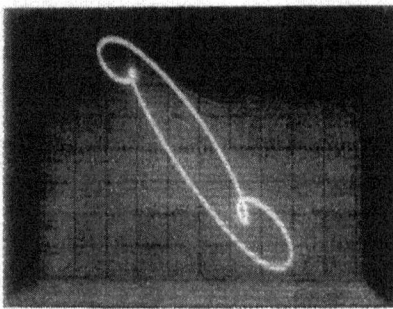

f(i)

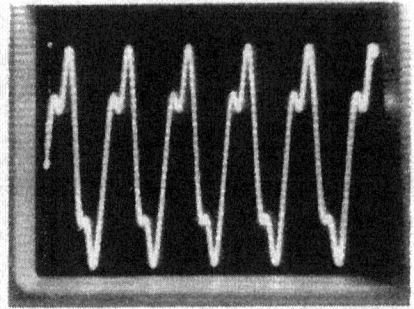

f(ii)

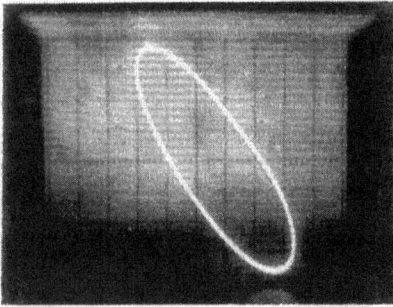

g(i)

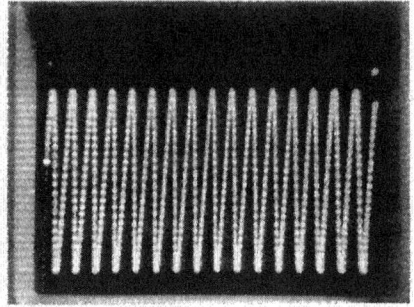

g(ii)

Fig. 7.22. (*Continued*)

Table 7.4. Summary of bifurcation phenomena of Eq. (7.10) for the parameters $\beta = 1.0, \nu = 0.015, a = -1.02, b = -0.55$ and $\omega = 0.75$ (Ref. [78]).

Amplitude (f)	Description of attractor	Fig. 7.23
$0 < f \leq 0.071$	Period-1 limit cycle	(a)
$0.071 < f \leq 0.089$	Period-2 limit cycle	(b)
$0.089 < f \leq 0.093$	Period-4 limit cycle	(c)
$0.093 < f \leq 0.19$	Chaos	(d, e)
$0.19 < f \leq 0.3425$	Period-3 window	(f)
$0.3425 < f \leq 0.499$	Chaos	(not included here)
$0.499 < f \leq 0.625$	Period-3 window	(not included here)
$0.625 < f$	Period-1 boundary	(g)

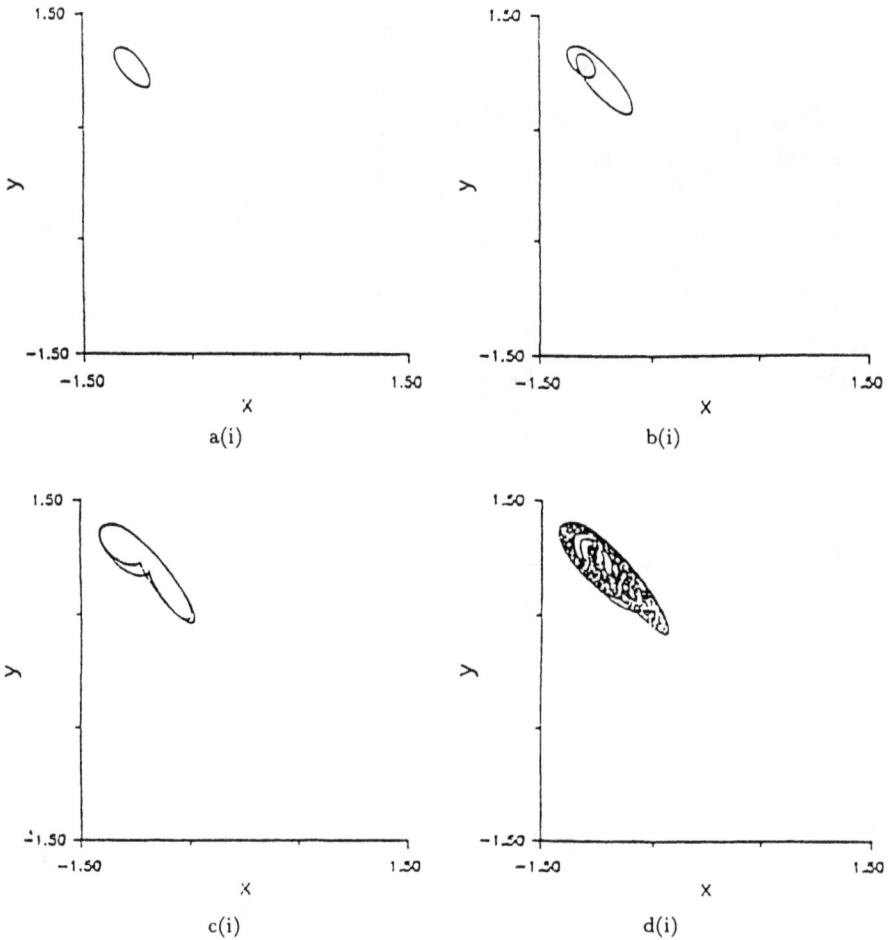

Fig. 7.23. (i) Phase portraits in $(x-y)$ plane of Eq. (7.10) for the parameters $\beta = 1.0$, $\nu = 0.015, a = -1.02, b = -0.55$, and $\omega = 0.75$: (a) $f = 0.065$ ($F_c = 0.046$ V$_{\rm rms}$), Period-1 limit cycle; (b) $f = 0.08$ ($F_c = 0.0565$V$_{\rm rms}$, Period-2 limit cycle; (c) $f = 0.091$($F_c = 0.06435$V$_{\rm rms}$), Period-4 limit cycle; (d) $f = 0.1$($F_c = 0.0707$V$_{\rm rms}$), One-band chaos; (Here F is the calculated value of the forcing amplitude using the relation $F = fB_P/\beta$, see Eq. (7.10)). (ii) Poincaré map plots in $(x-y)$ plane corresponding to Fig. 7.23(i) (e) Double-band chaotic attractor, $f = 0.15$($F_c = 0.106$V$_{\rm rms}$): (i) Phase portrait; (ii) Poincaré map; (iii) Power spectrum of the x signal. (f) Period-3 window, $f = 0.2$($F = 0.1414$V$_{\rm rms}$): (i) Phase portrait; (ii) Poincaré map of (i). (g) Period-1 boundary, $f = 0.7$($F = 0.495$V$_{\rm rms}$): (i) Phase portrait; (ii) Poincaré map of (i).

a(ii)

b(ii)

c(ii)

d(ii)

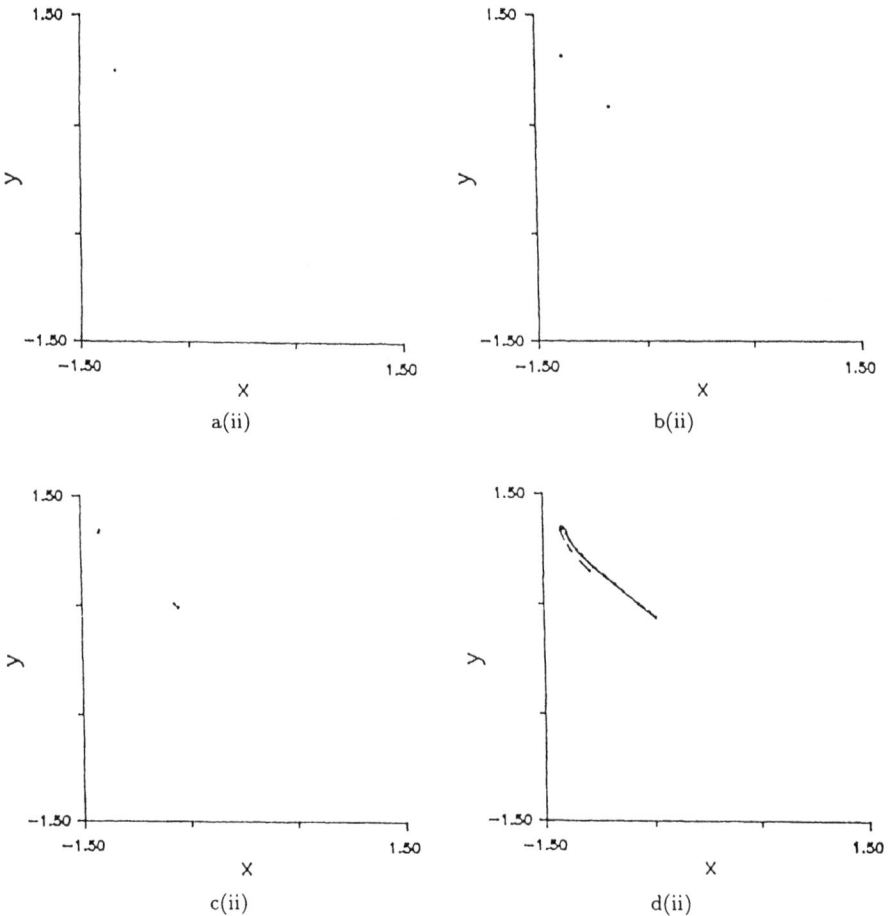

Fig. 7.23. (*Continued*)

sequences to Chua's one-band attractor, double-band attractor, periodic windows, boundary crisis, etc. These are illustrated in Figs. 7.22. Also by using the standard 4th-order Runge–Kutta integration routine we have carried out a numerical analysis of (7.10) with the rescaled circuit parameters of Fig. 7.21(b), $\beta = 1.0, \nu = 0.015, a = -1.02, b = -0.55$, and $\omega = 0.75$ (Ref. [17]), and with f as the control parameter. The results are summarized in Table 7.4, and some of them are also exhibited in Figs. 7.23. Figure 7.24(a) depicts the one-parameter bifurcation diagram in the $(f$–$x)$ plane, which clearly indicates the

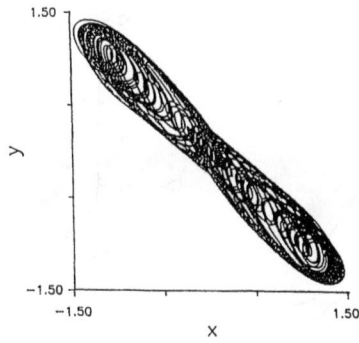

e(i)

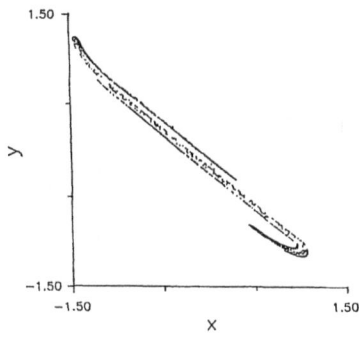

e(ii)

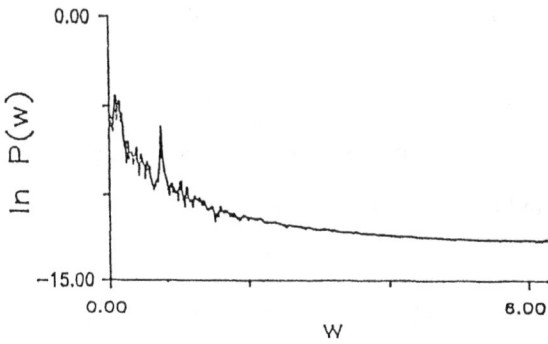

e(iii)

Fig. 7.23. (*Continued*)

f(i)

g(i)

f(ii)

g(ii)

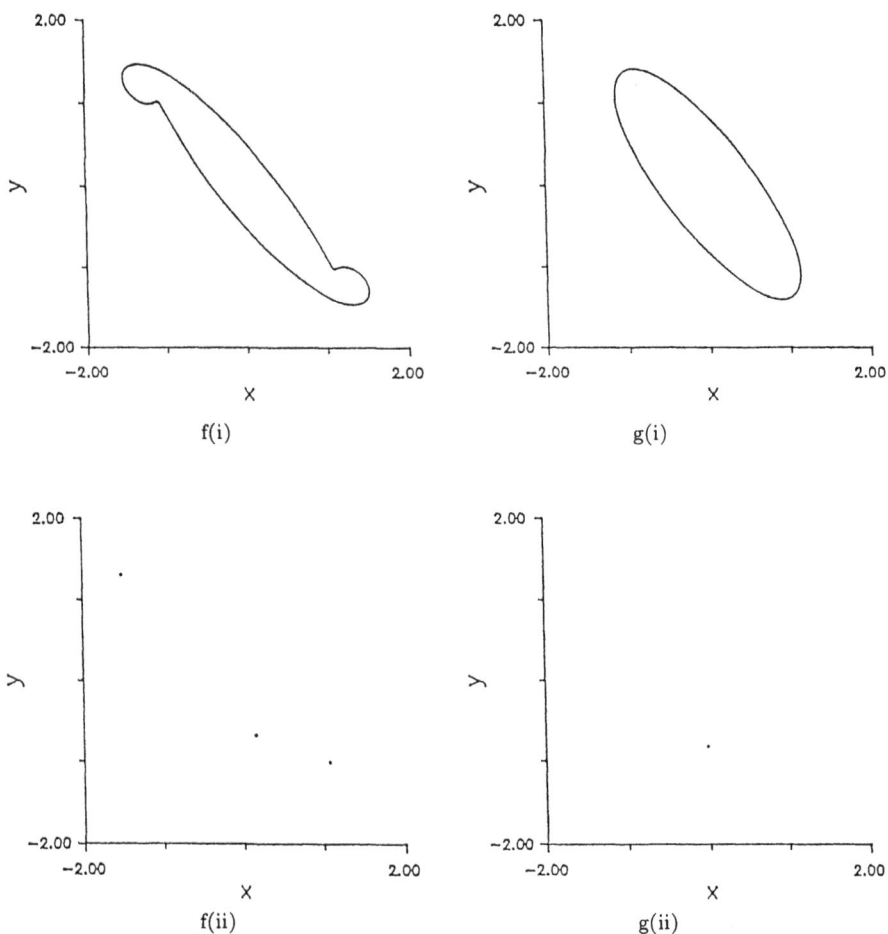

Fig. 7.23. (*Continued*)

familiar period-doubling bifurcation sequences, chaos, windows, etc. Also, in Fig. 7.24(b), the maximal Lyapunov spectrum in the $(f-\lambda_{\max})$ plane is plotted.

7.5.5. *Chua's Diode with Single Break-point and Spiral Chua's Attractor*

Figure 7.23(d)(i) shows the appearance of spiral Chua's attractor for $f = 0.1$. The invariance of Eq. (7.10) for $f = 0$ under the reflection $(x, y) \rightarrow (-x, -y)$ implies that there is another spiral-type chaotic attractor located

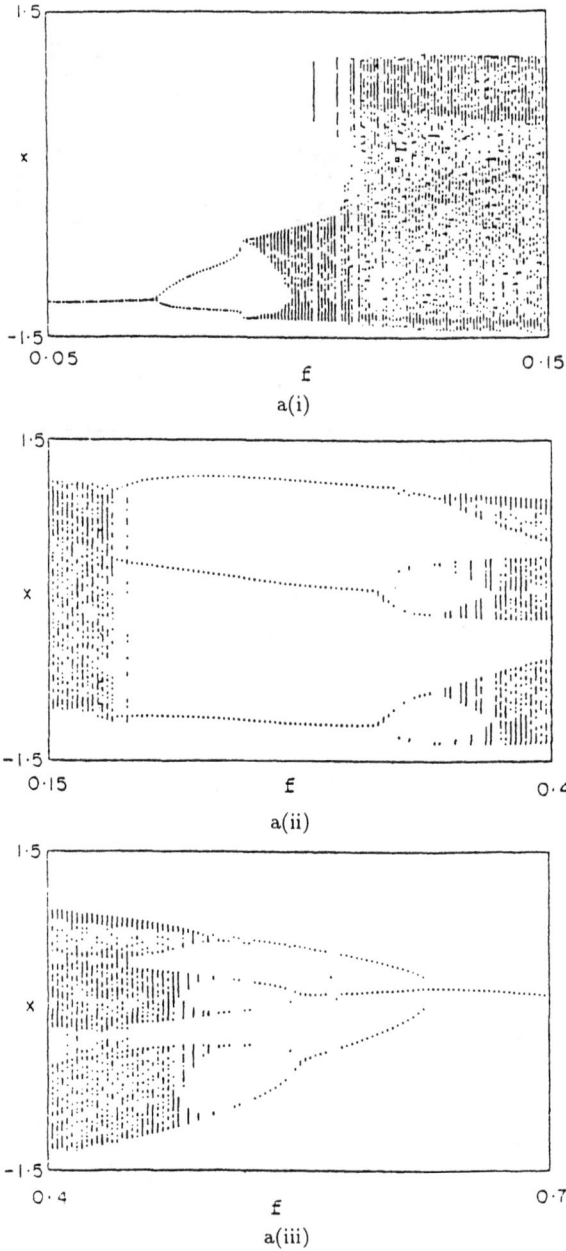

Fig. 7.24. (a) One-parameter bifurcation diagram in $(x-f)$ plane. (b) Plot of the maximal Lyapunov spectrum: (i) $0.05 \leq f \leq 0.15$; (ii) $0.15 \leq f \leq 0.4$; (iii) $0.4 \leq f \leq 0.7$.

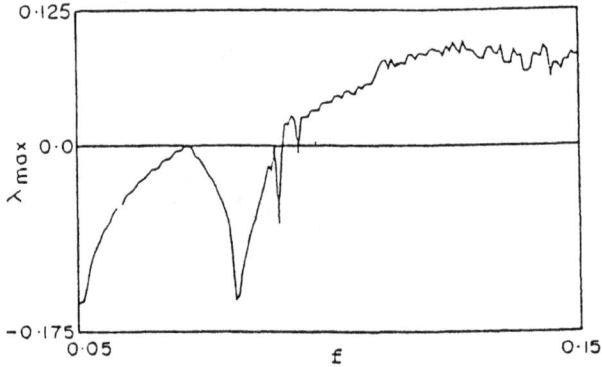

b(i)

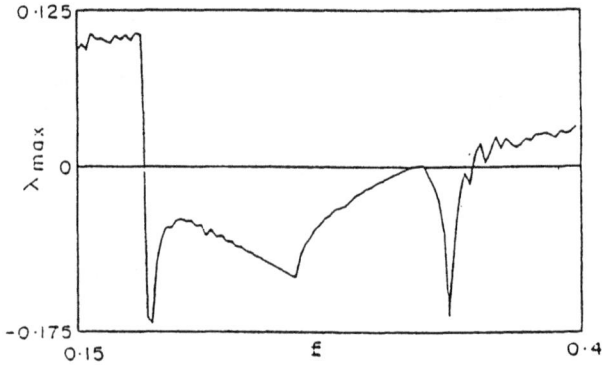

b(ii)

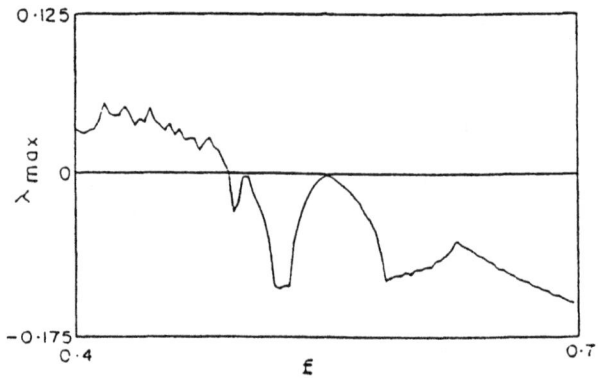

b(iii)

Fig. 7.24. (*Continued*)

symmetrically with respect to the origin, depending upon the initial conditions. This symmetry, in turn, stems from the symmetry of the function $h(.)$ in Eq. (7.11). This observation suggests that one attractor should still be present even if one replaces the 3-segment function $h(.)$ of Chua's diode with the 2-segment function as in Fig. 7.25. This conjecture is confirmed in Fig. 7.26, where the spiral-type chaotic attractor is observed with a piecewise-linear resistor having only one break point (Refs. [78, 147]). The mathematical representation of the characteristic curve of the Fig. 7.25 can be given (Ref. [49]) as

$$h(x) = 0.5(b-a)[|x-B_p| - |B_p|] + 0.5(a+b)x \,, \qquad (7.19)$$

or

$$h(x) = \begin{cases} bx + a - b & x \geq B_p \,, \\ ax & x \leq B_p \,. \end{cases} \qquad (7.20)$$

Here $B_p = 1.0, a = -1.02$ and $b = -0.55$. However, experimental realization of the nonlinear resistor with 2-segment characteristic curve is difficult but this can be attained by adding a bias battery in series with the sinusoidal source (Ref. [135]). Also one can obtain explicit solutions of various regions of operation of Eq. (7.10) along with Eq. (7.20) as in the 3-segment case.

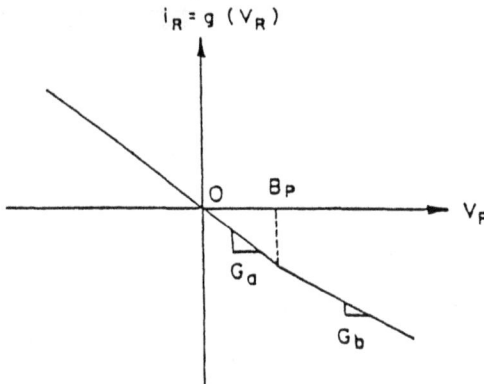

Fig. 7.25. Characteristic curve of Chua's diode with two segments.

7.6. Chaotic Dynamics of an Autonomous van Der Pol–Duffing (ADVP) Oscillator

In the circuit diagram of Fig. 7.5, N denotes a voltage-controlled nonlinear resistor described by $i = f(v_1)$. In our studies so far, this nonlinear resistor characteristic is represented by a piecewise-linear function of Chua's diode (Eq. 7.1). However this is not mandatory; it can as well be represented by any suitable smooth function. For example, one can choose a cubic nonlinear element of the form

$$i_N = f(v_1) = av_1 + bv_1^3 . \quad (a < 0, \ b > 0) \tag{7.21}$$

With such a nonlinear element, the circuit of Fig. 7.5 turns out to be an interesting physical model, namely, an autonomous van der Pol–Duffing (ADVP) oscillator. In the following we will consider some details of the dynamics of the circuits with nonlinear resistor represented by Eq. (7.21).

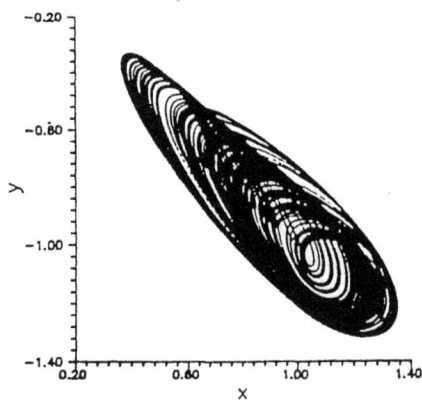

(a)

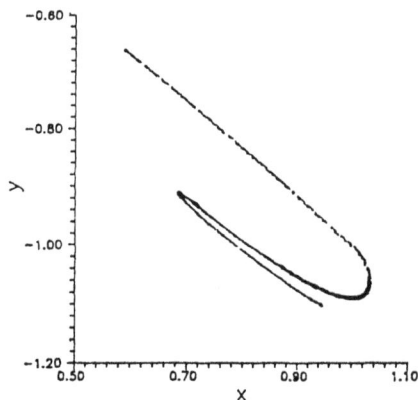

(b)

Fig. 7.26. (a) Phase portrait in (x–y) plane of Eq. (7.10) and Eq. (7.19). One-band chaotic attractor for $f = 0.1, a = -1.02, b = -0.55, \beta = 1.0, \nu = 0.015$, and $\omega = 0.75$. (b) Poincaré map of (a).

7.6.1. *Special Cases*

Now, the three nonlinear circuits (Ref. [148]) shown in Fig. 7.27 are obtained as limiting cases of Fig. 7.5.

(i) The circuit equation of Fig. 7.27(a) is

$$\frac{dv_1}{dt} = -(1/C_1)(bv_1^3 + av_1), \tag{7.22}$$

which may also be written as

$$\frac{dv_1}{dt} = -(1/C_1)\left[\frac{dF}{dv_1}\right],$$

where F is the double-well potential of the form

$$F(v_1) = \left(\frac{b}{4}\right)v_1^4 + \left(\frac{a}{2}\right)v_1^2. \qquad (a < 0, \ b > 0).$$

(ii) The circuit of Fig. 7.27(b) is a van der Pol oscillator (cf. Sec. 6.1) and may be obtained from Fig. 7.5 by setting either $C_1 = 0$ or $R = 0$. Assuming $R = 0$, one obtains the van der Pol's equation

$$\frac{dv_1}{dt} = -\left(\frac{1}{C_1}\right)(bv_1^3 + av_1 + i_L), \tag{7.23a}$$

$$\frac{di_L}{dt} = \frac{v_1}{L}, \tag{7.23b}$$

which in a more familiar form (cf. Appendix B) is

$$\ddot{v}_1 + \left(\frac{1}{C_1}\right)(a + bv_1^2)\dot{v}_1 + \frac{v_1}{LC_1} = 0. \quad (. = \frac{d}{dt}) \tag{7.24}$$

(iii) The circuit of Fig. 7.27(c) is a *second-order* force-free van der Pol–Duffing oscillator (DVP oscillator cf. Sec. 6.2). The circuit equations are given as

$$\dot{v}_1 = -(1/C_1)\left[bv_1^3 + av_1 + i_L\right], \tag{7.25a}$$
$$\dot{i}_L = (v_1/L) - (Ri_L/L). \tag{7.25b}$$

Equation (7.25) can be brought into a dimensionless form with the rescaling

$$x = \sqrt{bL/RC_1}\ v_1, \quad z = \sqrt{bLR/C_1}\ i_L, \quad \tau = Rt/L,$$

(a)

(b)

(c)

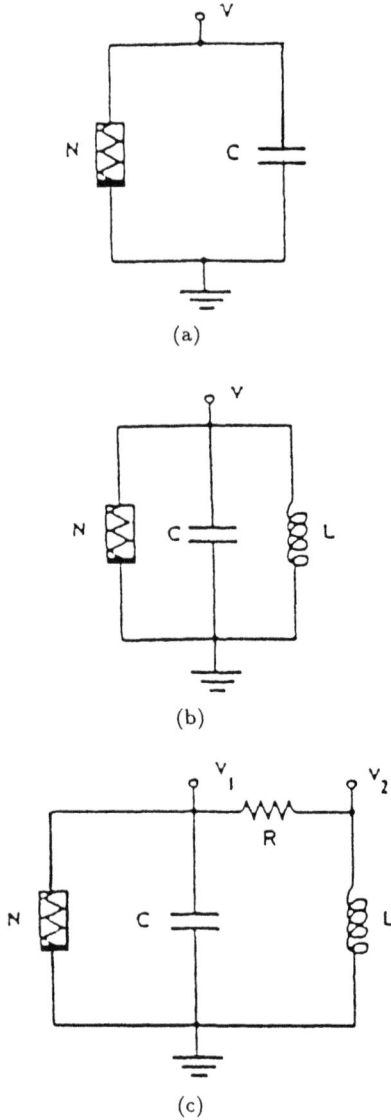

Fig. 7.27. Three nonlinear circuits embedded in the chaotic autonomous van der Pol–Duffing (ADVP) circuit: (a) Nonlinear RC circuit, (b) van der Pol oscillator, (c) van der Pol–Duffing oscillator (force free case).

which yields the two first-order systems of equations

$$\dot{x} = -(x^3 - \alpha x) - \Gamma z, \qquad (7.26a)$$

$$\dot{z} = x - z \,, \tag{7.26b}$$

where the differentiation is with respect to τ and the two parameters are given as $\alpha = -(La/RC_1)$ and $\Gamma = L/(C_1 R^2)$. Equation (7.26) can be rewritten in the form

$$\ddot{x} + [(1 - \alpha) + 3x^2]\dot{x} + \frac{dF}{dx} = 0 \,, \tag{7.27}$$

where F is the potential

$$F(x) = \frac{x^4}{4} - (\alpha - \Gamma)\frac{x^2}{2} \,.$$

Equation (7.27) is of course nothing but the force-free version of the DVP oscillator given by Eq. (6.22).

7.6.2. Third-order Autonomous DVP (ADVP) Oscillator

Now we consider the third-order autonomous van der Pol–Duffing oscillator (ADVP) model by including one more passive circuit element C_2 to that of Fig. 7.27(c). The resultant circuit can be easily seen to be equivalent to a Chua's autonomous circuit model (Fig. 7.5) but with a cubic nonlinear element (Refs. [148–150]). We now proceed with the chaotic dynamics of the ADVP oscillator. Applying the Kirchoff's laws to Fig. 7.5, we obtain the equations

$$\dot{v}_1 = -(1/C_1)[bv_1^3 + av_1 + (1/R)(v_1 - v_2)] \,, \tag{7.28a}$$
$$\dot{v}_2 = (1/C_2)[(1/R)(v_1 - v_2) - i_L] \,, \tag{7.28b}$$
$$\dot{i}_L = v_2/L \,. \tag{7.28c}$$

The dynamics of Fig. 7.5 can also be studied experimentally by realizing the nonlinear element (N) with suitable circuits with operational-amplifiers and a set of diodes (Refs. [148–150]). However, since it is difficult to realize this nonlinear function with precise parameters "a" and "b" as well as a large enough dynamic range, one has to take recourse to suitable numerical simulation algorithms or analog simulation as in Chapters 4 or 6. Here we study the dynamics of Equation (7.28) through numerical simulations. Eq. (7.28), after rescaling, can be written as

$$\dot{x} = -\nu[x^3 - \alpha x - y] \,, \tag{7.29a}$$
$$\dot{y} = x - y - z \,, \tag{7.29b}$$
$$\dot{z} = \beta y \,, \tag{7.29c}$$

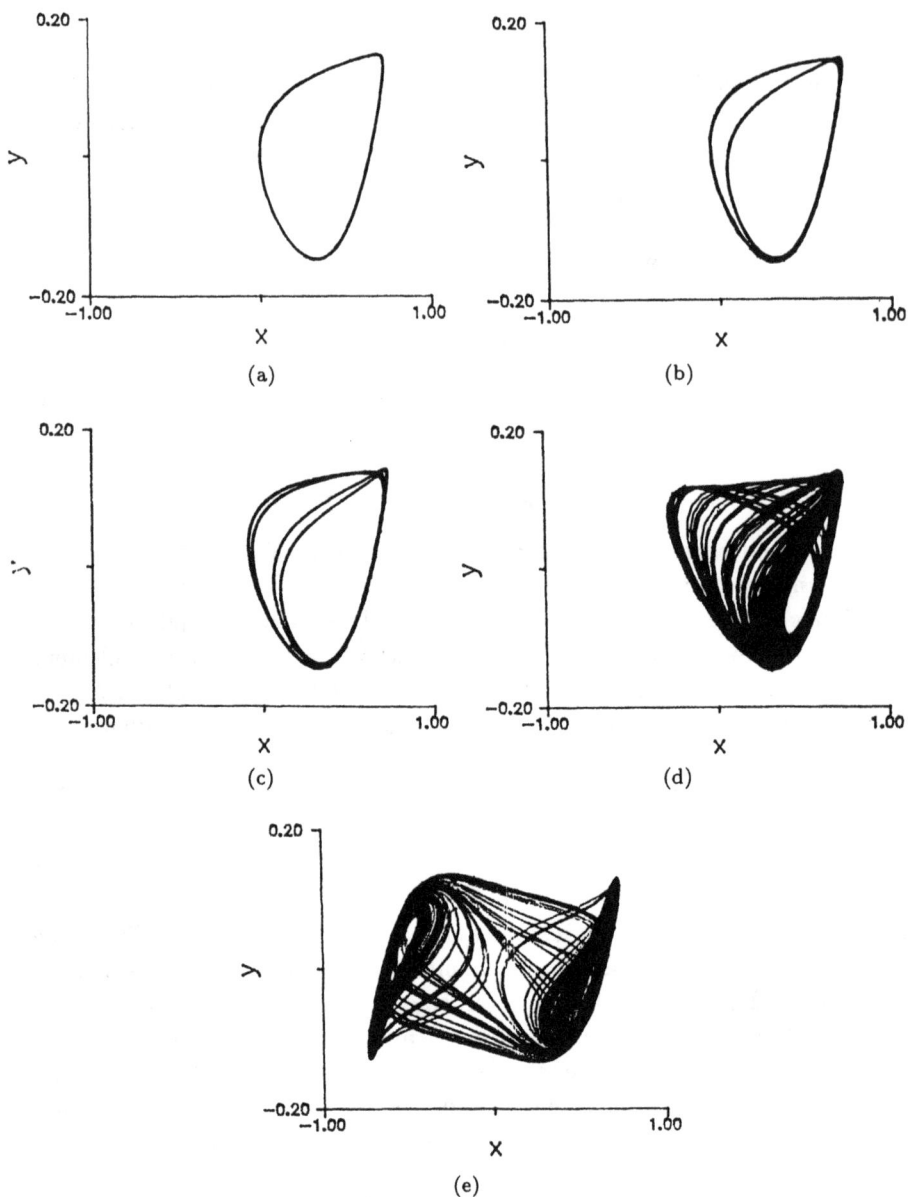

Fig. 7.28. Period-doubling bifurcation sequences to chaos of the third order ADVP oscillator Eq. (7.29) for $\alpha = 0.35, \nu = 100$: (a) Period-1 limit cycle; $\beta = 800$, (b) Period-2 limit cycle; $\beta = 750$, (c) Period-4 limit cycle; $\beta = 710$, (d) One-band chaos; $\beta = 600$, (e) Double-band chaos; $\beta = 300$.

where the scaled variables are related to the unscaled ones by

$$x = \sqrt{bR}\, v_1, \ y = \sqrt{bR}\, v_2, \ z = \sqrt{bR^3}\, i_L, \ \tau = t/RC_2,$$
$$\alpha = (1 + aR), \ \beta = C_2 R^2/L, \ \nu = C_2/C_1.$$

Equation (7.29) with fixed values of $\nu = 100$ and $\alpha = 0.35$ then exhibits period-doubling bifurcations leading to chaos as the parameter β is decreased from a large value. One observes period-1, period-2, period-4 limit cycles, one-band chaos and double-band chaos respectively at $\beta = 800, 750, 710, 600$ and 300 (see Fig. 7.28). In these simulations, the initial conditions were chosen as $x(0) = x(t = 0) = 0.5, y(0) = 0.1$ and $z(0) = 0.1$. Due to the rich dynamical phenomena the chaotic ADVP oscillator also becomes an important model from the circuit theoretic point of view, enabling one to use it for the study of synchronization and ensuing secure communication applications. These topics are covered in Chapter 9.

CHAPTER 8

CONTROLLING OF CHAOS

In the previous chapters we have studied the dynamics of several interesting nonlinear oscillators and circuits. Specifically, the nature of bifurcations and occurrence of chaotic behaviour have been investigated mostly under the influence of external periodic forces. Even though the presence of chaotic behaviour is generic and robust for suitable nonlinearities, ranges of parameters and external forces, there are practical situations where one wishes to avoid or control chaos so as to improve the performance of a dynamical system. Also even though sometimes chaos is useful, as in a mixing process or in heat transfer, often it is unwanted or undesirable. For example, increased drag in flow systems, erratic fibrillations of heart beating, extreme weather patterns, and complicated circuit oscillations are some situations where chaos is harmful. Clearly, the ability to control chaos, that is to convert chaotic oscillations into desired regular ones with a periodic time dependence, would be beneficial in working with a particular system. The possibility of purposeful selection and stabilization of particular orbits in a normally chaotic system using *minimal predetermined* efforts provides a unique opportunity to maximize the output of a dynamical system. It is thus of great practical importance to develop suitable control methods and to analyze their efficacy. Very recently much interest has been focussed on this type of problems (Refs. [79, 89, 151–184]).

8.1. Algorithms for Controlling of Chaos: Feedback and Non-Feedback Methods

Let us consider a general n-dimensional nonlinear dynamical system

$$\dot{X} = \frac{dX}{dt} = F(X; p; t),\qquad (8.1a)$$

where $X = (x_1, x_2, x_3, \ldots, x_n)$ represents the n state variables and p is a control or external parameter. Different control algorithms are essentially based on the fact that one would like to effect changes as minimal as possible to the original system so that it is not grossly deformed. From this point of view controlling methods or algorithms can be broadly classified into two categories:

(i) feedback methods and

(ii) non-feedback algorithms.

In the following, we briefly discuss a selection of control methods developed for controlling chaos from different viewpoints. Let $X(t)$ be a chaotic solution of Eq. (8.1a) with a specific value of the control parameter p. First, we briefly mention some of the *feedback* control methods.

8.1.1. *Feedback Controlling Algorithms*

Feedback methods essentially make use of the intrinsic properties of chaotic systems, including their sensitivity to initial conditions, to stabilize orbits already existing in the systems. The salient features of the prominent methods are as follows.

(a) *Adaptive control algorithm (ACA)*: In the method of Huberman and Lumer (Ref. [153]) and that of Sinha, Ramaswamy and Subba Rao (Ref. [154]), the system motion is set back to a desired state X_s by adding dynamics on the control parameter p through the evolution equation,

$$\dot{p} = \varepsilon G(X - X_s), \quad \varepsilon \ll 1, \quad (\cdot = d/dt) \qquad (8.1b)$$

where the function G is proportional to the difference between X_s and the actual output X, and ε indicates the stiffness of the control. The function G could be either linear or nonlinear. Specific applications and details are given in Sec. 8.2.1.

(b) *Ott–Grebogi–Yorke (OGY) method and Singer et al. method of stabilizing unstable periodic orbits*: This method for controlling of chaos was first proposed by Ott, Grebogi and Yorke (OGY) in 1990 (Ref. [152]). The key observation is that a chaotic attractor contains an infinite number of unstable periodic orbits. By applying small judiciously chosen temporal perturbations to a control parameter (p) of the system (8.1a) a chaotic trajectory can be stabilized around a desired periodic orbit. The determination of small nudges of the parameter $p(t)$ requires a knowledge of the eigenvalues and eigenvectors of the unstable orbit. A major advantage of the method is that it can be applied to experimental systems in which *a priori* knowledge of the system is usually not known. A time series found by measuring one of the system's dynamical variables in conjunction with the time-delay embedding method (Ref. [165]), which transforms a scalar time series into a trajectory in phase space, is sufficient to determine the desired unstable periodic orbits to be controlled and the relevant quantities required to compute the parameter perturbations. An added advantage of the OGY method is its flexibility in choosing the desired periodic orbit to be controlled. The method has attracted growing interest for controlling dynamical systems (Ref. [184]), Hamiltonian systems (Ref. [185]), control of transient chaos (Ref. [186]) and chaotic scattering (Ref. [180]). A variant of the OGY method applicable to high orbits of certain experimental systems has been proposed in Ref. [159].

Another method which is also able to stabilize the unstable orbits of the chaotic attractor is the small feedback control mechanism of Singer *et al.* (Ref. [156]). Introducing a small perturbation to an external parameter of a dynamical system, the sign of which depends on the sign of the deviation of actual output, and to a preset signal, Singer *et al.* demonstrated the conversion of a chaotic flow into a stationery (fixed point) solution in the Lorenz system and in an experimental thermal convection loop.

(c) *Control engineering approach*: Feedback controls, linear or otherwise, have been recognized to be very useful for stabilizing an unstable system while tracking a reference input and/or rejecting uncertain disturbances and so on. Recently Chen and Dong (Ref. [174]) have developed some new ideas and formalized successful techniques for controlling (or ordering) chaotic discrete-time and continuous-time systems using conventional engineering feedback controls, based essentially on the rigorous Lyapunov arguments. Using this approach they have shown (1) controlling of chaotic trajectories of some typical systems such as the discrete-time Lozi and Henon maps, the Duffing oscillator and Chua's circuit, (2) controlling chaotic trajectories of such systems to their unstable equilibria and/or unstable limit cycles (even multi-periodic orbits), and (3) that the feedback controller can be either nonlinear or linear.

8.1.2. *Non-feedback Methods*

In contrast to feedback control techniques, non-feedback methods (Refs. [89, 174, 176]) make use of a small perturbing external force such as a small driving force, a small noise term, a small constant bias, or a weak modulation to some system parameter. These methods modify the underlying chaotic dynamical system weakly so that stable solutions appear.

(i) *Parametric perturbation*: Aleixeev and Loskutov (Ref. [162]) and Lima and Pettini (Ref. [155]) have suggested that it is possible to bring a chaotic system into a regular regime by means of a small parametric perturbation of suitable frequency, that is replacing p in (8.1a) by $p + \eta \cos \Omega t$, with the Rössler model (Ref. [162]) and Duffing oscillator (Ref. [155]) as specific examples. In fact, Pikovsky observed earlier synchronization and stochastic motions in coupled oscillators by parametric perturbation (Ref. [163]). Suppression of chaos by resonant periodic parametric perturbations in an experimental set up of a Duffing oscillator has also been reported (Ref. [158]) some time ago.

(ii) *Addition of a weak periodic signal, constant bias or noise*: A few years ago, it was reported that with the addition of a second periodic force in a Duffing oscillator (Refs. [187, 188]) or a driven pendulum (Ref. [189]) a shift in period-doubling bifurcation occurs. Motivated by this, Braiman and Goldhirsch (Ref. [157]) have investigated the possibility of eliminating chaos in a driven pendulum with a second weak periodic force. In addition, the effect of an additional weak constant bias term to quench chaos has also been reported (Ref. [79]). Interestingly, there is another way of controlling chaos, that is by adding appropriate external noise (Ref. [176]), which can sometimes help to remove the strangeness of the chaotic attractor.

(iii) *"Entrainment" — open loop control*: Another representative approach for the control of chaos using a non-feedback method is the entrainment control method proposed by Jackson (Refs. [169–171]), Hubler (Ref. [172]), and others (Refs. [172, 173]). They have considered experimental models represented by $dx/dt = F(x)$, where $x \in R^n$, $F(x)$-differentiable, solutions existing for $\forall t \geq 0$ and the system possessing attracting limit sets for trajectories. Then the control problem is to find suitable perturbation of $dx/dt = F(x)$ such that the solutions $x(t)$ will be entrained to an arbitrarily chosen "goal dynamics" $g(t)$ such that $\lim_{t \to \infty} |x(t) - g(t)| = 0$. By applying a control of the form $E(g, \dot{g})$ to $F(x)$, the modified equation may be given as $dx/dt = F(x) + E(g, \dot{g})S(t)$, where $S(t) = 0$ for $t < 0$, $S(t) = 1$ for $t \geq 0$. Thus the control should have the form $E(g, \dot{g}) = \dot{g} - F(g)$. Following this, a control can be achieved for the initial conditions in some set $BE(g)$ called the basin of entrainment ($x_0 \in BE(g)$). This control method is applicable for entrainment and for elimination of multiple basins of attraction (in systems with many attractors).

(iv) *Oscillation absorber method*: Another effective non-feedback method for controlling chaos is by coupling a main chaotic system to a new but simple system (controller) with easily changeable parameters. This method has been developed by Kapitaniak *et al.* (Ref. [175]). Their method is based on the "oscillation absorber" concept and it has been successfully applied to both mechanical and electrical systems.

We should mention here that the above methods are only representative ones and are by no means exhaustive.

8.2. Controlling in the BVP Oscillator

In order to illustrate the applicability and usefulness of the various algorithms mentioned above, we now demonstrate the controlling of chaos in a BVP oscillator. The discussions closely follow the work of Rajasekar and Lakshmanan (Ref. [176]). As discussed in Chapter 6, the BVP oscillator can be represented by a set of odes,

$$\dot{x} = x - x^3/3 - y + A_0 + A_1 \cos \omega t \,, \tag{8.2a}$$
$$\dot{y} = c(x + a - by) \,, \tag{8.2b}$$

where a, b, c are parameters. The parameters are fixed at $a = 0.7, b = 0.8, c = 0.1$ and $\omega = 1$ as in Sec. 6.1 for our studies. Naturally the BVP oscillator with $A_1 = 0$ cannot show chaotic behaviour. As noted in Sec. 6.1, for $A_0 = 0$ and $A_1 > 0$, the system exhibits a variety of oscillatory (both periodic and chaotic) domains. Figure 8.1 shows the phase portrait of the typical chaotic orbit for $A_1 = 0.74, A_0 = 0$. The salient features of its dynamics is summarized in Table 8.1 for convenience.

Table 8.1. Complex behaviour of the system (8.2) as a function of the parameters A_0 and A_1 in the interval $(0,2)$ for $a = 0.7, b = 0.8, c = 0.1$ and $\omega = 1$ (for details, see Ref. [176]).

A_0	A_1	Behaviour
0–0.340	0	stable focus
0.341–1.41	0	limit cycle
1.420–2	0	stable focus
0	0–0.7181	period doublings
0	0.7182–1.289	complex oscillations (chaos, periodic windows, intermittencies)
0	1.290–2	reverse bifurcations
0–0.6	0.74	phase-locking and devil's staircase

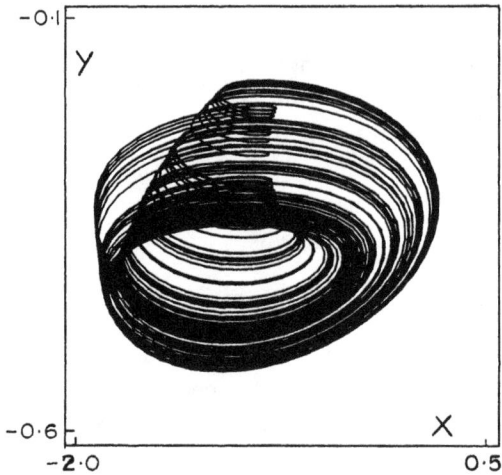

Fig. 8.1. Phase portrait of the BVP oscillator for $A_1 = 0.74$.

8.2.1. *Adaptive Control Algorithm (ACA) to the BVP Oscillator*

In this section we study the applicability of ACA to the BVP oscillator (8.2) with $A_1 = 0$ and then $A_0 = 0, A_1 \neq 0$. Let us first consider the case $A_1 = 0$ and $A_0 > 0$. In order to convert the dynamics of (8.2) from a limit cycle to a desired fixed point (x_s, y_s) we perturb the parameter A_0 as

$$\dot{A}_0 = -\varepsilon(x - x_s), \qquad (8.2c)$$

where we have assumed the control function G to be linear in $(x - x_s)$ to start with. More general forms can also be used as discussed below for the limit-cycle attractor. Now the fixed points $X_s = (x_s, y_s)$ of (8.2a,b) can be obtained from the relations

$$x_s^3 + 3x_s(1/b - 1) + 3a/b - 3A_0 = 0, \qquad (8.3a)$$

$$y_s = (x_s + a)/b. \qquad (8.3b)$$

Without (8.2c) we get a limit cycle for $A_0 = 0.4$ (Fig. 8.2(a)). When the control dynamics (8.2c) is included with $A_0(t = 0) = 0.4$ and given ε and x_s, the system evolves to reach the fixed point (x_s, y_s). This is shown in Fig. 8.2(b) for $\varepsilon = 0.1$ with $(x_s, y_s) = (-1.001, -0.376)$, which is the stable fixed point of (8.2a,b) for $A_0 = 0.2917$ and $A_1 = 0$.

(a)

(b)

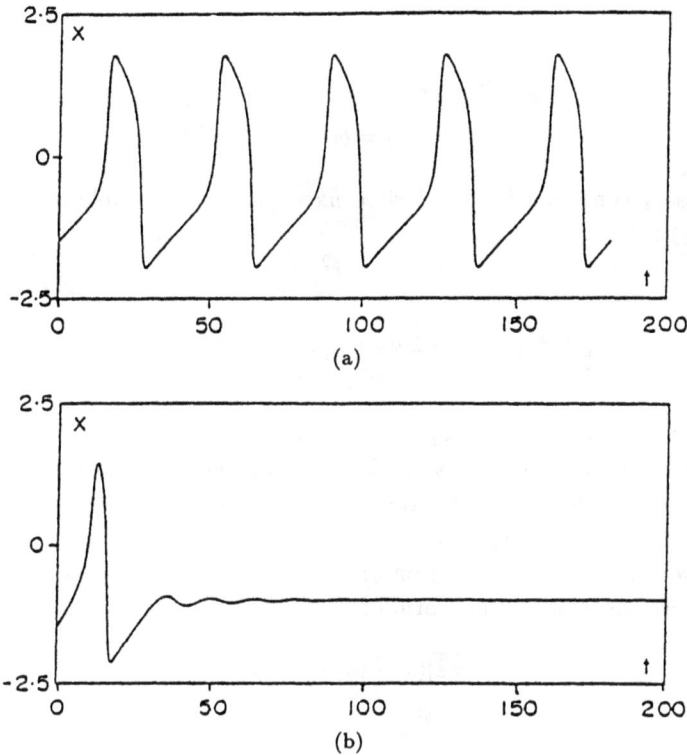

Fig. 8.2. Evolution of the variable x of the BVP oscillator. (a) Limit cycle motion without control; (b) Attraction to the fixed point $(x_s, y_s) = (-1.001, -0.376)$ under the control for $\varepsilon = 0.1$.

In general the control mechanism is sensitive to the ε value. The function in Eq. (8.1b) and the value ε which work for a particular (x_s, y_s) may not be suitable for some other (x_s, y_s). Thus it is essential to study the stability of the desired fixed point in the presence of (8.2c). To analyze its stability we note that the Jacobian of the right-hand side of Eqs. (8.2a–c) is

$$J = \begin{bmatrix} 1 - x_s^2 & -1 & 1 \\ c & -bc & 0 \\ -\varepsilon & 0 & 0 \end{bmatrix} . \tag{8.4}$$

Then the eigenvalues of (8.4), which determine the stability of (x_s, y_s), can be estimated from

$$\lambda^3 + a_1 \lambda^2 + a_2 \lambda + a_3 = 0, \tag{8.5a}$$

where

$$a_1 = bc - 1 + x_s^2, \quad a_2 = c + \varepsilon + bcx_s^2 - bc,$$
$$a_3 = bc\varepsilon. \tag{8.5b}$$

The necessary condition for the desired fixed point to be stable is $a_1 > 0$ and $(a_1 a_2 - a_3) > 0$, that is

$$(bc - 1 + x_s^2) > 0 \tag{8.6a}$$

and

$$\varepsilon > \frac{c[bx_s^4 + x_s^2(b^2 c - 2b + 1) + (bc - b^2 c + b - 1)]}{(1 - x_s^2)}. \tag{8.6b}$$

For fixed values of a, b, c and x_s as determined by Eq. (8.6a), (8.6b) gives a restriction on the value of ε so that the range of control stiffness is limited by stability considerations, provided the initial conditions are chosen in a suitable neighbourhood of $X_s \equiv (x_s, y_s)$.

Now we consider the conversion of a chaotic motion to a limit cycle. To start with we consider a linear form of G and let A_1 evolve as

$$\dot{A}_1 = \varepsilon[(x - x_s) - (y - y_s)]$$
$$\equiv \varepsilon\phi, \tag{8.2c'}$$

with $(x_s(t), y_s(t))$, as the desired solution. In the ACA the control strategy is the following. Without (8.2c') the motion of the system is chaotic. We choose x_s as a desired orbit to be stabilized. At time $t = t_0$, say, the control dynamics (8.2c') is switched on. The parameter A_1 evolves according to Eq. (8.2c') and adjusts its value until the desired state is reached. In the following we illustrate the above control mechanism.

For the choice $A_1 = 0.74$ and $A_0 = 0$, as noted in Fig. 8.1, the BVP oscillator exhibits chaotic motion without (8.2c'). Figure 8.3(a) shows the chaotic evolution in the $(x-t)$ plane. We choose (x_s, y_s) as the solution of Eqs. (8.2a, b) with $A_1 = 0.5$. For this choice of parameter value the long-time evolution is a period-T limit-cycle attractor. Figure 8.3(b) shows the evolution of the system under the control algorithm towards the limit-cycle motion for $\varepsilon = 0.1$. In Fig. 8.4 we have plotted A_1 as a function of t. As t increases from 0 the value of A_1 oscillates and finally settles to the constant value $A_{1f} = 0.5$. A useful quantity to characterize the dependence of the control process is the recovery time R_T, defined as the time taken to reach the desired state. We studied the dependence of R_T on ε with some assumed precision, say 10^{-4}.

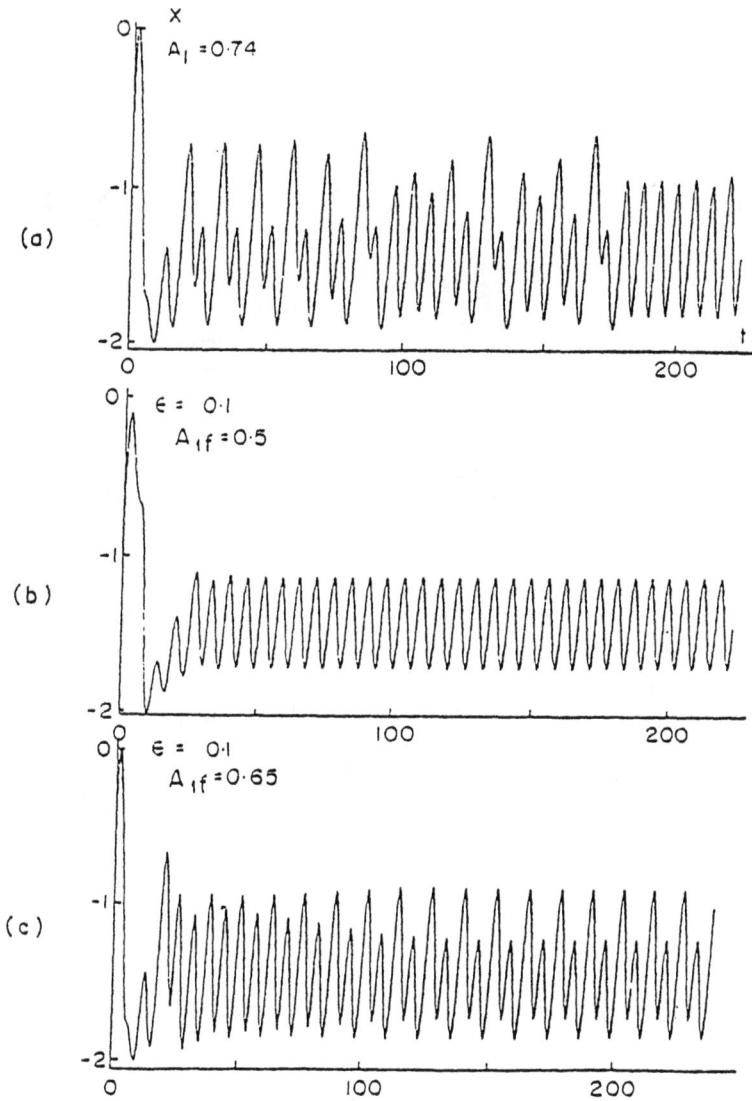

Fig. 8.3. Conversion of (a) chaotic motion to (b) a period-T limit cycle and (c) period-$2T$ limit cycle. A_{1f} is the value of A_1 associated with the desired solution.

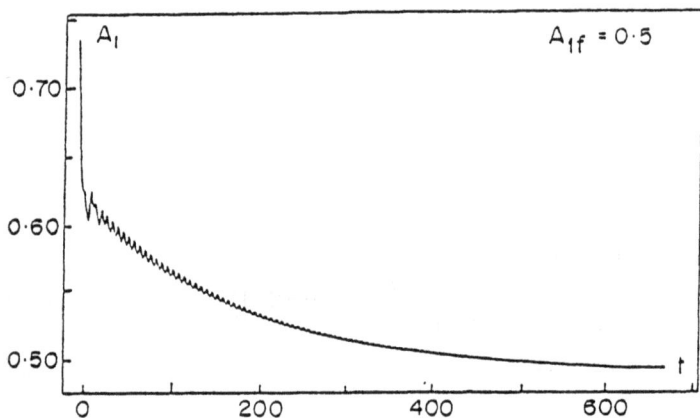

Fig. 8.4. Variation of A_1 according to the control (8.2c').

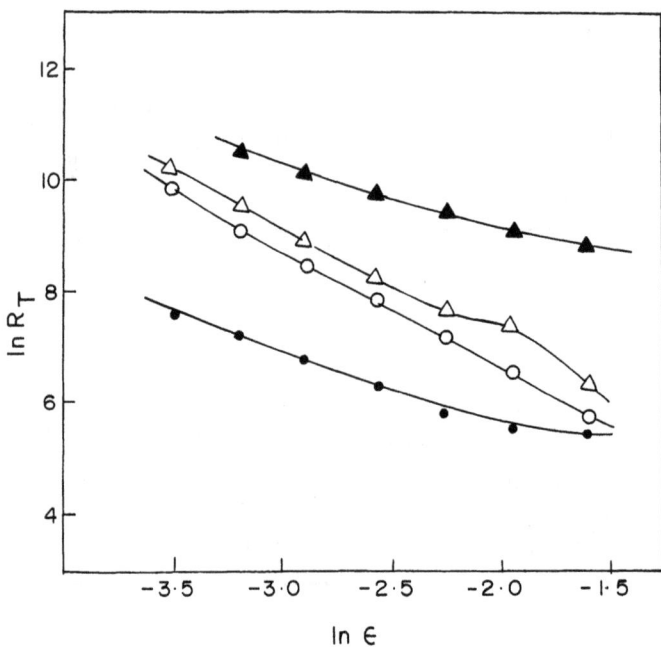

Fig. 8.5. Recovery time R_T versus the parameter ε for different control functions in Eq. (8.2c'). The control functions used are: ϕ (•), $\sin \phi$ (○), $\phi(1-\phi)$ (△), and $-(x-x_s)(y-y_s)$ (▲).

For ε in the interval 0.03–0.2 we estimated R_T for randomly chosen 100 initial conditions and averaged them. As expected, R_T decreases when ε increases.

The applicability of ACA has also been studied using nonlinear functions of the forms $\phi(1 - \phi)$, $\sin \phi$ and $-(x - x_s).(y - y_s)$ in (8.2c'). These nonlinear functions have also worked well in bringing the system back to the desired goal dynamics. The dependence of recovery time on the parameter ε for different control functions has also been determined and the results are shown in Fig. 8.5. For all the control functions used in the above analysis the recovery time is always inversely proportional to the stiffness of the control. Moreover, the recovery time for the linear form of ϕ in (8.2c') is relatively smaller than those of other nonlinear functions. That is, in the BVP system the linear control function seems to be a convenient choice. In the same manner, using the ACA we can also bring the chaotic evolution occurring for other parametric choices, including the case $A_0 > 0$, to a desired regular motion.

One might also ask the question whether it is possible to control the chaotic dynamics through ACA to a state which is not in the original dynamics (8.2a,b). For typical cases one finds that the system is not stable in this case and that the goal dynamics is not achieved in general.

8.2.2. *Suppression of Chaos by Parametric Perturbation*

In this section we study the effect of periodic parametric perturbation on the chaotic motion of the BVP oscillator. With periodic parametric perturbation of c, the BVP equation (8.2) can be written as

$$\dot{x} = x - \frac{1}{3}x^3 - y + A_1 \cos \omega t, \qquad (8.7a)$$

$$\dot{y} = c(1 + \eta \cos \Omega t)(x + a - by), \qquad (8.7b)$$

where η is the amplitude and Ω is the frequency of the parametric perturbation. For $\eta = 0$ we recover the usual BVP oscillator equation without the dc term. Starting from a chaotic region ($\eta = 0$), we explored the effect of the parametric perturbation on the dynamics by studying the behaviour of (8.7) in the (η, Ω) parameter space. While a full-fledged phase diagram would be invaluable, we present here the dynamics in a restricted region only, namely,

(i) Ω fixed at $0.7, 0.8, 1, 1.5$ and $\eta \in (0, 0.2)$ with interval $\Delta \eta = 0.005$;
(ii) η fixed at 0.12 and 0.2 and $\Omega \in (0, 2)$ with interval $\Delta \Omega = 0.01$.

As in Sec. 8.2 the parameters in Eq. (8.7) are fixed at $a = 0.7, b = 0.8, c = 0.1, \omega = 1$, and $A_1 = 0.74$ also.

Firstly, Table 8.2 summarizes the dynamical behaviour of the system (8.7) as a function of the parameter η for the above four fixed values of the parameter Ω. For $\Omega = 0.7$, chaotic behaviour persists up to $\eta \leq 0.075$. For $\eta > 0.075$

Table 8.2. Dynamical behaviour of the system (8.7) under parametric perturbation as a function of the parameters η and Ω for $a = 0.7, b = 0.8, c = 0.1, A_1 = 0.74$ and $\omega = 1$ (Ref. [176]).

Ω	$\eta(\Delta\eta = 0.005)$	Behaviour
0.7	0–0.075	chaos
	0.08–0.1	period-20
	0.105–0.2	period-10
0.8	0–0.12	chaos
	0.125–0.2	period-10
1	0–0.055	chaos
	0.06	period-16
	0.065-0.07	period-8
	0.075-0.105	period-4
	0.11-0.325	period-2
	0.33-0.4	period-1
1.5	0–0.055	chaos
	0.06–0.07	period-16
	0.075-0.08	period-8
	0.085-0.11	period-4
	0.115–0.155	period-2
	0.16-0.2	period-8

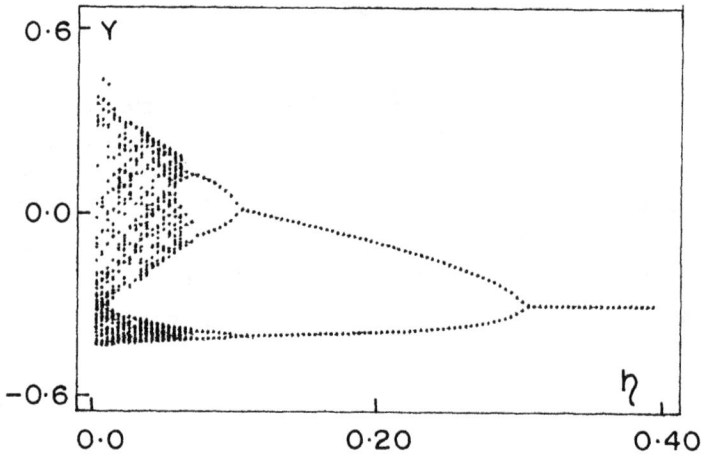

Fig. 8.6. Bifurcation diagram of the parametrically driven BVP system (8.7) showing the reverse bifurcation to period-T orbit.

mode-locked oscillations with periods 20 and 10 are observed. When the frequency of the perturbation matches the frequency of the driving force, namely for $\Omega = 1$, suppression of chaos by a period-halving phenomenon is observed as the value of η is increased from 0. Figure 8.6 shows reverse bifurcation to a period-T orbit for $\Omega = 1$. Regular behaviour is recovered for higher values of Ω also. For example, when $\Omega = 1.5$, regular dynamics is found for $\eta \geq 0.06$.

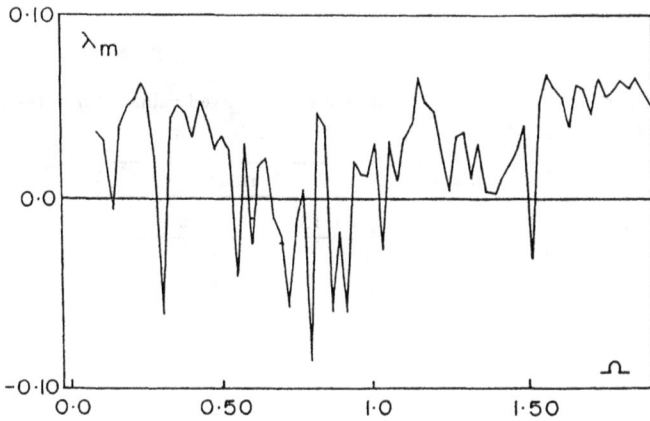

(a)

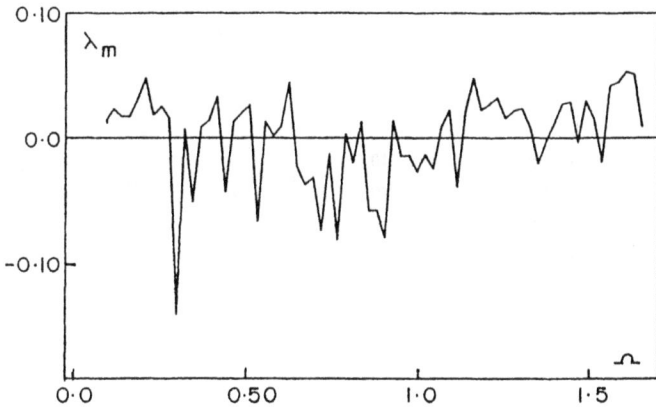

(b)

Fig. 8.7. Variation of the maximal Lyapunov exponent as a function of the parameter Ω for $\eta = 0.12$ (a), $\eta = 0.2$ (b).

Similarly, for fixed η the orbits for different Ω have been investigated (Table 8.3). We find many narrow regimes of regular motion interspersed with chaotic regimes. In Figs. 8.7(a) and 8.7(b) a detailed structure of the maximal Lyapunov exponent versus the frequency Ω is given for $\eta = 0.12$ and 0.2 respectively. Regular motion is recovered in the parameter interval where $\lambda_m < 0$. Here again the suppression of chaos is by mode-locking as in the earlier case. However, the mode-locked intervals are now very narrow. In Table 8.3 we list the values of Ω and the corresponding periods of the orbits for $\eta = 0.12$ and 0.2.

Table 8.3. List of periodic orbits as a function of Ω for fixed values of η in the parametric perturbation case (Ref. [176]).

$\Omega(\Delta\Omega = 0.005)$	Period	
	$\eta = 0.12$	$\eta = 0.2$
0.10	10	—
0.25	4	16
0.30	—	10
0.40	—	10
0.50	4	2
0.55	20	—
0.625	8	8
0.65	—	20
0.675	40	40
0.70	10	10
0.725	—	40
0.750	4	4
0.80	10	4
0.825	40	—
0.85	20	20
0.875	8	8
0.90	—	10
0.95	—	20
1	2	2
1.025	—	20
1.125	—	8
1.375	—	16
1.4	—	10
1.5	—	8

8.2.3. *Effect of Second Periodic Force*

One can also control system dynamics by the addition of an external weak force in the chaotic state. The BVP equation with a second periodic force $\eta \cos \Omega t$ can be written as

$$\dot{x} = x - x^3/3 - y + A_1 \cos \omega t + \eta \cos \Omega t \,, \tag{8.8a}$$

$$\dot{y} = c(x + a - by) \,. \tag{8.8b}$$

Equation (8.8) can be investigated by numerically solving it, with the remaining parameters fixed as in Sec. 8.2.2. Here again we present two sets of analysis, namely,

(i) Ω fixed, η varying,
(ii) η fixed, Ω varying.

Table 8.4. Details of the dynamical behaviour of the BVP system in the presence of a second periodic force for three fixed values of Ω as a function of η (Ref. [176]).

Ω	$\eta(\Delta\eta = 0.005)$	Dynamical behaviour
0.7	0–0.03	chaos
	0.035–0.04	period-20
	0.045–0.20	period-10
0.8	0–0.105	chaos
	0.11–0.20	period-5
1.5	0.0–0.075	chaos
	0.08–0.20	period-4

Table 8.4 summarizes numerical simulations of the system (8.8) as a function of the parameter η for three fixed values of Ω. From this table we note that in the η intervals regular behaviour is recovered for η values above a certain threshold value as in the previous case. On the other hand, to study the change in the dynamics of the system as a function of Ω for fixed η we have estimated the maximal Lyapunov exponent. When $\eta = 0$ and $A_1 = 0.74$ the maximal Lyapunov exponent is estimated as $\lambda_m \approx 0.056$. The graph λ_m versus Ω is plotted in Fig. 8.8 for $\eta = 0.015, 0.02$ and 0.2. For $\eta = 0.015$, λ_m takes negative values only at three narrow ranges of Ω in the interval we have studied. For example, attractors with period 50, 4 and 8 are found to occur at $\Omega = 0.34, 1.25$ and 1.5 respectively. However, as η is increased the size of the locked intervals increases. This is clear from Figs. 8.8(b) and 8.8(c). Figure 8.9 shows a period

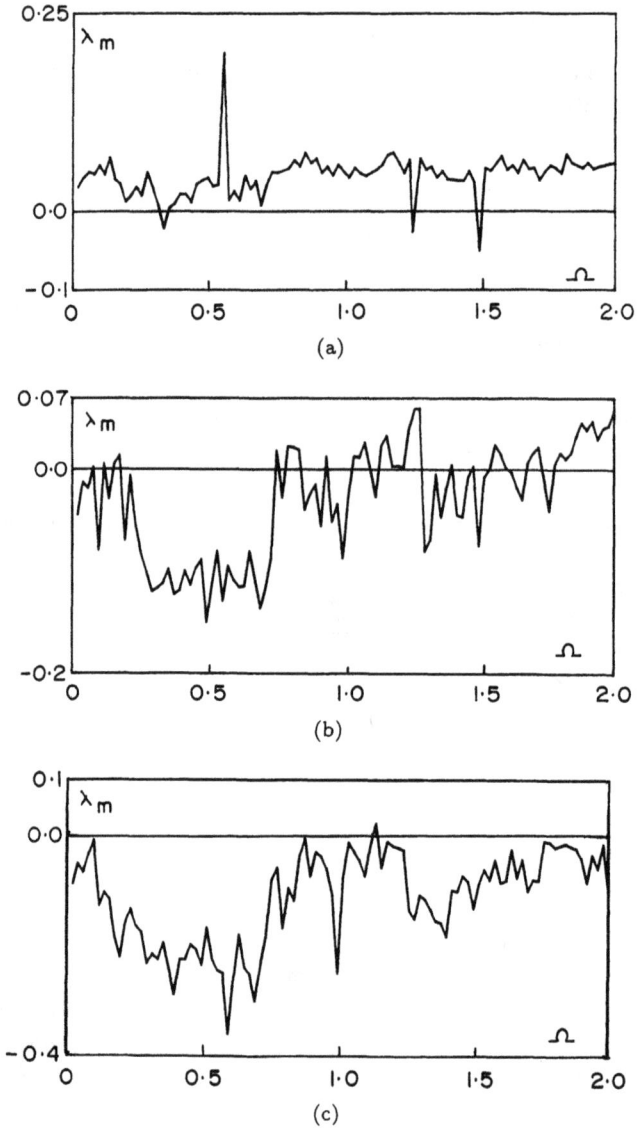

Fig. 8.8. Maximal Lyapunov exponent against the parameter Ω for the system (8.8) with $\eta = 0.015$ (a), $\eta = 0.02$ (b), and $\eta = 0.2$ (c).

diagram for $\eta = 0.2$. Between Ω intervals mode locking with periods 2, 4, 5, 10, 25 and 50 occurs. From Fig. 8.9 we also note that the locked intervals are dominated by the orbits with period 50. Similar results have been observed for larger values of η.

One thus notices that regular behaviour occurs over a wider range of Ω for larger η values. However, too large an η may mean a violent perturbation of the original system which is outside the gambit of the controlling. On the other hand we feel that perturbing through an external force is one of the easiest way of adjusting the system. So with an optimal choice of the strength of η, suitable regular motions can be achieved for appropriate intervals of Ω.

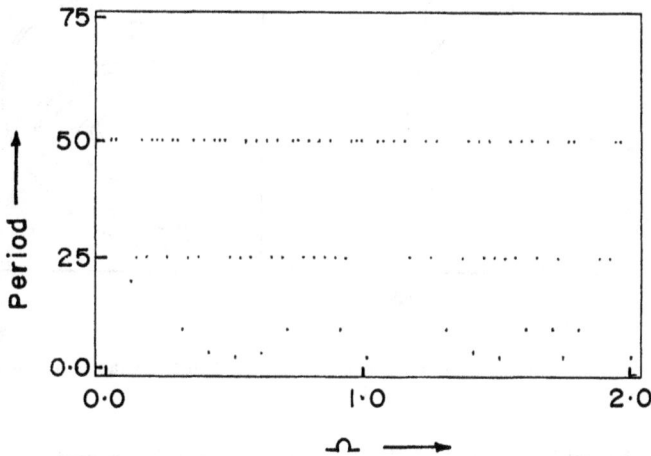

Fig. 8.9. Period versus Ω with $\eta = 0.2$ for the system (8.8).

8.2.4. *Stabilization of Unstable Periodic Orbits*

In the adaptive control and weak perturbation methods chaotic dynamics is converted into a new periodic orbit, the period of which depends on the applied perturbation. On the other hand, chaos can also be eliminated by stabilizing the unstable periodic orbits (Refs. [190, 191]) which are embedded in the chaotic orbit itself by making small adjustments to a suitable external parameter. In this section we investigate the effectiveness of such methods by applying them to the BVP oscillator.

(a) *Ott–Grebogi–Yorke (OGY) method*: In the OGY technique (Ref. [152]), stabilization of an unstable periodic orbit in a suitably defined Poincaré surface

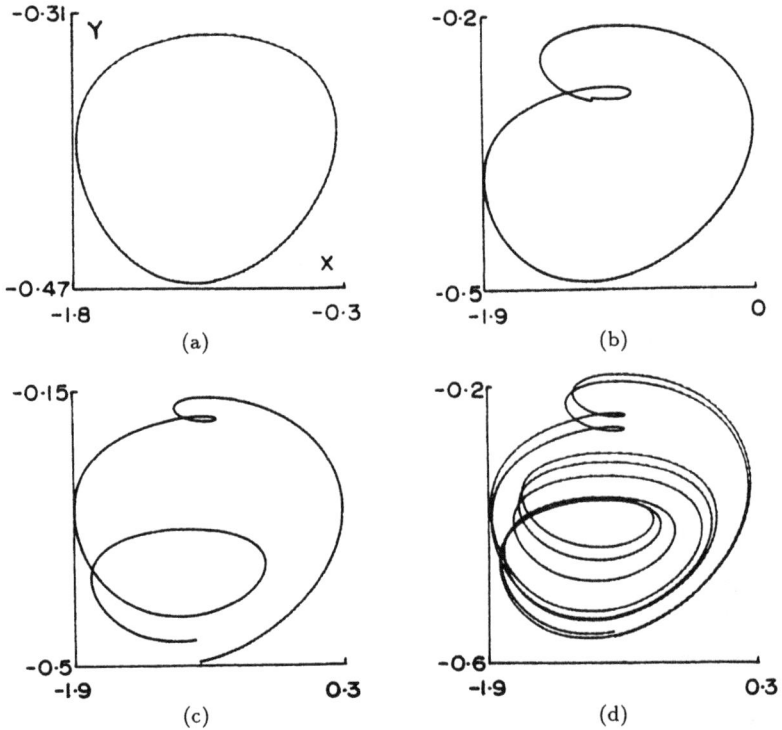

Fig. 8.10. Some of the unstable period-k orbits embedded in the chaotic attractor of the BVP oscillator (Fig.8.1).

of section can be achieved by slightly adjusting a control parameter. To apply this technique to the BVP equation (8.2), we consider its Poincaré map and stabilize an unstable periodic orbit embedded in the chaotic attractor. To start with, we first recall the salient features of the OGY technique with reference to a two-dimensional discrete map.

Suppose we wish to stabilize an unstable period-1 orbit contained in the chaotic attractor (Fig. 8.1), which occurs for $A_0 = 0$ and $A_1 = 0.74$. As the control parameter, say A_0, in Eq. (8.2) is varied slightly from $A_0 = 0$ to some $\bar{A}_0 (\neq 0)$, the period-1 fixed point will now shift from $X_F(A_0 = 0)$ to $X_F(A_0 = \bar{A}_0)$ where $X_F = (x_F, y_F)$. We define a vector g as

$$g \equiv \left. \frac{\partial X_F(A_0)}{\partial A_0} \right|_{A_0=0} \approx \frac{X_F(A_0) - X_F(0)}{A_0}. \tag{8.9}$$

Near the period-1 orbit $X_F(0)$ and for small values of the control parameter one may write

$$\delta X_{n+1} \approx M \cdot \delta X_n \,, \tag{8.10}$$

where M is a 2×2 matrix, $\delta X_n = X_n - X_F(\bar{A}_0)$, and X_n is the state of the system at the nth iteration. Let λ_u and λ_s denote the unstable and stable eigenvalues of the matrix M respectively, with $|\lambda_u| > 1$ and $|\lambda_s| < 1$. Then $M \cdot e_u = \lambda_u e_u$ and $M \cdot e_s = \lambda_s e_s$, where e_u and e_s are the unstable and stable eigenvectors of M respectively. Further, the contravariant basis vectors are defined by $f_s \cdot e_s = f_u \cdot e_u = 1, f_s \cdot e_u = f_u \cdot e_s = 0$. When the nth iteration X_n is close to X_F, the value of A_0 is changed slightly such that X_{n+1} falls on the stable manifold of X_F. That is, we choose the control δA_{0n} so that

$$f_u \cdot \delta X_{n+1} = 0 \,, \tag{8.11}$$

which yields the control formula

$$\delta A_{0n} = \frac{\lambda_u f_u \cdot \delta X_n}{(\lambda_u - 1)f_u \cdot g} \,. \tag{8.12}$$

The control is switched on only if δA_{0n} is less than some preassigned maximal allowed perturbation $\delta A_{0\,\max}$; otherwise δA_{0n} is zero.

For the BVP oscillator (8.2) with chaotic motion, first we determine some of the low-period unstable periodic orbits as follows (Refs. [43, 44]). Let the chaotic attractor in the Poincaré map be the set of points $X_1, X_2, \ldots, X_n$, where $X = (x, y)$. Then the points $(X_n, X_{n+1}, \ldots, X_{n+k})$ constitute an unstable period-k orbit with accuracy ε, if for $k > 0$

$$\|X_{n+k} - X_n\| < \varepsilon \tag{8.13}$$

is satisfied. For example, Fig. 8.10 shows four such different orbits associated with the chaotic attractor of Fig. 8.1 with $\varepsilon = 0.05$. Let us denote $X_F(A_0 = 0)$ as an unstable period-1 orbit to be stabilized. Then we find all pairs of iterates falling within the circle of radius 0.01 about the point $X_F(0)$. To this set of data we have fitted the approximate linear map (8.10) and then determined the form of M and its eigenvalues. The approximate location of the period-1 orbit is calculated as $X_F \equiv (x_F, y_F) \simeq (-0.9560, -0.4685)$. We have obtained $\lambda_u \simeq 1.46$, $\lambda_s \simeq -0.68$, $g \simeq (0.105, 0.97)$. To stabilize the orbit X_F we have chosen the maximum allowed perturbation as $|\delta A_{\max}| = 0.01$. Figure 8.11(a) illustrates the control of period-1 orbit. The period-1 orbit gets stabilized as long as the control is included. Once the control is removed the system evolves chaotically. Figure 8.11(b) shows the perturbations δA_0

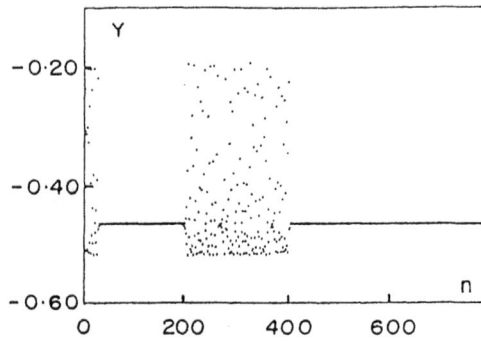

(a)

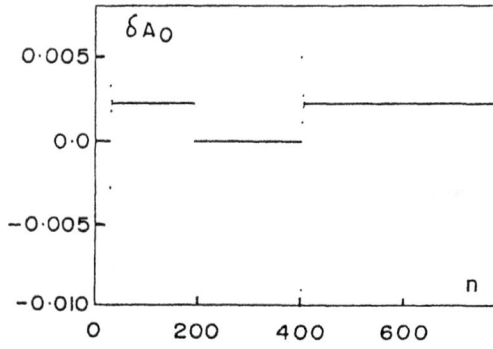

(b)

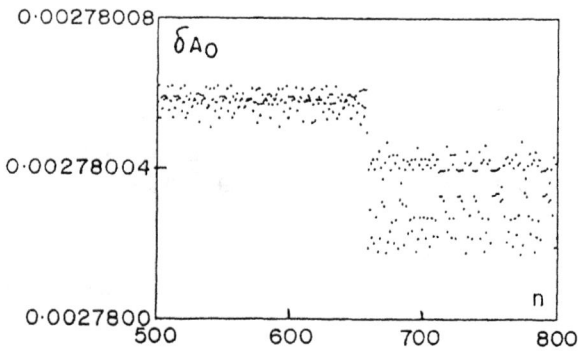

(c)

versus n used to stabilize the desired orbit. As the required perturbation δA_0 is quite small the evolution of the control parameter in Fig. 8.11(c) appears to be constant. For clarity, in Fig. 8.11(c) we have plotted δA_0 against the iteration number n in the interval 500–800. From Fig. 8.11(a) we note that before reaching the desired controlled orbit, for a given set of initial conditions, the system trajectory exhibits a transient evolution. To study the dependence of the recovery time R_T, one can estimate R_T for randomly chosen 100 initial conditions and average them for several values of $\delta A_{0\,max}$. As expected the average length of the transient $\langle R_T \rangle$ increased with a decrease in $\delta A_{0\,max}$.

One can also test the control strategy by adjusting the parameter A_1. Using the method described earlier, the vector g and other quantities can be calculated to estimate the perturbation δA_1. Figure 8.12 shows the stabilization of period-1 orbit. Figures 8.11 and 8.12 clearly illustrate the flexibility offered by

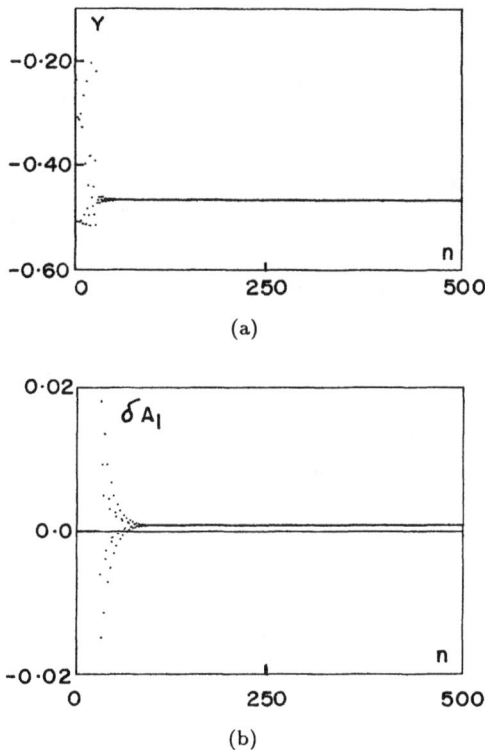

Fig. 8.12. Control of period-1 orbit with A_1 as the control parameter.

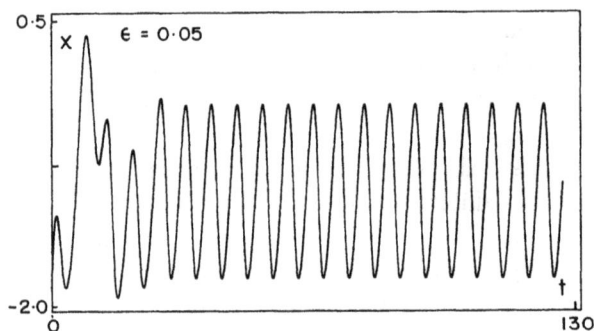

(a)

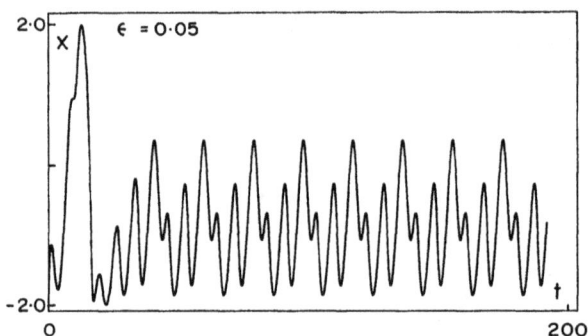

(b)

Fig. 8.13. Time-series plot illustrating the control of an unstable period-T (a) and $3T$ (b) orbits.

the OGY method in stabilizing a specific orbit by means of different control parameters.

Even though in the above we have only discussed the stabilization of the period-1 orbit in Fig. 8.10 by the OGY method, the same procedure can be extended, in principle, to stabilize other periodic orbits too.

(b) *Small feedback control*: In this subsection we illustrate the stabilization of unstable periodic orbits by the method of Singer *et al.* (Ref. [156]). To stabilize a desired unstable periodic orbit, assume that the parameter A_0 in Eq. (8.2) is available for small perturbation and the BVP equation with the controller is given by

$$\dot{x} = x - x^3/3 - y + A_1 \cos \omega t + [A_0 - \varepsilon \operatorname{sgn}(x - \tilde{x})], \qquad (8.14a)$$

$$\dot{y} = c(x + a - by). \qquad (8.14b)$$

In Eq. (8.14), $\tilde{x}$ is the x-component of the unstable period-k orbit, ε represents the stiffness of the control signal and sgn corresponds to the sign of $x - \tilde{x}$. As noted earlier, for $A_0 = 0$, and $\varepsilon = 0$, the system exhibits chaotic motion (Fig. 8.1) for $A_1 = 0.74$. When the control is switched on, the controller reacts to the deviation of $x(t)$ from $\tilde{x}(t)$ and modifies the magnitude of the dc component A_0. In this way, by making small adjustments by an amount $\pm\varepsilon$, $\varepsilon \ll 1$, to the parameter A_0, it is possible to suppress the chaotic behaviour and thereby stabilize a chosen unstable periodic orbit $\tilde{X}$ contained in the chaotic orbit.

In Figs. 8.13(a) and 8.13(b) we show the stabilization of unstable period-T and period-$3T$ orbits respectively for $\varepsilon = 0.05$. The initial condition used is $(x(0), y(0)) = (-1.8, -0.3)$ and the control is initiated at $t = 0$. From Fig. 8.13 it is clear that when the controller is applied to a chaotic system it takes some time, which is the recovery time R_T, for converting the actual motion to a desired orbit $\tilde{X}$. Detailed estimation of R_T for $0 < \varepsilon < 0.5$ shows that it is inversely proportional to the stiffness of the control.

8.2.5. *Controlling of Chaos by Noise*

In the various control methods discussed so far, a system parameter or an external forcing parameter is suitably adjusted so that ultimately periodic motion sets in. In the present section we point out a rather simple approach without disturbing any of the system parameters, namely the addition of an external noise to suppress chaotic behaviour.

We now present the effect of a Gaussian noise $\eta(t)$ on the chaotic dynamics of the BVP equation. We add $\eta(t)$ to the right-hand side of Eq. (8.2a) for the chaotic case $A_1 = 0.74$ with $A_0 = 0$ and the other parameters fixed as before. Gaussian random numbers with standard deviation σ is added to the system at every 0.01 time step. In order to know whether the motion is regular or chaotic, the maximal Lyapunov exponent is estimated. In Fig. 8.14 λ_m is plotted against the parameter σ. It is seen that for $\sigma < 0.072$, λ_m is positive, that is the motion is chaotic. For $\sigma \geq 0.072$, λ_m takes negative values which confirms the regular behaviour of the system. We can illustrate the influence of noise more clearly by looking at the distance $S(t)$ between two orbits starting from two nearby initial conditions. The distance $S(t)$ between two trajectories starting from $X(0)$ and $X'(0)$ is given by

$$S(t) = ||X(t) - X'(t)|| . \tag{8.15}$$

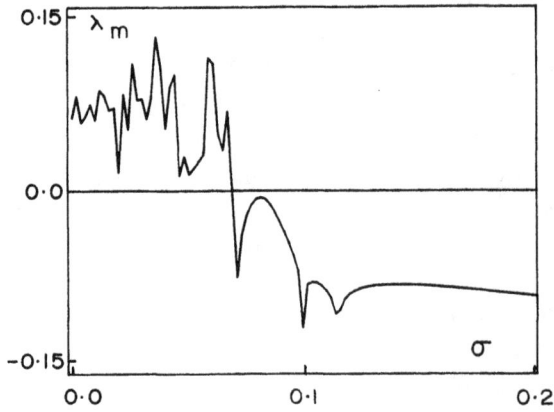

Fig. 8.14. Maximal Lyapunov exponent as a function of standard deviation σ of the external noise.

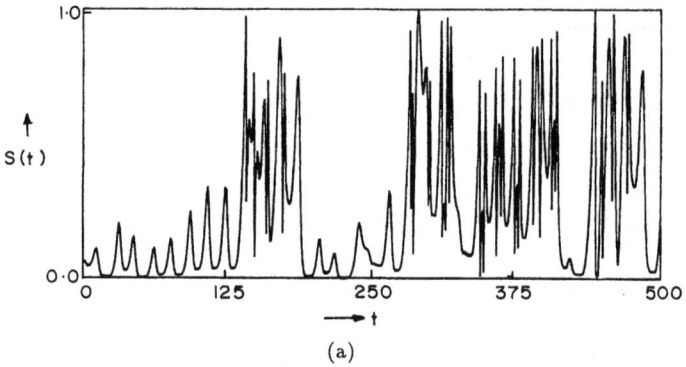

(a)

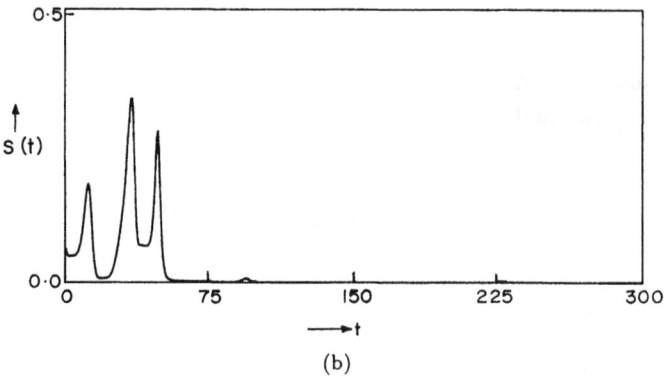

(b)

Fig. 8.15. Separation distance plot for (a) $\sigma = 0.06$ and (b) $\sigma = 0.09$.

For regular behaviour $S(t)$ decays to zero in the limit $t \to \infty$. On the other hand, it will vary irregularly with time for chaotic motion. For $\sigma < 0.072$, $S(t)$ shows irregular variation while for $\sigma \geq 0.072$ it diminishes with time after an irregular initial-stage transient. That is, chaotic behaviour persists for $\sigma < 0.072$ but is suppressed for $\sigma \geq 0.072$. In Fig. 8.15 we report the calculated $S(t)$ for two noise levels. In calculating the separation distance the same noise is added for both the initial conditions, $X(0)$ and $X'(0)$. The irregular variation of $S(t)$ in Fig. 8.15(a) for $\sigma = 0.06$ indicates sensitive dependence on the initial states of the system. For $\sigma = 0.09$ (Fig. 8.15(b)), $S(t)$ diminishes to zero after some short-lived irregular variation. This shows what is perhaps the most dramatic effect of noise in the above studies: the loss of sensitive dependence on the initial conditions when the standard deviation exceeds a certain critical value.

In controlling chaos it is necessary to study the ultimate motion of the system. Figure 8.16 shows the Poincaré map of the attractor for $\sigma = 0.09$. The λ_m of the attractor is nearly -0.026 while its correlation dimension is found to be 1.31. That is, the attractor is strange but nonchaotic. The addition of noise not only suppresses the chaotic nature of the attractor, but it has substantially changed the pattern of the attractor in the above case. This is quite apparent from Figs. 6.4 and 8.16. The striking result is that while the attractor of Fig. 6.4 is chaotic, the attractor shown in Fig. 8.16 is nonchaotic. We have found similar strange nonchaotic attractors for noise levels $\sigma > \sigma_c$. Of course, strange nonchaotic attractors are not unknown in the literature and they have been observed in certain quasiperiodically forced systems (Refs. [192–195]). We also wish to note that in the long Josephson-junction oscillator under the influence of Gaussian random noise (Ref. [196]), for a range of noise level chaotic motion becomes converted to a limit cycle. Whether such regular periodic attractors are possible under the influence of external noise in the BVP oscillator also requires a more detailed analysis in other parameter regions.

Finally, it is important to know whether the noise effect is sensitive to the time step Δt used for the numerical calculation of trajectories. On analyzing the suppression of chaos by adding noise with different time steps, one obtains the interesting result that for all the time steps chosen in Ref. [176] the maximal Lyapunov exponent λ_m is found to be positive for $\sigma > \sigma_c(\Delta t)$ and negative for $\sigma < \sigma_c(\Delta t)$. In Fig. 8.17, σ_c is plotted against the time step Δt and we find that it is inversely proportional to Δt, leading to the relation

$$\sigma_c(\Delta t) = -0.4667\Delta t + 0.07667 . \qquad (8.16)$$

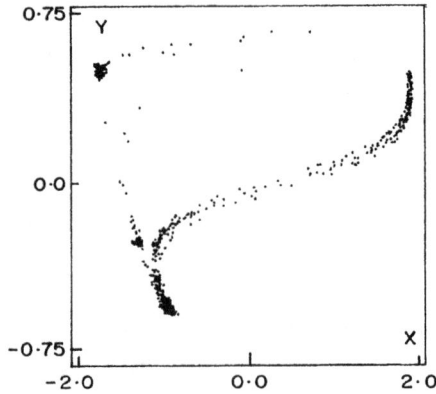

Fig. 8.16. Strange nonchaotic attractor for $\sigma = 0.09$ (Poincaré map).

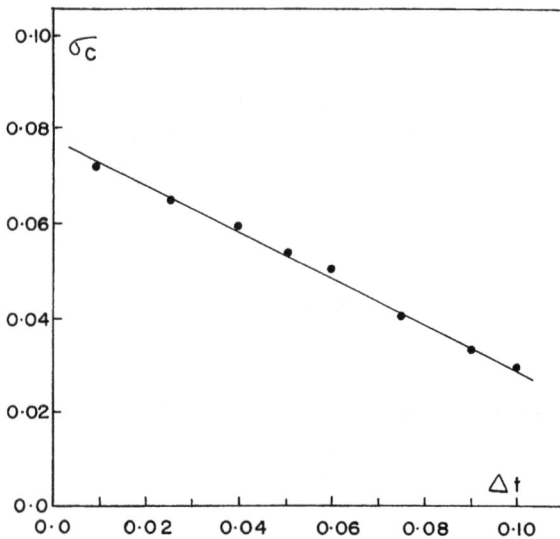

Fig. 8.17. Data showing the σ_c plotted against the time step Δt after which noise is added. The solid line is obtained by straight-line fitting of the numerical data.

Thus one finds that in the BVP equation the interaction between the dynamic force and stochastic (noise) force gives rise to a negative maximal Lyapunov exponent and strange nonchaotic attractors. For $\sigma < \sigma_c$ the noise term is relatively weak compared to the dynamic force and the system behaviour is still chaotic. However, for $\sigma \geq \sigma_c$ the noise term is able to overcome the effect of chaos giving rise to motions free from sensitive dependence on initial conditions.

8.3. Controlling in Other Oscillators

We now wish to present a brief account of the applicability of the control algorithms to control chaos in Chua's circuit, the driven Chua's circuit, Murali–Lakshmanan–Chua (MLC) circuit discussed in Chapter 7, and the Duffing oscillator in order to demonstrate further the feasibility of controlling.

8.3.1. *Controlling of Chaos in Chua's Circuit*

(a) *Oscillator absorber concept*: As noted earlier the method here is to add a supplementary dynamical system to quench the chaotic oscillations of the original system. Kapitaniak *et al.* (Ref. [175]) proposed such a "chaotic oscillation absorber" for Chua's circuit. As seen in Fig. 8.18 the "absorber" is a parallel RLC circuit coupled with the original Chua's circuit via the resistor R_x. The equations describing the dynamics of this augmented system can be given in a dimensionless form (Ref. [175]) (see also Chapter 7, Sec. 7.3):

$$\dot{x} = \alpha[y - x - h(x)], \tag{8.17a}$$
$$\dot{y} = x - y + z - \varepsilon(y - y'), \tag{8.17b}$$
$$\dot{z} = -\beta y, \tag{8.17c}$$
$$\dot{y}' = \alpha'[-\gamma'y' + z' + \varepsilon(y - y')], \tag{8.17d}$$
$$\dot{z}' = -\beta'y'. \tag{8.17e}$$

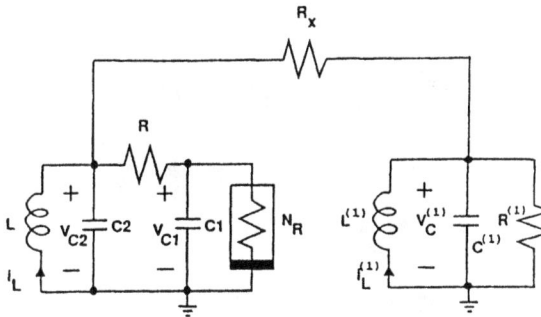

Fig. 8.18. Chua's circuit coupled to a two-dimensional linear system (Ref. [175]).

In terms of the circuit equations we have the additional set of equations for the "absorber" variables (y', z') and weak mutual coupling term $(\varepsilon(y' - y))$ via which the original equations of Chua's circuit are modified. Here ε denotes the coupling stiffness. Depending on the choice of ε various stable asymptotic states can be obtained in the original chaotic circuit. Figure 8.19 depicts the results of experimental simulations confirming the desired action of the "absorber". Specifically, Fig. 8.19(a) corresponds to Chua's double-scroll attractor observed when there is no absorber, while Figs. 8.19(b)–(g) show different types of periodic orbits (depending on the magnitude of ε).

(b) *Control engineering approach*: We now demonstrate the applicability of the control engineering approach suggested by Chen and Dong (Ref. [174]) to the Chua's circuit. The control problem for the present case can be stated as follows. Denoting by $(\tilde{x}(t), \tilde{y}(t), \tilde{z}(t))$ any solution of Chua's circuit, which exists for a given parameter set (e.g., an unstable periodic orbit), the goal is to control the actual system trajectory such that for any given $\varepsilon > 0$, there exists a $T_\varepsilon > t_0$ for which

$$
\begin{aligned}
|x(t) - \tilde{x}(t)| &\leq \varepsilon, \\
|y(t) - \tilde{y}(t)| &\leq \varepsilon, \\
|z(t) - \tilde{z}(t)| &\leq \varepsilon, \quad \text{for all } t \geq T_\varepsilon.
\end{aligned}
\tag{8.18}
$$

The following result has been proved in (Ref. [174]).

Theorem: Let $(\tilde{x}(t), \tilde{y}(t), \tilde{z}(t))$ be the unstable limit cycle of Chua's circuit. Then, the chaotic trajectory $(x(t), y(t), z(t))$ of the circuit can be driven to reach this limit cycle by a linear feedback control of the form

$$
\begin{bmatrix} u_1 \\ u_2 \\ u_3 \end{bmatrix} = -K \begin{bmatrix} x - \tilde{x} \\ y - \tilde{y} \\ z - \tilde{z} \end{bmatrix} \equiv - \begin{bmatrix} 0 & 0 & 0 \\ 0 & K_{22} & 0 \\ 0 & 0 & 0 \end{bmatrix} \begin{bmatrix} x - \tilde{x} \\ y - \tilde{y} \\ z - \tilde{z} \end{bmatrix},
\tag{8.19}
$$

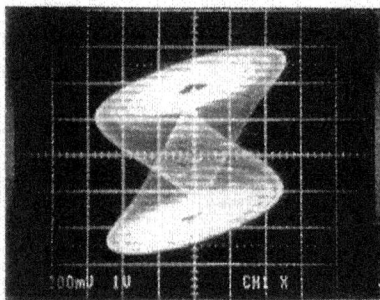

(a)

Fig. 8.19. Effect of experimental controlling procedure using "oscillator absorber" concept (Ref. [175]). (a) The double-scroll chaotic attractor ($\varepsilon = 0$) in (v_1-v_2) plane. (b) period-1 orbit ($\varepsilon = 0.322$). (c) period-2 orbit ($\varepsilon = 0.302$). (d) period-4 orbit ($\varepsilon = 0.148$). (e) period-5 orbit ($\varepsilon = 0.105$). (f) period-3 orbit ($\varepsilon = 0.097$). (g) fixed point ($\varepsilon = 0.322$).

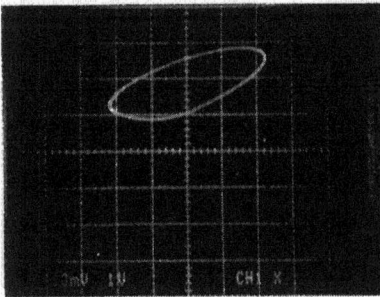

(b)

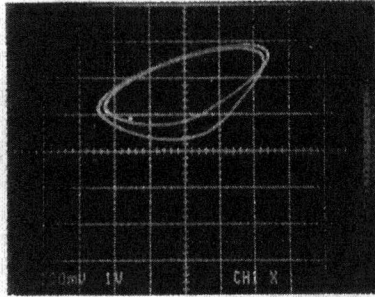

(c)

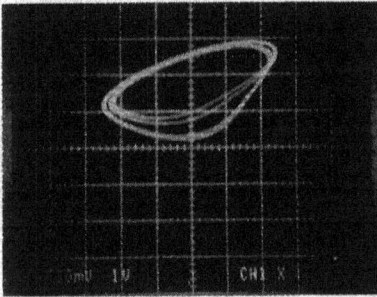

(d)

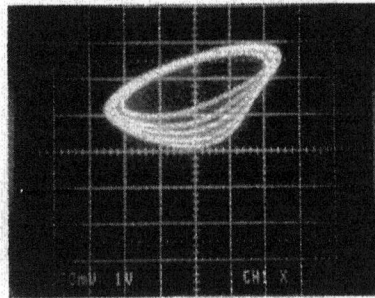

(e)

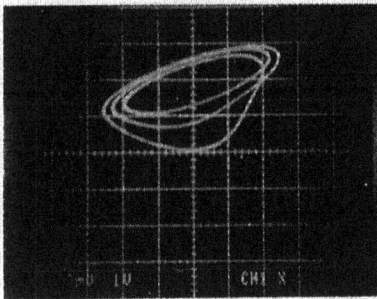

(f)

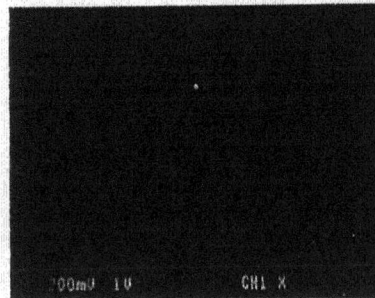

(g)

Fig. 8.19. (*Continued*)

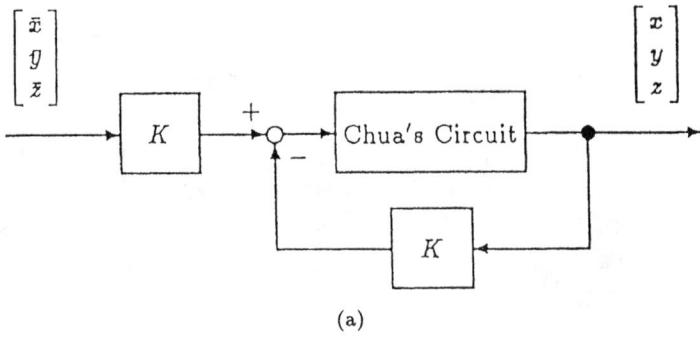

(a)

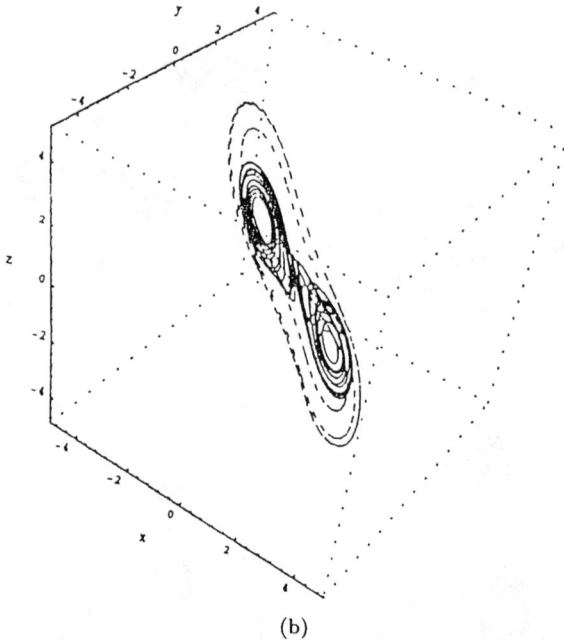

(b)

Fig. 8.20. (a) Block diagram of the feedback control of a chaotic system using classical linear controller (Chen–Dong approach) (Ref. [174]). (b) A Chua's circuit solution trajectory (before and when a feedback control is applied) in the $x - y - z$ space.

provided that

$$0 < \frac{1}{(G_b + 1)} \leq K_{22} \leq \frac{-1}{(G_a + 1)}, \tag{8.20}$$

where G_a and G_b are the middle and outer slopes of the characteristic curve of Chua's diode (cf. Sec. 7.3). The block diagram of the control scheme is shown in Fig. 8.20(a). The dimensionless equations of the controlled circuit are

$$\dot{x} = \alpha[y - x - h(x)], \tag{8.21a}$$

$$\dot{y} = x - y + z - K_{22}(y - \tilde{y}), \tag{8.21b}$$

$$\dot{z} = -\beta y. \tag{8.21c}$$

Using the above method of Chen and Dong (Ref. [174]) one is able to stabilize the chaotic trajectories by directing them towards the saddle-node unstable limit cycle coexisting with the chaotic attractor (see Fig. 8.20(b)).

(c) *Occasional proportional feedback control (OPF)*:   Johnson and Hunt (Ref. [159]) have applied the OPF method to Chua's circuit and showed that both periodic and chaotic orbits can be controlled. This method is to perturb the magnitude of the negative resistance as it does not alter the natural frequency of the system in a first approximation. The overall negative resistance is obtained by letting a fixed nonlinear resistance be in parallel with a voltage-variable resistance (VVR) which is provided by the field effect transistor. Control of the system is achieved by modulating this VVR. Since the system is autonomous, an external synchronization signal (with frequency ν_{sync}) is used and is supplied by a separate signal generator. The circuit implementation is shown in Fig. 8.21.

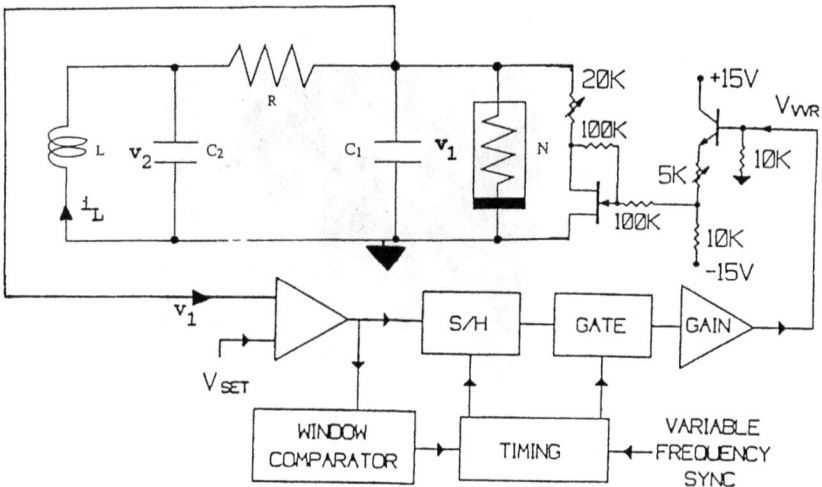

Fig. 8.21. Block diagram of the OPF control circuit (Ref. [159]). Chua's circuit is shown with the addition of the voltage-variable resistor in parallel with the nonlinear resistor labeled N.

(a)

(b)

(c)

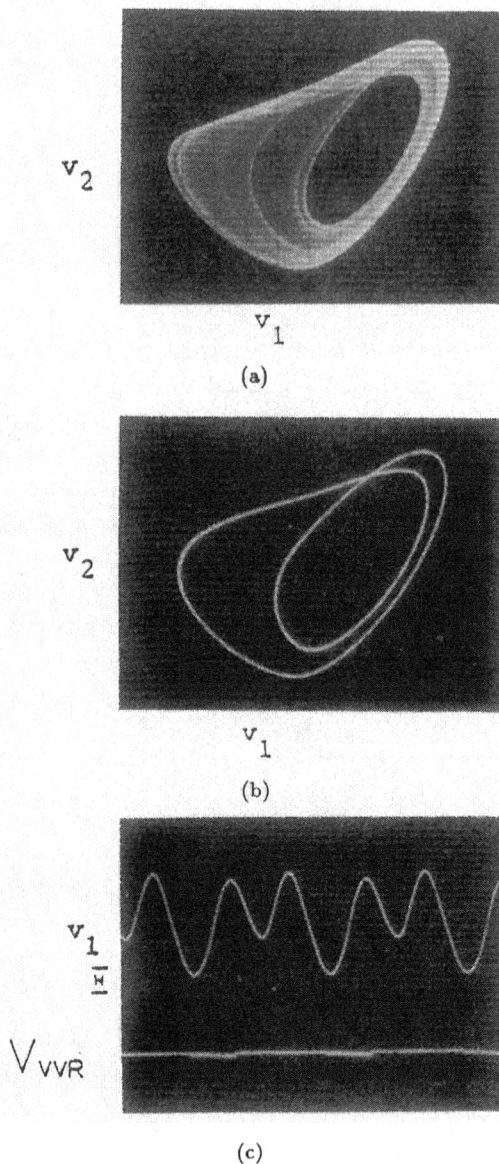

Fig. 8.22. A single-correction, low-period orbit. Period-2 is stabilized by applying a single correction lasting less than one oscillation. A plot of v_2 versus $v_R(= v_1)$ shows the system (a) before and (b) after the control circuit is turned on. (c) The upper trace is the voltage across the nonlinear resistor, $v_R(= v_1)$, and the lower trace is the feedback signal V_{VVR}. The window edges are shown as indicated by the label "W" (Ref. [159]).

The voltage V_R is sampled and fed back to the window comparator where it is compared to the offset control voltage V_{set}. When the voltage peak falls within the comparator, the gate switches a signal proportional to ($V_{R,peak}$ − V_{set}), which is then amplified to become the feedback control voltage signal V_{VVR}. In the experiment, the circuit oscillates chaotically until it visits the neighbourhood of the fixed point (i.e., inside the window), then the proper feedback control signal ensures that on the next cycle the circuit solution trajectory approaches the fixed point. When control is to be attempted in a single basin, the system is set up so that oscillations take place in the chaotic regime, which continue to one of the two Chua's spiral-type attractors. Control is switched on, and the control variables are adjusted until the system locks in on a stabilized orbit. Periods 1, 2, and 4, remnants of period-doubling cascade, are easily stabilized, and the control signal is very small. Figure 8.22(a) shows Chua's spiral-type chaotic attractor, and 8.22(b) the stabilized period-2 orbit. Figure 8.22(c) shows the lower peak of V_R in the window producing the control signal in the lower trace. We also note that recently the OPF method or similar approaches were employed to control chaos in a globally-coupled, multimode, autonomous laser system, in fiber lasers, and in diode resonators system (Ref. [182]).

We may also mention that other control methods like the OGY method (Refs. [89, 152]), the classical regulators and state space techniques (Ref. [167]), the distortion control techniques (Ref. [166]), and so on, are also applicable to Chua's circuit, details of which are given in the reviews of Chen and Dong (Ref. [174]) and Ogorzalek (Ref. [89]).

8.3.2. Controlling of Chaos in Driven Chua's Circuit: Experimental Results

In order to study the controlling of the chaotic aspect of the driven Chua's circuit (Sec. 7.4.2) experimentally, we add a weak second periodic signal $f_2 \sin \omega_2 t$ in series with the periodic signal $f_1 \sin \omega_2 t$ as in Fig. 8.23. Then the system becomes a quasiperiodically driven one. Correspondingly, the defining equations for the circuit becomes

$$C_1 \frac{dv_1}{dt} = (1/R)(v_2 - v_1) - f(v_1)\,, \tag{8.22a}$$

$$C_2 \frac{dv_2}{dt} = (1/R)(v_1 - v_2) + i_{L2} - i_{L1}\,, \tag{8.22b}$$

$$L_1 \frac{di_{L1}}{dt} = v_2 - (f_1 \sin \omega_1 t + f_2 \sin \omega_2 t)\,, \tag{8.22c}$$

$$L_2 \frac{di_{L2}}{dt} = -v_2 . \qquad (8.22d)$$

As noted in Chapter 7, for the chosen parametric values given in the circuit (Fig. 8.23) and $f_2 = 0$, the system exhibits a variety of bifurcation sequences. For the present analysis, we set the forcing amplitude f_1 and frequency ($\nu_1 = \omega_1/2\pi$) at values for which chaos is initially observed. When the amplitude f_2 of the second periodic signal is slowly varied ($f_2 > 0$), we find a remarkable suppression of chaos to take place giving rise to ordered motions. The bifurcation diagram in the f_1-ν_1 plane (Fig. 7.11) shows that there are at least three prominent chaotic regimes, Ch_1, Ch_2 and Ch_3.

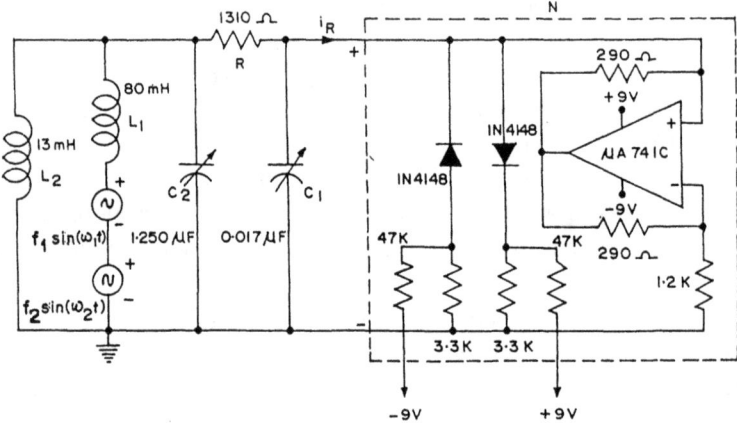

Fig. 8.23. The driven Chua's circuit with a second periodic signal $f_2 \sin \omega_2 t$.

(1) *Ch_1 region*: When $f_2 = 0$, a double-band chaotic attractor of Fig. 8.24(a) is observed for $f_1 = 364.5$mV ($\nu_1 = 1200$Hz). When the amplitude of the second sinusoidal perturbation is slowly and continuously increased from 0mV (frequency $\nu_2 = (\omega_2/2\pi) = 10$KHz), the following scenario is observed:

  (i) The double-band chaos of Fig. 8.24(a) is slowly controlled or suppressed to a single-band chaos of Fig. 8.24(b) when $f_2 = 50.1$mV, which persists within the region 50.1mV$< F_2 < 123.2$mV.

 (ii) When f_2 reaches the value 123.2mV, the single-band chaos of Fig. 8.24(b) is converted into the period-4 limit cycle of Fig. 8.24(c). This periodic orbit continues to persist within the region 123.2mV $< f_2 < 237.5$mV.

(iii) When f_2 has the value 237.5mV, the period-4 attractor of Fig. 8.24(c) is again controlled to the period-2 limit cycle of Fig. 8.24(d) and it continuously appears within the region 237.5mV $< f_2 < 415.6$mV.

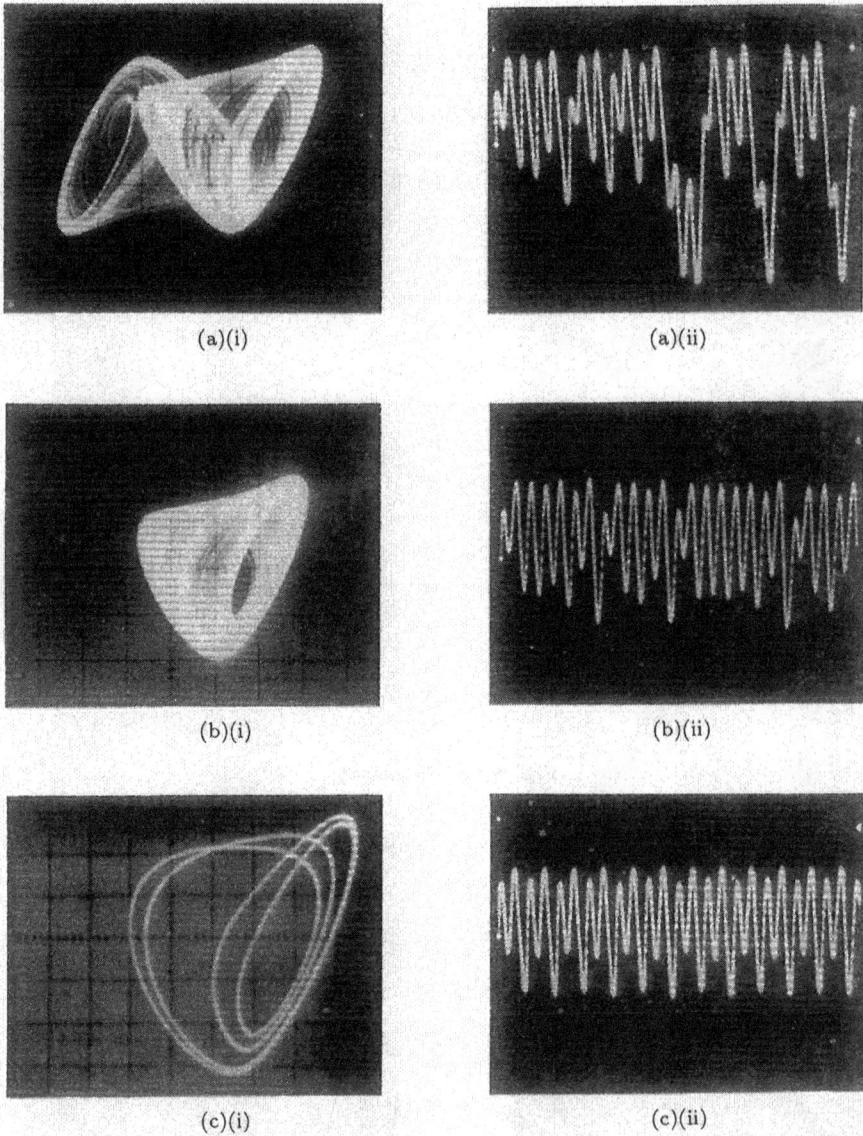

Fig. 8.24. Controlling of chaos in the Ch_1 region of Fig. 7.11(a) ($f_1 = 364.5$ mV, $\nu_1 = 1200$Hz, and $\nu_2 = 10$ KHz): (i) Phase portrait in v_1-v_2 plane; (ii) wave form v_1. (a) Double-band chaos ($f_2 = 0$ mV); (b) Single-band chaos ($f_2 = 50.1$mV); (c) Period-4 limit cycle ($f_2 = 123.2$mV); (d) Period-2 limit cycle ($f_2 = 237.5$mV); (e) Period-1 limit cycle ($f_2 = 415.6$mV).

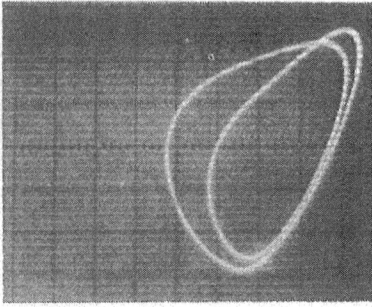

(d)(i)

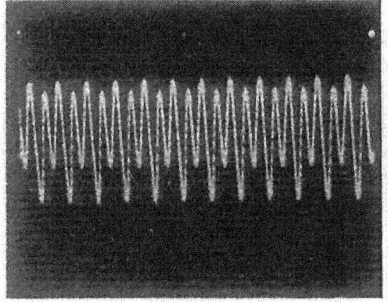

(d)(ii)

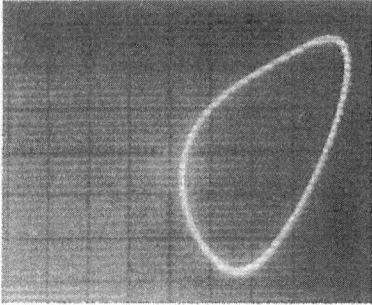

(e)(i)

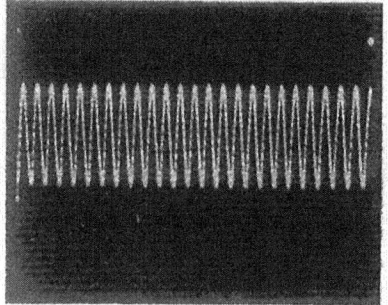

(e)(ii)

Fig. 8.24 (*Continued*)

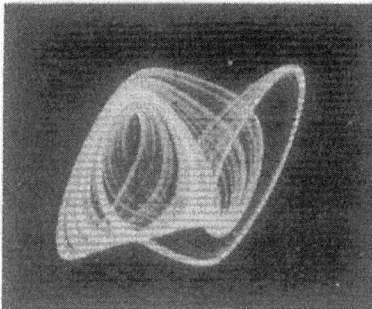

(a)(i)

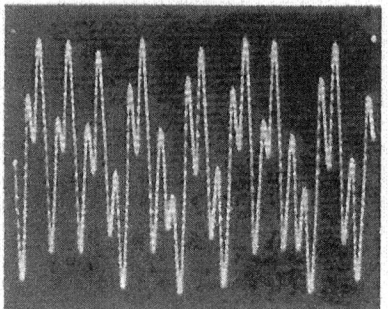

(a)(ii)

Fig. 8.25. Controlling of chaos in the Ch_2 region of Fig.7.11(a): ($f_1 = 584.7$ mV, $\nu_1 =$ 1200 Hz, and $\nu_2 = 10$ KHz): (i) Phase portrait in $v_1 - v_2$ plane; (ii) wave form v_1. (a) Double-band chaos ($f_2 = 0$ mV). (b) Period-3 attractor ($f_2 = 212.2$ mV).

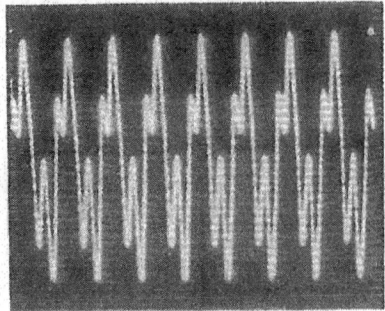

(b)(i)

(b)(ii)

Fig. 8.25 (*Continued*)

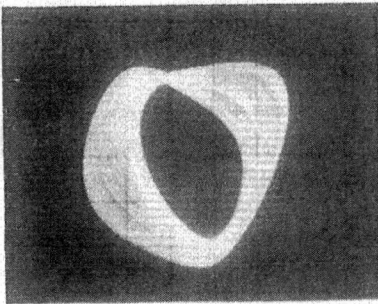

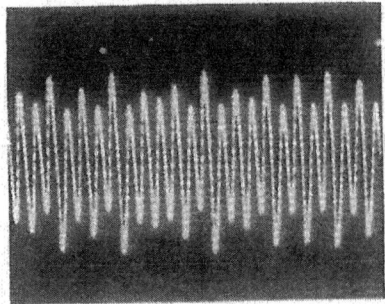

(a)(i)

(a)(ii)

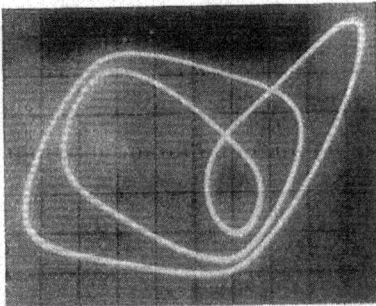

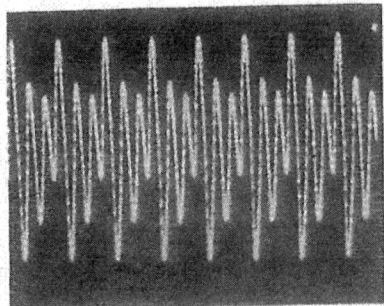

(b)(i)

(b)(ii)

Fig. 8.26. Controlling of chaos in the Ch$_3$ region of Fig. 7.11(a) ($f_1 = 630.4$ mV, $\nu_1 = 1200$ Hz, and $\nu_2 = 10$ KHz): (i) Phase portrait in $v_1 - v_2$ plane; (ii) wave form v_1. (a) Chaotic attractor ($f_2 = 0$ mV). (b) Period-3 window ($f_2 = 116.2$ mV).

(iv) The period-2 limit cycle of Fig. 8.24(d) is further transformed into the period-1 attractor of Fig. 8.24(e) when f_2 reaches the value 415.6mV. Generally speaking, in this regime the variation of the second periodic perturbation seems to make the system follow reverse period-doubling bifurcations from the original chaotic oscillations to periodic behaviour.

(2) Ch_2 *region*: When $f_2 = 0$, a chaotic attractor of Fig. 8.25(a) is observed for $f_1 = 584.7$mV ($\nu_1 = 1200$ Hz) in the Ch_2 chaotic region of Fig. 7.11. Then as f_2 is increased to a value of 212.2mV ($\nu_2 = 10$KHz), the chaotic attractor of Fig. 8.25(a) is controlled and converted into a period-3 window of Fig. 8.25(b).

(3) Ch_3 *region*: Another chaotic attractor of Fig. 8.26(a) is observed in the Ch_3 region of Fig. 7.11 for $f_1 = 630.4$mV ($\nu_1 = 1200$Hz and $f_2 = 0$mV). Then due to the inclusion of the second periodic perturbation f_2 (frequency $\nu_2 = 10$KHz), this chaotic attractor of Fig. 8.26(a) is suppressed to a period-3 window of Fig. 8.26(b) when f_2 reaches the value 116.2mV.

In general, due to the effect of the second periodic perturbation f_2, the chaotic behaviour of the circuit is controlled to nearby periodic attractors (Fig. 7.11). These results indicate how a second periodic signal can change the dynamics of this non-autonomous circuit in a drastic fashion so that even strong chaos can be completely controlled or altered to periodic behaviour. One possible mechanism to explain this phenomenon is that one of the infinite unstable limit cycles embedded in the chaotic attractor (Refs. [17, 152]) gets stabilized by the addition of the second external periodic perturbation (Ref. [157]).

8.3.3. *Controlling of Chaos in Duffing Oscillator and MLC circuit*

A. *Addition of second weak periodic force*: We now apply the method of quenching chaos by the addition of a second weak periodic signal to the Duffing oscillator (Chapter 4) and the simplest dissipative non-autonomous MLC circuit (Chapter 7) numerically.

(1) Considering the Duffing oscillator with a second periodic force, the equation of motion becomes (with specific choice of parameters)

$$\ddot{x} + 0.5\dot{x} - x + x^3 = f_1 \sin \omega_1 t + f_2 \sin \omega_2 t. \tag{8.23}$$

This system indeed exhibits chaotic behaviour as shown in Fig. 8.27(a) for $f_1 = 0.42$, $\omega_1 = 1$ (c.f. Chapter 4), in the absence of the second periodic signal

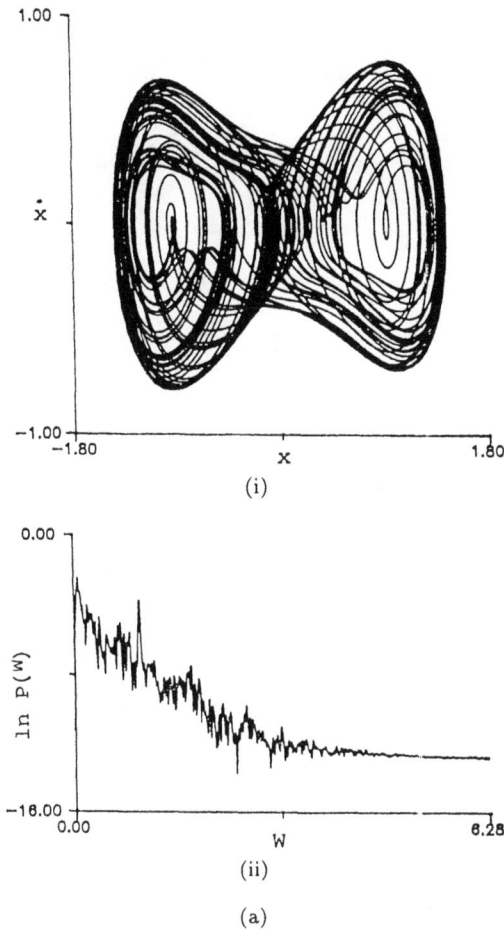

Fig. 8.27. (a): (i) Chaotic attractor of Eq. (8.23): $f_1 = 0.42, f_2 = 0, \omega_1 = \omega_2 = 1$, (ii) Power spectrum of x signal. (b) Bifurcation diagram in $(f_2 - x)$ plane: (i) $0 \leq f_2 \leq 0.1$, (ii) $0.1 \leq f_2 \leq 0.2$, (iii) $0.2 \leq f_2 \leq 0.3$. (c): (i) Controlled period-5 attractor: $f_1 = 0.42, f_2 = 0.04, \omega_1 = \omega_2 = 1$, (ii) Power spectrum of x signal.

$(f_2 = 0)$. However, if we consider the effect of the second periodic signal $(f_2 > 0)$, for certain ranges of values of f_2 and ω_2, the initial chaotic motion of Eq. (8.23) (Fig. 8.27(a)) is completely eliminated or controlled into a periodic one. With the further choice $\omega_2 = 1$, Fig. 8.27(b) shows the bifurcation diagrams of Eq. (8.23) in the $(x-f_2)$ plane covering the range $0 \leq f_2 \leq 0.3$. As discussed in Chapter 4, there appears "reverse period-5 bubble" and "period-3 bubble" structures in Fig. 8.27(b)(i) and Fig. 8.27(b)(ii) respectively as the

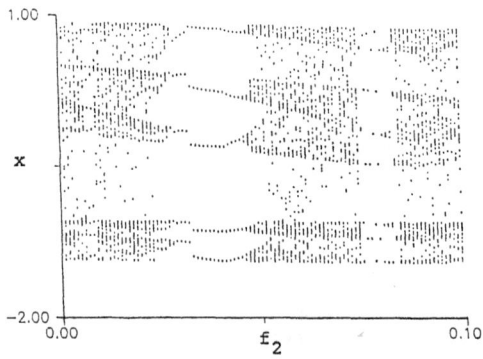

(i)

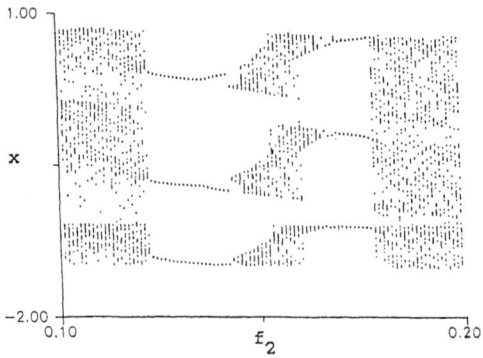

(ii)

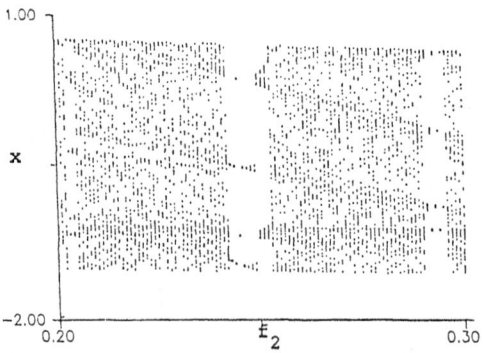

(iii)

(b)

Fig. 8.27. (*Continued*)

(i)

(ii)

(c)

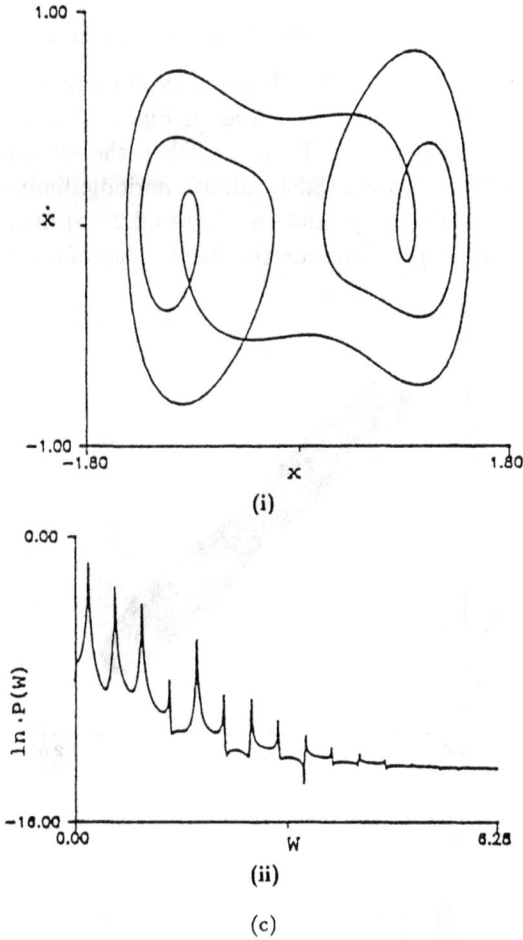

Fig. 8.27. (*Continued*)

control parameter f_2 is changed. The existence of the patterns in the bi-furcation diagram immediately implies that this value of f_2 is antimonotone (Refs. [50, 51] (see Sec. 4.2.1(D)). A typical sample plot for $f_2 = 0.04$, $\omega_2 = 1$ indicating the controlled period-5 limit cycle is shown in Fig. 8.27(c).

(2) Next we consider the MLC circuit equations (Ref. [79]), whose chaotic dynamics has been well discussed in the previous Chapter 7. The governing equations of motion of this circuit along with the second periodic signal is given as

$$\dot{x} = y - h(x), \qquad (8.24a)$$

$$\dot{y} = -\beta(1+\nu)y - \beta x + f_1 \sin \omega_1 t + f_2 \sin \omega_2 t . \qquad (8.24b)$$

Numerical simulation exhibits a double-scroll type chaotic attractor for $\beta = 1$, $\nu = 0.015$, $f_1 = 0.15$ and $\omega_1 = 0.75$ as in Fig. 8.28(a) in the absence of a second periodic signal ($f_2 = 0$). However, when the second periodic signal is included ($f_2 > 0$) this system (8.24) admits periodic limit-cycle behaviour for certain ranges of values of f_2 and ω_2. Figure 8.28(b) depicts the bifurcation diagrams in the (x–f_2) plane for the fixed value of $\omega_2 = 0.75$, covering

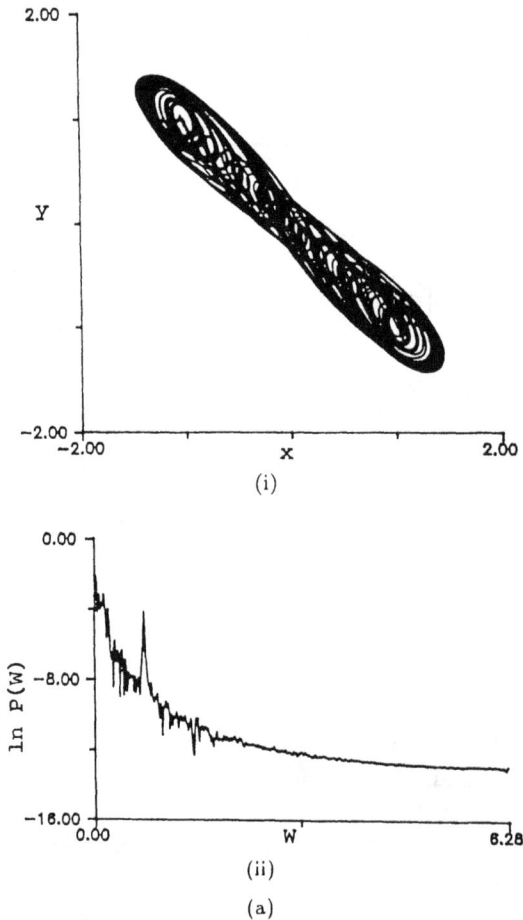

(i)

(ii)

(a)

Fig. 8.28. (a): (i) Chaotic attractor of Eq. (8.24): $f_1 = 0.15$, $f_2 = 0$, $\omega_1 = \omega_2 = 0.75$, $\beta = 1$, and $\nu = 0.015$, (ii) Power spectrum of x signal. (b) Bifurcation diagram in ($f_2 - x$) plane: (i) $0 \le f_2 \le 0.1$, (ii) $0.1 \le f_2 \le 0.2$, (iii) $0.2 \le f_2 \le 0.3$. (c): (i) Controlled period-3 attractor: $f_1 = 0.15$, $f_2 = 0.05$, $\omega_1 = \omega_2 = 0.75$, (ii) Power spectrum of x signal.

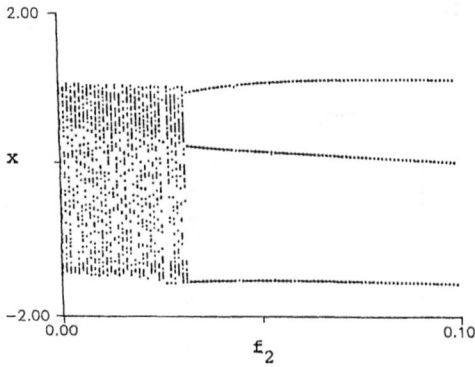

(i)

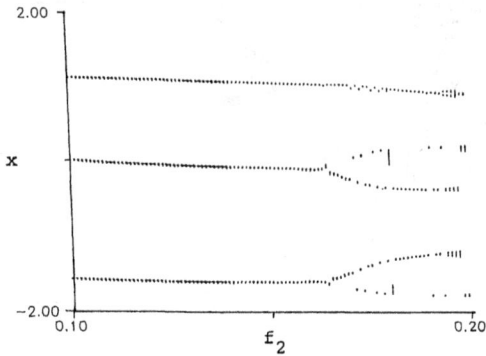

(ii)

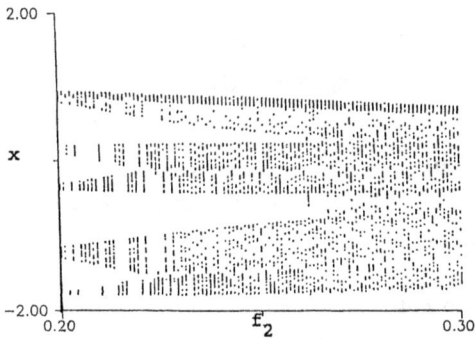

(iii)

(b)

Fig. 8.28. (*Continued*)

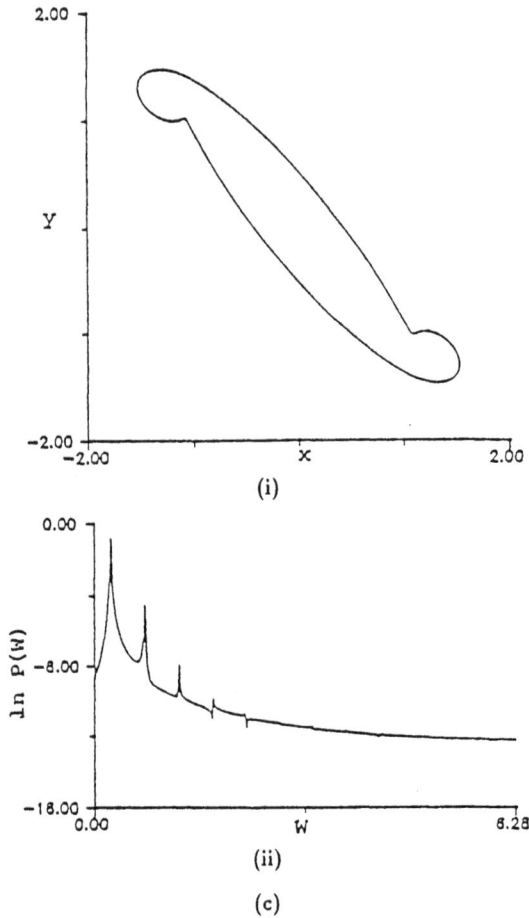

(i)

(ii)

(c)

Fig. 8.28. (*Continued*)

the range $0 \leq f_2 \leq 0.3$. As the parameter f_2 is varied, the original chaotic motions are altered/controlled in the form of periodic windows. For example for $f_2 = 0.05$ and $\omega_2 = 0.75$ a controlled period-3 limit cycle appears, as shown in Fig. 8.28(c).

It is interesting to note that the above non-feedback method of controlling chaos, namely the addition of a second periodic force, is simpler to implement effectively in the field of chaotic circuits and chaotic lasers. Moreover, this method requires no prior knowledge of the structure of the chaotic attractor as a function of the system parameters.

B. *Controlling of chaos by addition of small constant bias*: Interestingly, one can even add just a small constant biasing term to control or quench the chaotic attractor to a desired periodic one in typical nonlinear non-autonomous circuits and systems (Ref. [79]). In some sense, this is effectively inducing phase or mode-locking into the system dynamics; however, it ensures effective controlling in a very simple way. In order to understand this simple controlling approach in a better way, in the present subsection, we apply it to the MLC circuit (Ref. [79]) and the driven Duffing oscillator (Ref. [53]), both experimentally and numerically.

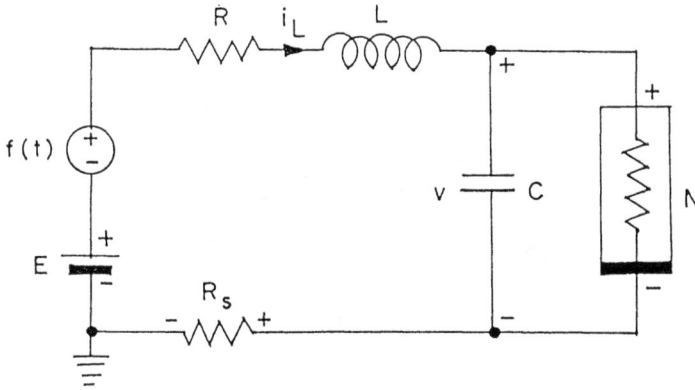

Fig. 8.29. The simplest dissipative non-autonomous circuit augmented with a constant bias voltage source E. Here, N is the Chua's diode. $L = 18$ mH, $C = 10$nF, $R = 1340\Omega$, $f = 0.107V_{rms}$, and frequency of the sinusoidal source $= 8890$Hz.

(1) MLC circuit with constant bias:

To the standard non-autonomous MLC circuit (Fig. 7.21 of Chapter 7), a constant bias voltage (E) is inserted in series with the periodic signal source (Fig. 8.29). The corresponding governing equations are

$$C\frac{dv}{dt} = i_L - f(v), \tag{8.25a}$$

$$L\frac{di_L}{dt} = -Ri_L - R_s i_L - v + f\sin\Omega t + E. \tag{8.25b}$$

In the absence of the constant bias voltage $(E = 0)$ this circuit exhibits chaotic behaviour for a range of forcing parameters (c.f. Chapter 7). The circuit parameters used are $C = 10$nF, $L = 18$mH, $R = 1340\Omega$, $R_s = 20\Omega$, and the frequency $(= \Omega/2\pi)$ of the external forcing source is 8890Hz. Figure 8.30(a)

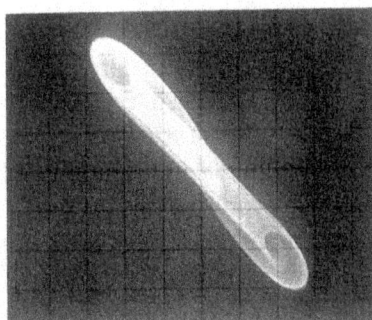

(a)

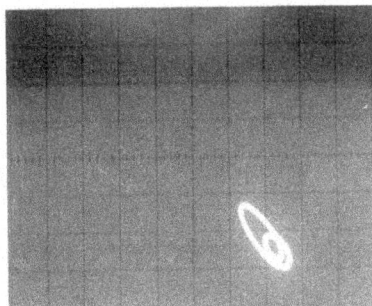

(b)

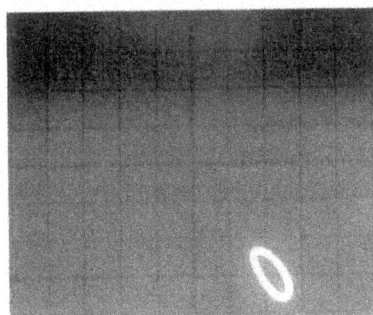

(c)

Fig. 8.30. (a) Chaotic attractor in v versus $v_s(= i_L R_s)$ plane for $E = 0$ V. (b) Controlled period-2 limit cycle for $E = 30$ mV. (c) Controlled period-1 limit cycle for $E = 50$ mV

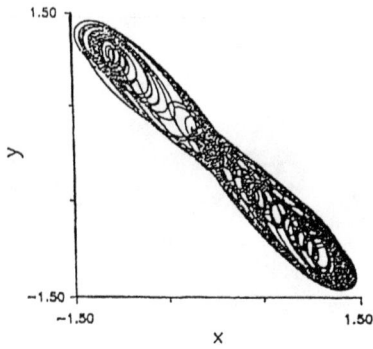

(a)

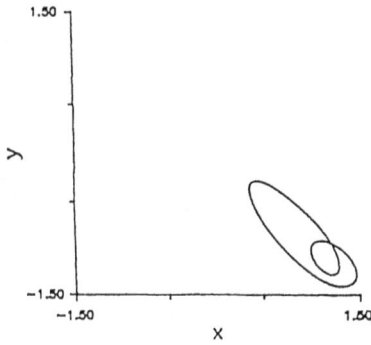

(b)

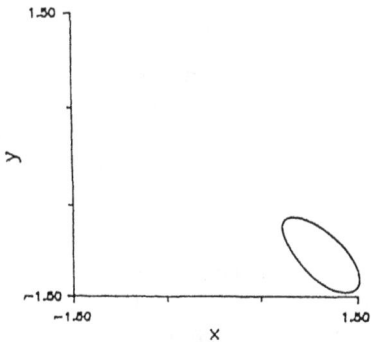

(c)

Fig. 8.31. Numerical confirmation results of Fig. 8.30 using Eq. (8.25) with rescaled parameters. (a) Double-band chaotic attractor in $(x-y)$ plane for $f = 0.15, \omega = 0.75$, and $E' = 0$. (b) Controlled period-2 limit cycle for $E' = 0.03$. (c) Controlled period-1 limit cycle for $E' = 0.05$.

(a)

(b)

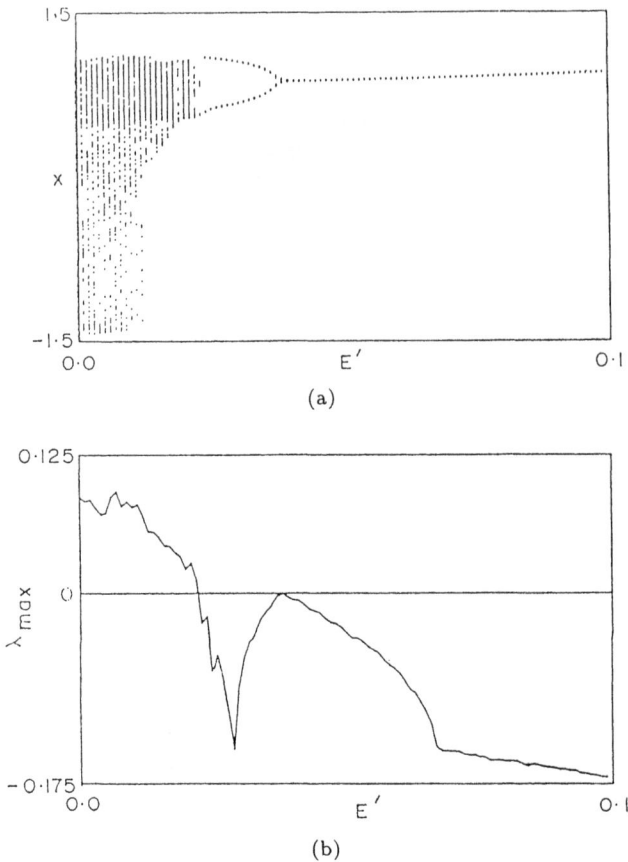

Fig. 8.32. (a) One-parameter bifurcation diagram in $(x - E')$ plane using rescaled equations of Eq. (8.25). Here, E' is the rescaled value of E in Fig.8.30. (b) Maximal Lyapunov exponent spectrum.

shows the phase portrait of the double-scroll Chua's attractor in the v–i_L plane for $f = 0.107\text{Vrms}$ and $E = 0.0\text{V}$.

Now if we consider the effect of the constant bias voltage E by increasing it from zero upwards $(E > 0)$, the chaotic behaviour of Fig. 8.30(a) is altered to the period-2 attractor of Fig. 8.30(b) for $E = 0.03\text{V}$, and to the period-1 attractor of Fig. 8.30(c) for $E = 0.05\text{V}$ (Ref. [79]). The corresponding numerical confirmation is given in Figs. 8.31, where the x and y variables represent the rescaled current and voltage variables (cf. Chapter 7). In Figs. 8.32, the associated one-parameter bifurcation diagram and the maximal Lyapunov exponent λ_{max} are also shown.

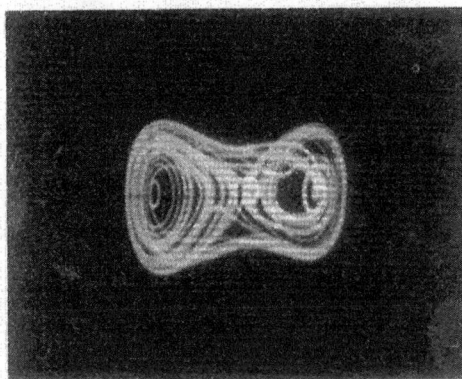

(a)

(b)

Fig. 8.33. (a) Chaotic attractor of the Duffing oscillator for $f = 0.42$ V and $E = 0$ V (Analog circuit simulation result). (b) Controlled period-1 limit cycle for $f = 0.42$ V and $E = 0.15$ V.

(2) Duffing oscillator with constant bias:
With a periodic driving signal $f \sin(t)$ along with a constant bias voltage E, the Duffing oscillator (for fixed parameters) is represented as

$$\ddot{x} + 0.5\dot{x} - x + x^3 = f \sin(t) + E. \qquad (8.26)$$

The chaotic dynamics of this system in the absence of constant bias ($E = 0$) is already extensively discussed through both experimental and numerical simulations in Chapter 4. Now we include the constant bias E and study its effect on the Duffing oscillator (Eq. 8.26). Carrying out both experimental analog simulation and numerical simulation for $f = 0.42$, a typical double-band chaotic

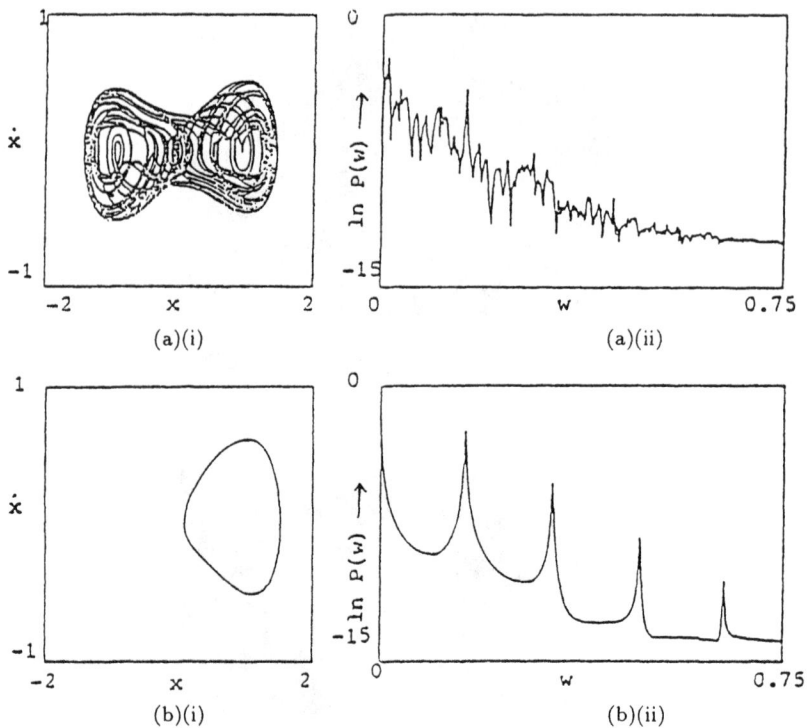

Fig. 8.34. (a) (i) Chaotic attractor in $(x - y)$ plane of Eq. (8.26) for $f = 0.42$ and $E = 0$ V (Numerical simulation result); (ii) Power spectrum of x signal. (b) (i) Controlled period-1 limit cycle for $f = 0.42$ and $E = 0.15$; (ii) Power spectrum of x signal.

attractor is noted as shown in Figs. 8.33(a) and Fig. 8.34(a), in the absence of constant bias $(E = 0)$. Now if we consider the effect of the constant bias voltage E by increasing it from zero upwards $(E > 0)$, the chaotic behaviour of Figs. 8.33(a) and 8.34(a) is altered to the period-1 attractor of Figs. 8.33(b) and 8.34(b) respectively for $E = 0.15$. So by including a small constant bias in series with the existing periodic driving source one can control the chaos behaviour in this oscillator also. Figure 8.35(a) shows the one-parameter bifurcation diagram of Eq. (8.26) for the value of $f = 0.42$ in the $(E-x)$ plane, while Fig. 8.35(b) represents the corresponding Lyapunov spectrum. This bifurcation diagram clearly demonstrates the controlling of chaos behaviour as the constant bias term E is increased.

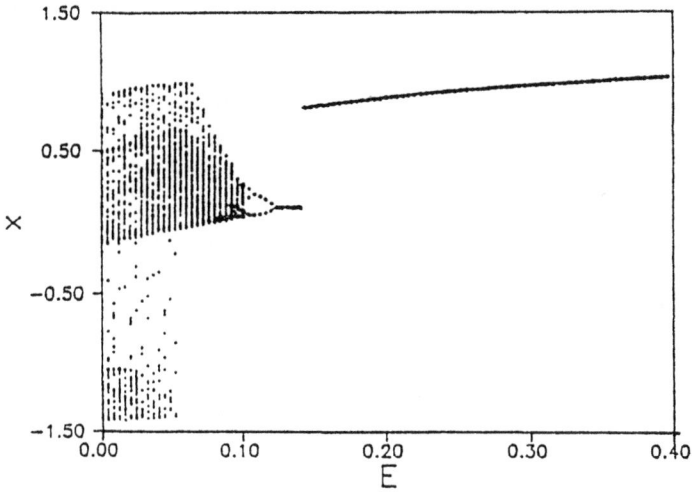

(a)

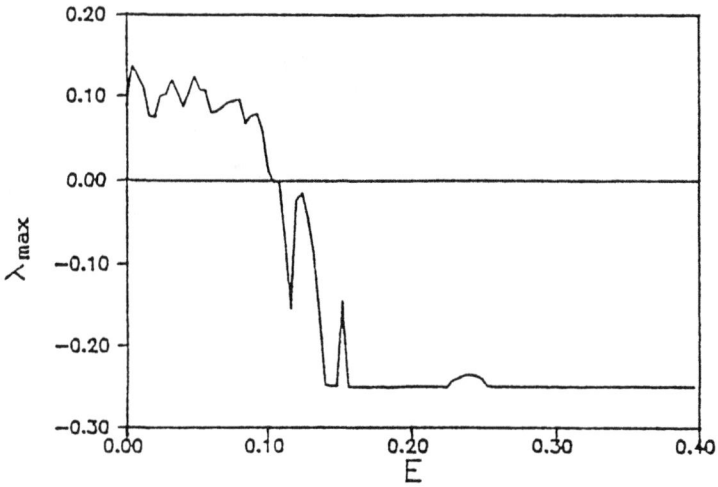

(b)

Fig. 8.35. (a) One-parameter bifurcation diagram of Eq. (8.26) in $(E-x)$ plane for $f = 0.42$. (b) Maximal Lyapunov exponent spectrum.

CHAPTER 9

SYNCHRONIZED CHAOTIC SYSTEMS AND SECURE COMMUNICATION

Having studied the bifurcation, chaos and suppression of chaos aspects in the previous chapters, we will now discuss another fascinating concept, namely, synchronization of chaos, which is attracting considerable interest among chaos researchers in recent times (Refs. [198–233]). The possibility of two or more chaotic systems oscillating in a coherent and synchronized way is not an obvious one. One of the main features often associated with the definition of chaotic behaviour is the sensitive dependence on initial conditions. Then one may conclude that synchronization is not feasible in chaotic systems because it is not possible in real systems either to reproduce exactly identical initial conditions or exact specification of system parameters for two similar systems. We may be able to build "nearly" identical systems, but there is always an inevitable technological mismatch and noise, impeding exact reproduction of all the parameters and initial conditions. Thus, even an infinitesimal deviation in any one of the parameters or initial conditions will eventually result in the divergence of nearby starting orbits.

In this connection, the recent suggestion of Pecora and Carroll (Ref. [198]) that it is possible to synchronize even chaotic systems by introducing appropriate coupling between them has revolutionized our understanding. They have shown that if a reproduced part (response or subsystem) of a chaotic system responses to a chaotic signal from the full (drive) system, under some conditions the signals in the response part would converge to the corresponding signals in the drive system. Such a possibility is known as *synchronization* of chaotic systems. Further, this idea has been generalized by cascading (Ref. [198]) the reproduced parts or subsystems. Here the considered chaotic system may be divided into two subsystems, each of which will synchronize with the full system when driven by the appropriate chaotic signal. By arranging one of these synchronized subsystems to drive the other subsystem, it is possible to produce synchronized signals in the subsystems for every signal in the full drive system.

Another innovative method which has been developed recently for chaos synchronization is the method of one way coupling between two identical chaotic systems (Refs. [206, 208, 210, 226–229]). In this configuration, the response chaotic system variables follow identically the drive chaotic system variables for the appropriate one-way coupling strength. Moreover, the behaviour of the response system is dependent on the behaviour of the drive system, but the drive system is not influenced by the response system's behaviour.

Thus by reproducing all the signals at the receiver under the influence of a single chaotic signal from the drive, the synchronized chaotic systems provide a rich mechanism for signal design and generation, with potential applications to communications and signal processing. Further, one can use the pair of cascading driven subsystems or the identical response system with one-way coupling term as attractor-recognition devices for chaotic attractors. Hence, a single input signal will allow recognition and reconstruction of the driving system and/or rejection of non-matching device systems. Another interesting area of application would be *cryptography*. For example, the non-driven part of the chaotic system produces signals, which are typically broad-band, noise-like, and difficult to predict, and so they can be ideally utilized in various contexts of masking information-bearing waveforms. They can also be further used as modulating waveforms in spread spectrum techniques. In the following, we shall discuss the various methods of chaos synchronization (Secs. 9.1 and 9.2), the possibility of secure transmission of signals (Sec. 9.3), analog signal transmission (Sec. 9.4), and digital signal transmission (Sec. 9.5). We shall also give a brief account of some further recent developments of chaos synchronization in Sec. 9.6.

9.1. Pecora and Carroll Method

In this section, we will introduce the method of Pecora and Carroll (Ref. [198]) for chaos synchronization and its modification by cascading. Further, the method is also applied to typical dynamical systems discussed in the previous chapters.

9.1.1. *Drive-response Concept*

In this method, Pecora and Carroll have considered an n-dimensional system governed by a state equation of the form

$$dX/dt = f[X(t)], \quad X = (x_1, x_2, \ldots, x_n)^T. \tag{9.1}$$

They divide the system into two parts in an arbitrary way as $X = (x_D, x_R)^T$. The D part is referred to as the driving subsystem variables and the R part as the response subsystem variables. Then Eq. (9.1) can be rewritten as

$$\dot{x}_D = g(x_D, x_R), \qquad (m\text{-dimensional}) \tag{9.2a}$$

$$\dot{x}_R = h(x_D, x_R), \qquad (k\text{-dimensional}) \tag{9.2b}$$

where $x_D = (x_1, x_2, \ldots, x_m)^T$, $x_R = (x_{m+1}, x_{m+2}, \ldots, x_n)^T$, $g = [f_1(X), \ldots, f_m(X)]^T$, $h = [f_{m+1}(X), \ldots, f_n(X)]^T$, and $m + k = n$.

Pecora and Carroll suggested building an identical copy of the response subsystem with variables x'_R and drive it with the x_D variables coming from the original system. In such a model, we have the following compound system of equations:

$$\dot{x}_D = g(x_D, x_R), \qquad (m\text{-dimensional}) \left.\begin{array}{c}\\\\\end{array}\right\}\text{-drive,} \tag{9.3a}$$

$$\dot{x}_R = h(x_D, x_R), \qquad (k\text{-dimensional}) \tag{9.3b}$$

$$\dot{x}'_R = h(x_D, x'_R), \qquad (k\text{-dimensional})\text{-response.} \tag{9.3c}$$

Under the right conditions, as time elapses the x'_R variables will converge asymptotically to the x_R variables and continue to remain in step with the instantaneous values of $x_R(t)$. Here, the drive or master system controls the response, or slave system, through the x_D component. The other component x'_R is allowed to have different initial conditions from that of x_R. The generalized schematic set-up for this type of synchronization is shown in Fig. 9.1.

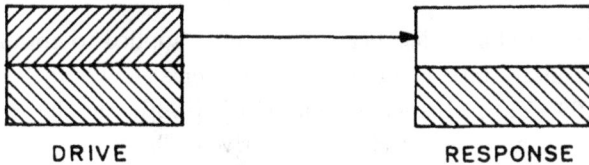

DRIVE   RESPONSE

Fig. 9.1. Schematic of drive-response scenario of chaos synchronization. ⬜ Variables used to drive the response. ⬜ The replica part of the drive system used in response. ⬜ The drive variables which are fed directly into the response.

One can easily see that if all the Lyapunov exponents of the response system consisting of Eq. (9.3c) are less than zero, then after the decay of the initial transients, $x'_R(t)$ will be exactly in step with $x_R(t)$ (Ref. [198]). The Lyapunov exponents for Eq. (9.3c) are called conditional Lyapunov exponents(CLE's). Also, the response system with negative conditional Lyapunov exponents is called a *stable subsystem or a stable response system* (Ref. [198]). If at least one of the conditional Lyapunov exponents is positive, then synchronization will not occur between x'_R and x_R variables. If the x'_R and x_R subsystems are under perfect synchronization then the difference between x'_R and x_R variables, $x^*_R = x'_R - x_R$, will tend to zero asymptotically, that is, $\lim_{t\to\infty} x^*_R(t) \to 0$. Then we have

$$\dot{x}^*_R(t) = \dot{x}'_R(t) - \dot{x}_R(t) = h(x_D, x'_R) - h(x_D, x_R). \qquad (9.4)$$

If the subsystem is linear, we have

$$\dot{x}^*_R(t) = A\, x^*_R(t), \qquad (9.5)$$

where A is a $(k \times k)$ constant matrix. Let the eigenvalues of A be $(\lambda_1, \lambda_2, \ldots, \lambda_k)$. The real parts of these eigenvalues are by definition the CLE's we seek.

If all the CLE's are negative then $\lim_{t\to\infty} x^*_R(t) = 0$ and the subsystems will synchronize; if there is a positive CLE the subsystems will grow further apart as $t \to \infty$ and they will never synchronize. Interestingly, an intermediate case occurs if one or more of the CLE's are zero but none are positive; in this case, as $t \to \infty$ the trajectories of the subsystems will be separated by a fixed distance depending upon the initial conditions (Ref. [198]).

If the response system is nonlinear in nature, then the conditional Lyapunov exponents (CLE) are not so easily determined, and we must resort to numerical procedures to calculate them. But if the response system is linear (for example, liner circuit with passive elements) it is trivial to calculate the CLE in *certain cases*.

The above methodology has been successfully applied to obtain chaos synchronization in many important nonlinear systems including Lorenz system (Refs. [198, 199, 213], Rössler system (Ref. [198]), the hysteretic circuit (Ref. [198]), Chua's circuit (Ref. [200]), driven Chua's circuit (Ref. [215]), DVP oscillator (Refs. [150, 207]), phase-locked loops (Refs. [201, 202]), and so on. We will illustrate the method for a couple of dynamical systems in the following.

9.1.2. *Chaos Synchronization in Chua's Circuit*

Using the Pecora and Carroll approach, Chua *et al.* (Refs. [200, 217]) have recently demonstrated chaos synchronization in the Chua's circuit both experimentally and numerically. The rescaled and dimensionless version of the Chua's circuit equations (Ref. [217]) can be written as (cf. Chapter 7, Sec. 7.3)

$$\dot{x} = \alpha(y - x - h(x)), \qquad (9.6a)$$

$$\dot{y} = x - y + z, \qquad (9.6b)$$

$$\dot{z} = -\beta y, \qquad (9.6c)$$

where $h(x) = bx + 0.5(a - b)[|x + 1| - |x - 1|]$. Here α, β, a and b are rescaled circuit parameters, which are fixed at $\alpha = 10, \beta = 14.87$, $a = -1.27$ and $b = -0.68$. As we have seen earlier the necessary condition for chaos synchronization is that the CLE's should be negative. This can be checked for the systems (9.6) by a relatively simple method provided the subsystem is linear.

In our present case (9.6), there will be three kinds of drive-response systems:

Table 9.1.

Case	Drive	Response	Subsystem
(a)	x	(y, z)	linear
(b)	y	(x, z)	nonlinear
(c)	z	(x, y)	nonlinear

We discuss each of these in turn:

(a) *x-drive configuration*: Figure 9.2 shows the experimental circuit realization. The state equations become (after rescaling as in Sec. 7.3)

drive:

$$\dot{x} = \alpha(y - x - h(x)), \qquad (9.7a)$$

$$\dot{y} = x - y + z, \qquad (9.7b)$$

$$\dot{z} = -\beta y, \qquad (9.7c)$$

response:

$$\dot{y}' = x - y' + z', \qquad (9.7d)$$

$$\dot{z}' = -\beta y', \qquad (9.7e)$$

and the difference system for $y^* = (y - y'), z^* = (z - z')$ in matrix form is

$$\begin{bmatrix} \dot{y}^* \\ \dot{z}^* \end{bmatrix} = \begin{bmatrix} -1 & 1 \\ -\beta & 0 \end{bmatrix} \begin{bmatrix} y^* \\ z^* \end{bmatrix}. \qquad (9.8)$$

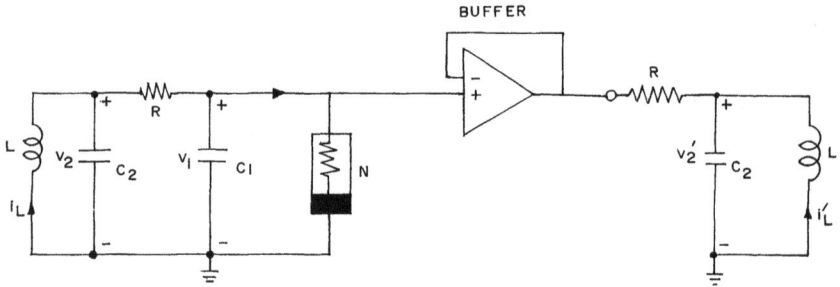

Fig. 9.2. x-drive configuration; op-amp from AD712 chip (Ref. [217]).

The eigenvalues are $-0.5 \pm 0.5i\sqrt{4\beta - 1}$ giving the solution

$$y^*(t) = z^*(t) = e^{-t/2}\left(C\cos(\sqrt{4\beta - 1}/2)t + D\sin(\sqrt{4\beta - 1}/2)t\right), \quad (9.9)$$

where C and D are constants of integration. The CLE's are $(-0.5, -0.5)$ and, as expected, $\lim_{t\to\infty} y^*(t) = 0$ and $\lim_{t\to\infty} z^*(t) = 0$. Then the response system synchronizes; see Figs. 9.3(a) and 9.3(b).

(b) *y-drive configuration*: Figure 9.4 shows the experimental set-up. The state equations are

    drive:

$$\dot{x} = \alpha(y - x - h(x)), \quad\quad\quad (9.10a)$$
$$\dot{y} = x - y + z, \quad\quad\quad (9.10b)$$
$$\dot{z} = -\beta y, \quad\quad\quad (9.10c)$$

    response:

$$\dot{x}' = \alpha(y - x' - h(x')), \quad\quad\quad (9.10d)$$
$$\dot{z}' = -\beta y. \quad\quad\quad (9.10e)$$

Using numerical procedures (for example, INSITE (Ref. [234]) Software Package) the CLE's of Eqs. (9.10(d)–(e)) are found to be $(-2.5 \pm 0.05, 0)$. As expected the response system synchronizes. Because the second eigenvalue is 0, $z(t)$ and $z'(t)$ will remain apart by a constant distance $B = |z(0) - z'(0)|$ (see Figs. 9.5(a) and 9.5(b)).

X-DRIVE CIRCUIT

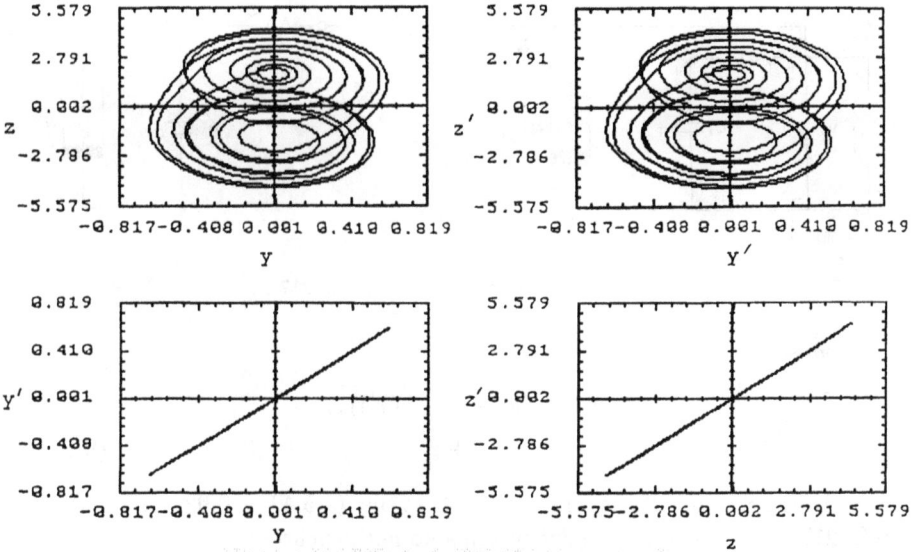

(a)

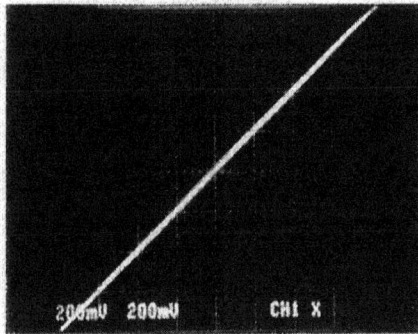

(b)

Fig. 9.3. (a) x-drive synchronization; initial conditions $x = -2$, $y = 0.02$, $z = 4.0$, $y' = 0.4$, $z' = -0.8$. (b) Experimental confirmation: v_1-drive synchronization (Ref. [217]).

(c) *z-drive configuration*: The state equations for this case becomes the following:

drive:

$$\dot{x} = \alpha(y - x - h(x)), \qquad (9.11a)$$

$$\dot{y} = x - y + z, \qquad (9.11b)$$

$$\dot{z} = -\beta y, \qquad (9.11c)$$

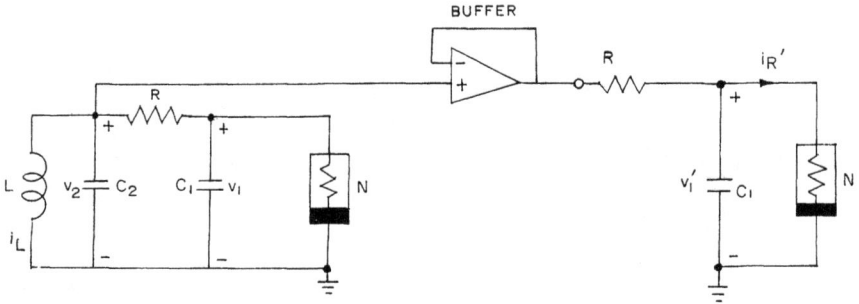

Fig. 9.4. *y*-drive configuration (Ref. [217]).

response:

$$\dot{x}' = \alpha(y' - x' - h(x')), \tag{9.11d}$$

$$\dot{y}' = x' - y' + z. \tag{9.11e}$$

Using INSITE, the CLE's are reported to be $(-5.42 \pm 0.02, 1.23 \pm 0.03)$ (Ref. [217]). As expected the subsystems do not synchronize.

9.1.3. *The Synchronized ADVP Oscillator*

As our second example, in the following we will consider the ADVP oscillator (Ref. [150]) (cf. Chapter 7, Sec. 7.6). By following the approach of chaos synchronization discussed above, the drive-response system for the x-drive is given as

drive:

$$\dot{x} = -\nu[x^3 - \alpha x - y], \tag{9.12a}$$

$$\dot{y} = x - y - z, \tag{9.12b}$$

$$\dot{z} = \beta y, \tag{9.12c}$$

response:

$$\dot{y}' = x - y' - z', \tag{9.12d}$$

$$\dot{z}' = \beta y'. \tag{9.12e}$$

Now by considering the combined system of equations with the parameter values $\alpha = 0.35$, $\nu = 100$ and $\beta = 300$, we find that in spite of the differences in initial conditions of (y', z') and (y, z) variables, the primed and unprimed systems do synchronize so that for $t \to \infty$, $(y' - y) \to 0$ and $(z' - z) \to 0$ (see Fig. 9.6). Similar synchronization was observed with a "y" feedback from the drive system, while the "z" feedback failed to show synchronization.

Y-DRIVE CIRCUIT

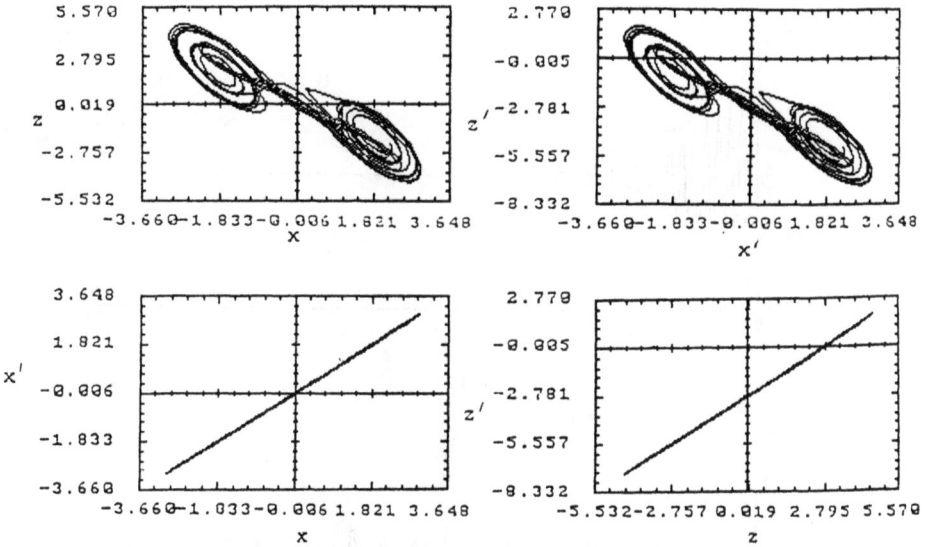

(a)

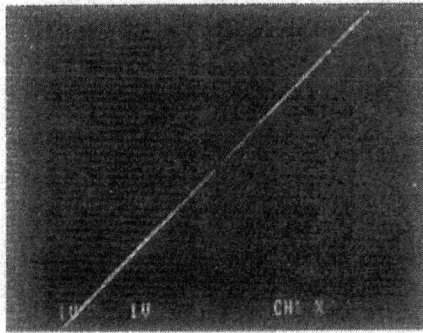

(b)

Fig. 9.5. (a) y-drive synchronization; initial conditions $x = -2, y = 0.02, z = 2.0, x' = 0.7, z' = -0.8$. Note the difference in $z - z'$ offset due to initial conditions. (b) Experimental confirmation: v_2-drive synchronization (Ref. [217]).

Considering the set of Eqs. (9.12) for x-drive, we note that the response system is linear in nature. The associated difference system can obviously be expressed in the form (see also Eq. (9.8))

$$\begin{bmatrix} \dot{y}^* \\ \dot{z}^* \end{bmatrix} = \begin{bmatrix} -1 & -1 \\ \beta & 0 \end{bmatrix} \begin{bmatrix} y^* \\ z^* \end{bmatrix}. \tag{9.13}$$

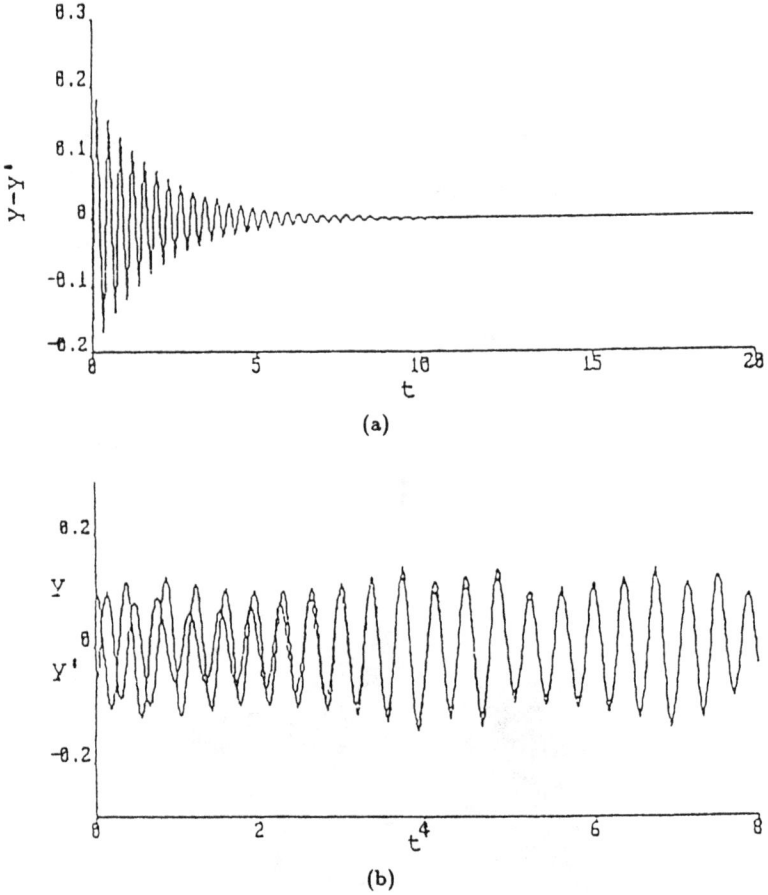

Fig. 9.6. (a) Trajectory $(y - y')$ versus t plot of Eq. (9.12) for $\alpha = 0.35, \nu = 100$, and $\beta = 300$. (b) Trajectory plot in the $(y$–$t)$ and $(y'$–$t)$ planes of (a).

The eigenvalues of Eq. (9.13) are $[-0.5 \pm 0.5i(\sqrt{4\beta - 1})]$, giving solution of the form

$$y^*(t) = z^*(t) = e^{-t/2}(C \cos((\sqrt{4\beta - 1})/2)t + D \sin((\sqrt{4\beta - 1})/2)t), \quad (9.14)$$

where C and D are constants of integration. Here also the CLE's are $(-0.5, -0.5)$ and then the response system with x-drive configuration does synchronize.

9.1.4. *Asymptotic Stability and Lyapunov Function*

A criterion based on the asymptotic stability of dynamical systems has been developed by He and Vaidya (Ref. [199]) as a necessary and sufficient condition for the synchronization of periodic and chaotic systems. Here asymptotic stability refers to the condition for a given synchronized chaotic system with drive-response configuration to reach the same eventual state at a fixed (but sufficiently far enough) time, no matter what the initial conditions were. One of the practical ways to establish this asymptotic stability of the response system is to find an appropriate Lyapunov function. Its use can be shown by first considering the Lyapunov function in connection with the difference system of equations.

For example, first let us consider the difference system of Chua's circuit given by Eq. (9.8). Now if we consider the Lyapunov function

$$E = (1/2)[(\beta y^* - z^*)^2 + \beta y^{*2} + (1+\beta)z^{*2}], \qquad (9.15)$$

then the time rate of change of E along the trajectories is given by

$$\dot{E} = (\beta \dot{y}^* - \dot{z}^*)(\beta y^* - z^*) + \beta \dot{y}^* y^* + (1+\beta)\dot{z}^* z^* ,$$
$$= -\beta(y^{*2} + z^{*2}) \le 0 \qquad (\beta > 0).$$

Since E is positive definite and $\dot{E}$ is negative definite, Lyapunov's theorem (Refs. [235, 236]) implies that $y^*, z^* \to 0$ as $t \to \infty$. Therefore, synchronization occurs as $t \to \infty$.

Similarly for the ADVP oscillator case the difference system dynamics (9.13) is globally asymptotically stable at the origin, provided $\beta > 0$. This result follows by considering the Lyapunov function defined by

$$E = (1/2)[(\beta y^* + z^*)^2 + \beta y^{*2} + (1+\beta)z^{*2}], \qquad (9.16)$$

and so

$$\dot{E} = (\beta \dot{y}^* + \dot{z}^*)(\beta y^* + z^*) + \beta \dot{y}^* y^* + (1+\beta)\dot{z}^* z^*$$
$$= -\beta(y^{*2} + z^{*2}) \le 0 \qquad (\beta > 0).$$

The equality sign applies only at the origin; therefore the subsystem (9.12d–e) is globally asymptotically stable. Thus the drive [Eqs. (9.12a–c)] and response [Eqs. (9.12d–e)] systems eventually synchronize.

9.1.5. *Cascading Synchronized Systems*

Now we proceed one step further with synchronized chaotic systems by having a cascade of response systems. For example, for the general system (9.3), let us use Eq. (9.3a) with a new variable x''_D (in place of the variable x_D) as our *second response subsystem*. We then substitute the variable x'_R for the variable x_R in this new response subsystem, leading to a new set of equations containing (9.3a–c) and

$$\dot{x}''_D = g(x''_D, x'_R).\tag{9.17}$$

As before, if all the conditional Lyapunov exponents in Eq. (9.17) are negative, then $x''_D \to x_D$ and, as before, $x'_R \to x_R$, asymptotically. Using Eqs. (9.3c) and (9.17) as a drive-response system, all the variables in the drive dynamical system variables can be effectively reproduced by driving only one variable x_D (see schematic diagram of Fig. 9.7).

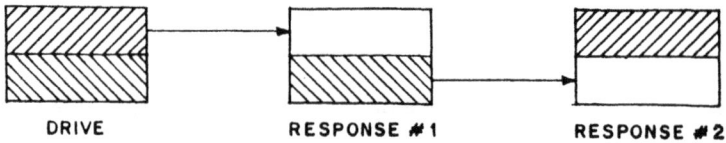

DRIVE   RESPONSE #1   RESPONSE #2

Fig. 9.7. Schematic representation of cascading synchronization. Drive & response #1: ▨ Variables used to drive the response #1. ▨ The replica part of drive system used in response. ▢ Drive variables which are fed directly into the response #1. Response #2: ▢ Variables which are fed directly into the response #2 from response #1. ▨ The replica part of the drive system.

If the parameters in the response system of Eqs. (9.3c) and (9.17) do not match the corresponding equations in the drive system of Eqs. (9.3a–b), then $x''_D(t)$ will not equal $x_D(t)$ and $x'_R(t)$ will not equal $x_R(t)$. If the response system is stable (has negative Lyapunov exponents), then the response system variables remain close to the drive system variables (Ref. [198]).

The cascading approach can again be used for synchronization in typical dynamical systems, such as the Chua's circuit and ADVP oscillator considered in the previous subsection. For example for the Chua's circuit, let us consider Eqs. (9.7a–e) and a second response system equation which is represented by *response 2*:

$$\dot{x}'' = \alpha(y' - x'' - f(x'')).\tag{9.7a'}$$

A synchronized chaotic behaviour of Eqs. (9.7a–e, 9.7a') is shown in Fig. 9.8. The initial conditions are fixed as $[x(0) = 0.1, y(0) = 0.1, z(0) = 0.2, y'(0) = 0.15, z'(0) = 0.22$ and $x''(0) = 0.15]$.

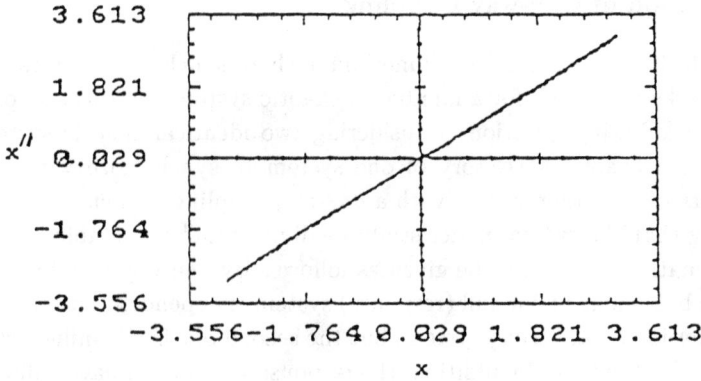

3.613

1.821

x'' 0.029

−1.764

−3.556

−3.556 −1.764 0.029 1.821 3.613

x

Fig. 9.8. Synchronized chaotic behaviour of Eqs. (9.7a–e, 9.7a′) in the $(x–x'')$ plane.

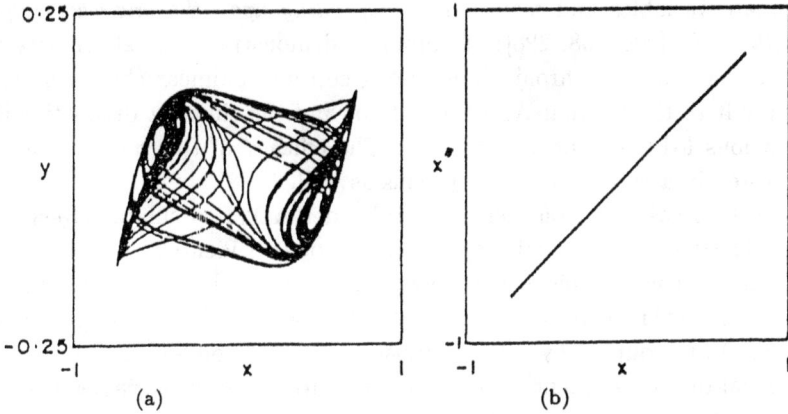

0·25

y

−0·25

−1   x   1

(a)

1

x''

−1

−1   x   1

(b)

Fig. 9.9. (a) Chaotic attractor projected on the $(x–y)$ plane for $\alpha = 0.35, \nu = 100$, and $\beta = 300$. (b) Synchronization of chaos between $x - x''$ through cascading (see Eqs. (9.12)).

For the second example of the ADVP oscillator represented by Eqs. (9.12), the second response system is taken as (Ref. [150]).

response 2:

$$\dot{x}'' = -\nu[(x'')^3 - \alpha x'' - y'].\qquad(9.12a')$$

A synchronized chaotic behaviour which exists between x and x'' for $\alpha = 0.35$, $\nu = 100$, and $\beta = 300$ is shown in Fig. 9.9 through a numerical simulation of Eqs. (9.12a–f and 9.12a′). Similar kind of cascaded synchronized chaotic systems have also been reported by a number of authors (Refs. [150, 200, 207, 213]) for various dynamical systems.

9.2. Method of One-way Coupling

Although the Pecora and Carroll method of chaos synchronization as discussed above works fairly well for a number of chaotic systems, in this section we address the following question: Considering two identical chaotic systems, can one make a chaotic trajectory of one system to synchronize with a chaotic trajectory of the other system with a *one-way* coupling element alone, without requiring that the system under study be divided into two stable subsystems? An affirmative answer can be given as follows. By one-way coupling we mean that the behaviour of one full (response) system is dependent on the behaviour of another identical (drive) system, but the second one is not influenced by the behaviour of the first. In addition, the response system can have a different set of initial conditions other than that of the drive system. As time progresses, the two identical chaotic systems can achieve a perfect synchronization among their state variables and maintain it, *depending upon the one-way coupling strength* (Refs. [206–208, 228]). In order to demonstrate this alternative way of achieving chaos synchronization among certain nonlinear chaotic systems, we apply it to the familiar ADVP oscillator and the simplest dissipative non-autonomous MLC circuit as examples. The efficacy of this method has also been tested in a variety of other systems as well.

As a first case, we consider the third-order ADVP oscillator (Refs. [206, 207]). The schematic circuit representation of two identical circuits with a homogeneous coupling element is shown in Fig. 9.10. In this circuit, the two ADVP systems, namely, the drive and response (circuits within the broken line boxes) are coupled by a linear resistor R_c and a buffer. The buffer acts as a signal-driving element that isolates the drive system variables from the response system variables, thereby providing a one-way coupling. In the absence of the buffer the system represents two identical oscillators coupled by a common resistor R_c when both the drive and response systems will mutually affect each other. The rescaled dynamical equations of the circuit model of Fig. 9.10 can be represented as

drive:

$$\dot{x} = -\nu[x^3 - \alpha x - y]\,, \tag{9.18a}$$

$$\dot{y} = x - y - z\,, \tag{9.18b}$$

$$\dot{z} = \beta y\,, \tag{9.18c}$$

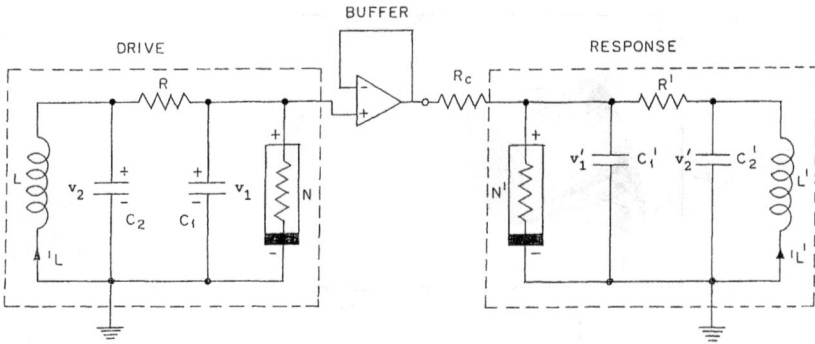

Fig. 9.10. Schematic circuit realization of two ADVP oscillators with one-way coupling resistor R_c. Here, the buffer acts as a signal driving element.

response:

$$\dot{x}' = -\nu[(x')^3 - \alpha x' - y'] + \nu\varepsilon(x - x'),\qquad(9.18\text{d})$$

$$\dot{y}' = x' - y' - z',\qquad(9.18\text{e})$$

$$\dot{z}' = \beta y'.\qquad(9.18\text{f})$$

Here α, ν and β are the rescaled parameters and $\varepsilon(= R'/R_c)$ is the one-way coupling parameter. Again for the parameters $\alpha = 0.35, \nu = 100$ and $\beta = 300$ the drive system exhibits a double-band chaotic attractor as shown in Fig. 9.11(a). Now the differential equations for the difference variables $x^* = x - x', y^* = y - y', z^* = z - z'$ can be written as

$$\dot{x}^* = -\nu[x^3 - (x')^3 - \alpha x^* - y^*] - \nu\varepsilon x^*,$$

$$= -\nu[(x^2 + xx' + (x')^2)x^* - \alpha x^* - y^*] - \nu\varepsilon x^*,$$

$$= [-a\nu x^* - \nu x^* + \nu y^*],\qquad(9.19\text{a})$$

$$\dot{y}^* = x^* - y^* - z^*,\qquad(9.19\text{b})$$

$$\dot{z}^* = \beta y^*,\qquad(9.19\text{c})$$

where $a = (x^2 + xx' + (x')^2) = (x - x')^2 + 3xx' \geq 0$ and $\varepsilon = (1 + \alpha)$.

By following the criterion based on the asymptotic stability as discussed previously, we consider the Lyapunov function for Eqs. (9.19) in the form

$$E = (\beta/2)x^{*2} + (\nu\beta/2)y^{*2} + (\nu/2)z^{*2} \geq 0,\qquad(9.20)$$

then

$$\dot{E} = -\nu\beta a x^{*2} - \nu\beta(x^* - y^*)^2 \leq 0 \qquad (\beta, \nu > 0, \quad a \geq 0).\qquad(9.21)$$

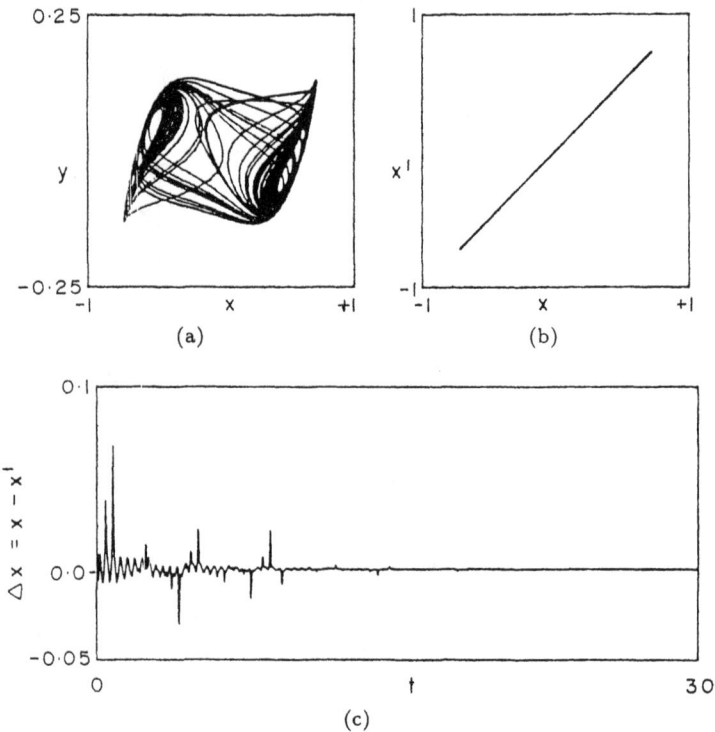

Fig. 9.11. (a) Chaotic attractor in $(x-y)$ plane. (b) Projection of trajectories in $(x-x')$ plane for $\varepsilon = 1.35, \alpha = 0.35, \nu = 100, \beta = 300$ (*synchronization by one-way coupling*). (c) Trajectory $(x-x')$ versus t plot.

If $\dot{E}$ is to vanish identically for $t > t_1$, then x^* and y^* must be zero for all $t > t_1$. From Eq. (9.19b) this requires that $\dot{y}^* = 0$ for all $t > t_1$ and so z^* must also be equal to zero (Refs. [206, 236] for all $t > t_1$. Thus $\dot{E}$ vanishes identically only at the origin. Therefore the solutions of Eqs. (9.18d–f) are globally asymptotically stable for the specific value $\varepsilon = (1 + \alpha)$. So both the drive (9.18a–c) and response (9.18d–f) systems eventually synchronize for all $t > t_1$ for the above choice of the coupling parameter. The synchronized chaotic behaviour of Eq. (9.18) between x and x' variables is shown in Fig. 9.11(b). The initial conditions are fixed as $x(0) = 0.1, y(0) = 0.1, z(0) = 0.2, x'(0) = 0.15, y'(0) = 0.2$ and $z'(0) = 0.3$ (see also Fig. 9.11(c)). Furthermore, a detailed numerical simulation also shows that the system (9.18) eventually synchronizes for all values of $\varepsilon > 0.8$, and it should therefore be possible to choose an appropriate Lyapunov function E for other sets of ε values (different from $\varepsilon = (1 + \alpha)$) also.

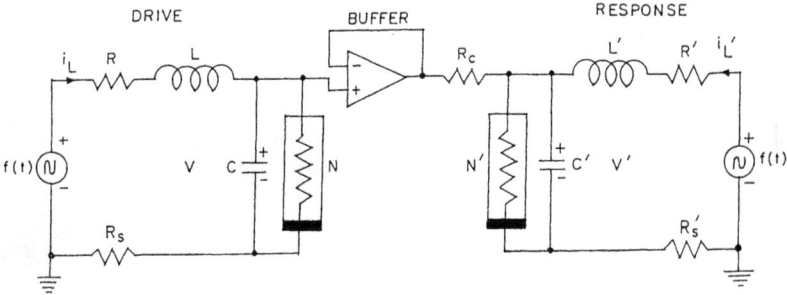

Fig. 9.12. Schematic circuit realization of two identical MLC circuits with one-way coupling resistor R_c.

The applicability of this method of chaos synchronization is not restricted to third-order autonomous systems alone, but is equally well suited for second-order non-autonomous nonlinear systems as well. To illustrate this, we consider now the simplest dissipative non-autonomous MLC circuit (Refs. [77, 79, 237]) discussed in Sec. 7.5. The schematic representation of two identical MLC circuits with one-way coupling element is shown in Fig. 9.12. The normalized state equations of Fig. 9.12 (cf. Chapter 7, Sec. 7.5) are represented as

drive:

$$\dot{x} = y - h(x), \tag{9.22a}$$

$$\dot{y} = -\beta(1+\nu)y - \beta x + F\sin(\omega t), \tag{9.22b}$$

response:

$$\dot{x}' = y' - h(x') + \varepsilon(x - x'), \tag{9.22c}$$

$$\dot{y}' = -\beta(1+\nu)y' - \beta x' + F\sin(\omega t), \tag{9.22d}$$

where $\varepsilon = (R'/R_c)$ is the coupling parameter. The difference system of Eqs. (9.22) is

$$\dot{x}^* = y^* - (h(x) - h(x')) - \varepsilon x^*,$$

$$\dot{y}^* = -\beta(1+\nu)y^* - \beta x^*, \tag{9.23}$$

where $x^* = (x - x')$ and $y^* = (y - y')$. Here $h(x) - h(x') = h'(\eta)(x - x') = h'(\eta)x^*$, and $h'(\eta)$ takes two values 'a' and 'b' (due to the piecewise-linear characteristics of $h(x)$ and $h(x')$, (cf. Sec. 7.5 for details)) (Refs. [79, 217]) depending upon the region of operation. Then,

$$\dot{x}^* = y^* - s_i x^* - \varepsilon x^*,$$

$$\dot{y}^* = -\beta(1+\nu)y^* - \beta x^*, \tag{9.24}$$

(a)

(b)

(c)

(d)

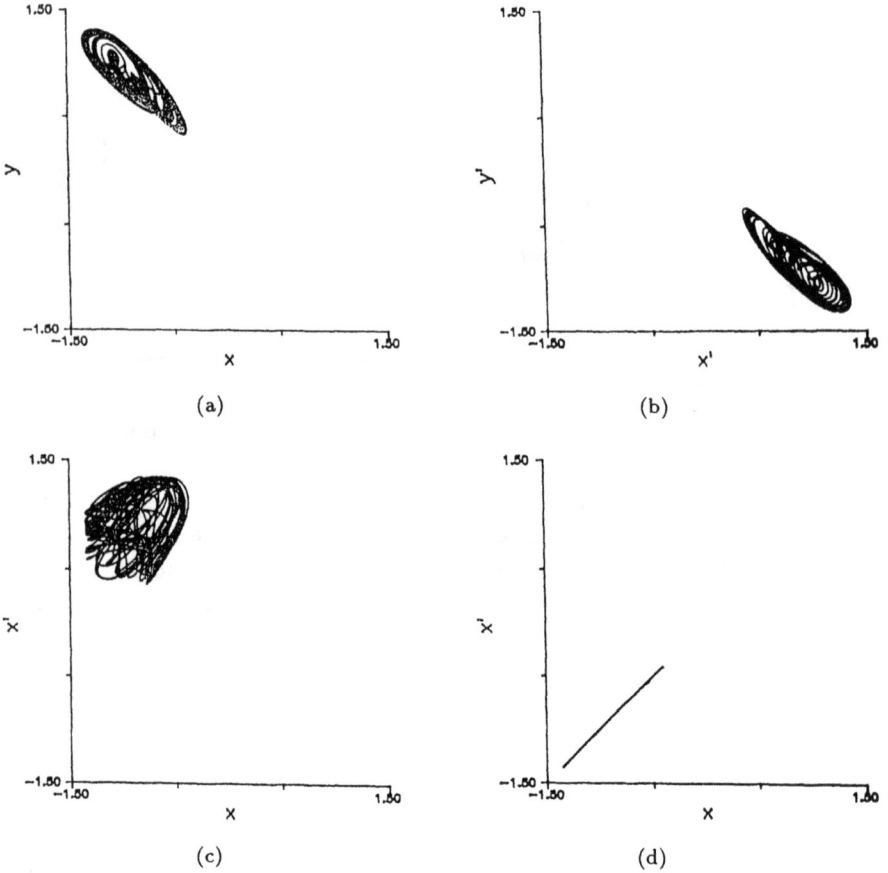

Fig. 9.13. (a) One-band chaotic attractor of Eqs. (9.22a–b) for $\varepsilon = 0, f = 0.1$ [$x(0) = -0.5, y(0) = 0.1$]. (b) One-band chaotic attractor of Eqs. (9.22c–d) for $\varepsilon = 0, f = 0.1$ [$x'(0) = 0.5, y'(0) = 0.11$]. (c) Unsynchronized motion in (x–x') plane for $\varepsilon = 0$. (d) Synchronized motion for $\varepsilon = 1$. (e): (i) Waveform $x^*(t) = (x-x')$ for $\varepsilon = 0$; (ii) Waveform $x^*(t)$ for $\varepsilon = 1$. (f): (i) One-band chaotic attractor of Eqs. (9.22a–b) for $\varepsilon = 0, f = 0.1$ [$x(0) = 0.5, y(0) = 0.11$]; (ii) One-band chaotic attractor of Eqs. (9.22c–d) for $\varepsilon = 0, f = 0.1[x'(0) = -0.5, y'(0) = 0.1]$; (iii) Unsynchronized motion in ($x - x'$) plane for $\varepsilon = 0$; (iv) Synchronized motion for $\varepsilon = 1$.

or

$$\begin{bmatrix} \dot{x}^* \\ \dot{y}^* \end{bmatrix} = \begin{bmatrix} -s_i - \varepsilon & 1 \\ -\beta & -\beta(1+\nu) \end{bmatrix} \begin{bmatrix} x^* \\ y^* \end{bmatrix}, \tag{9.25}$$

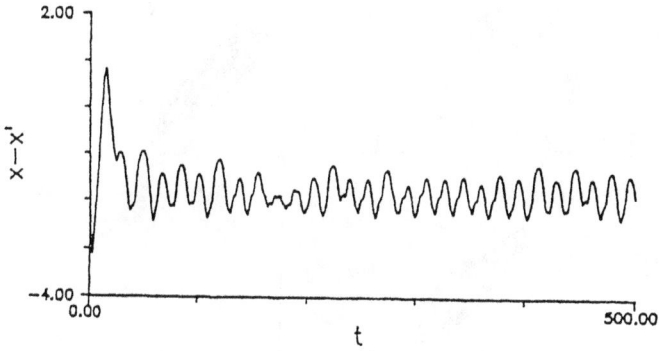

(e)(i)

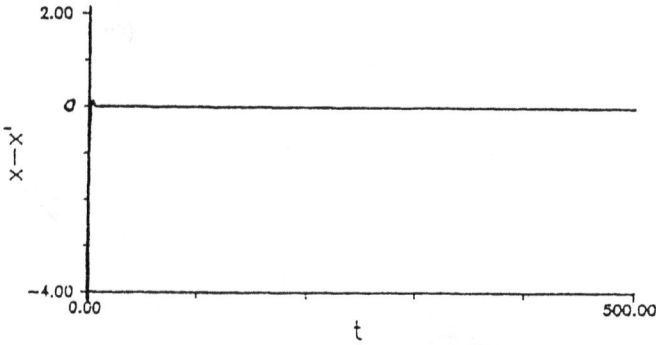

(e)(ii)

Fig. 9.13. (*Continued*)

where $s_i = a, b$; $i = 1, 2$. The characteristic equation is

$$\lambda^2 + \mu\lambda + \xi = 0, \tag{9.26}$$

where $\mu = (\varepsilon + s_i + \beta + \beta\nu)$ and $\xi = (\varepsilon\beta + \varepsilon\beta\nu + s_i\beta + s_i\beta\nu + \beta)$. If $\mu > 0$ and $\xi > 0$, then $x^* = y^* = 0$ is a stable point and the two systems (9.22a–b and 9.22c–d) will eventually synchronize. In the present case where $\beta = 1, \nu = 0.015, a = -1.02$, and $b = -0.55$ (Ref. [79]), the value of ε turns out to be 0.0348, and thus for all $\varepsilon > 0.0348$ the two systems (9.22a–b) and (9.22c–d) will synchronize asymptotically. In order to confirm this, one can numerically integrate Eqs. (9.22) simultaneously for different values of ε. Figures 9.13(a) and 9.13(b) show the chaotic attractor of Eqs. (9.22a–b) for $\varepsilon = 0, \beta = 1, \nu = 0.015, \omega = 0.75$ and $F = 0.1$ for different sets of initial conditions. Due to the symmetry of the system (9.22a–b) (Ref. [79]), it has

(f)(i)

(f)(ii)

(f)(iii)

(f)(iv)

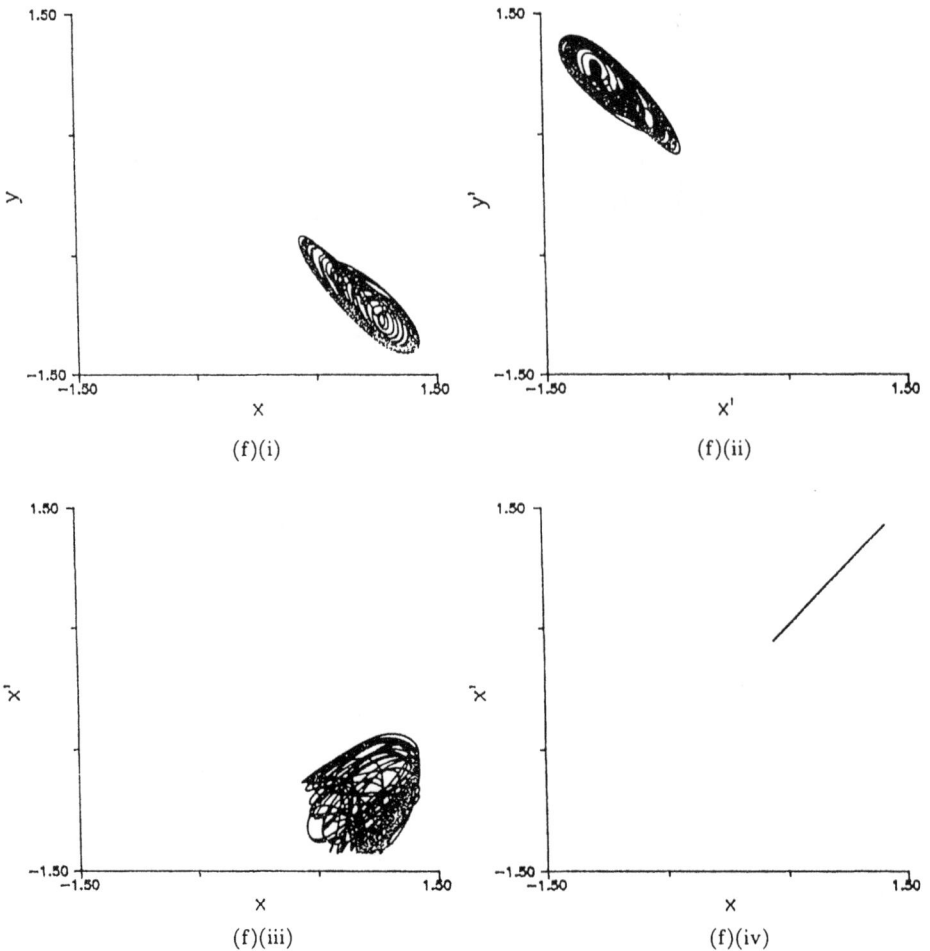

Fig. 9.13. (*Continued*)

two basins of attraction depending upon the initial conditions. Figure 9.13(a) shows a one-band chaotic attractor in the $(x-y)$ plane at one basin of attraction for the initial conditions $x(t = 0) = x(0) = -0.5$ and $y(0) = 0.1$ for Eq. (9.22a–b). Figure 9.13(b) shows the chaotic attractor in the $(x'-y')$ plane in the other basin of attraction of Eq. (9.22c–d) for the initial conditions $x'(t = 0) = x'(0) = 0.5$ and $y'(0) = 0.11$ for $\varepsilon = 0$. Since for $\varepsilon = 0$ the two systems (9.22a–b) and (9.22c–d) are uncoupled, the trajectories of these systems will not synchronize as shown in Fig. 9.13(c). However for $\varepsilon = 1$,

(a)

(b)(i)

b(ii)

(c)(i)

c(ii)

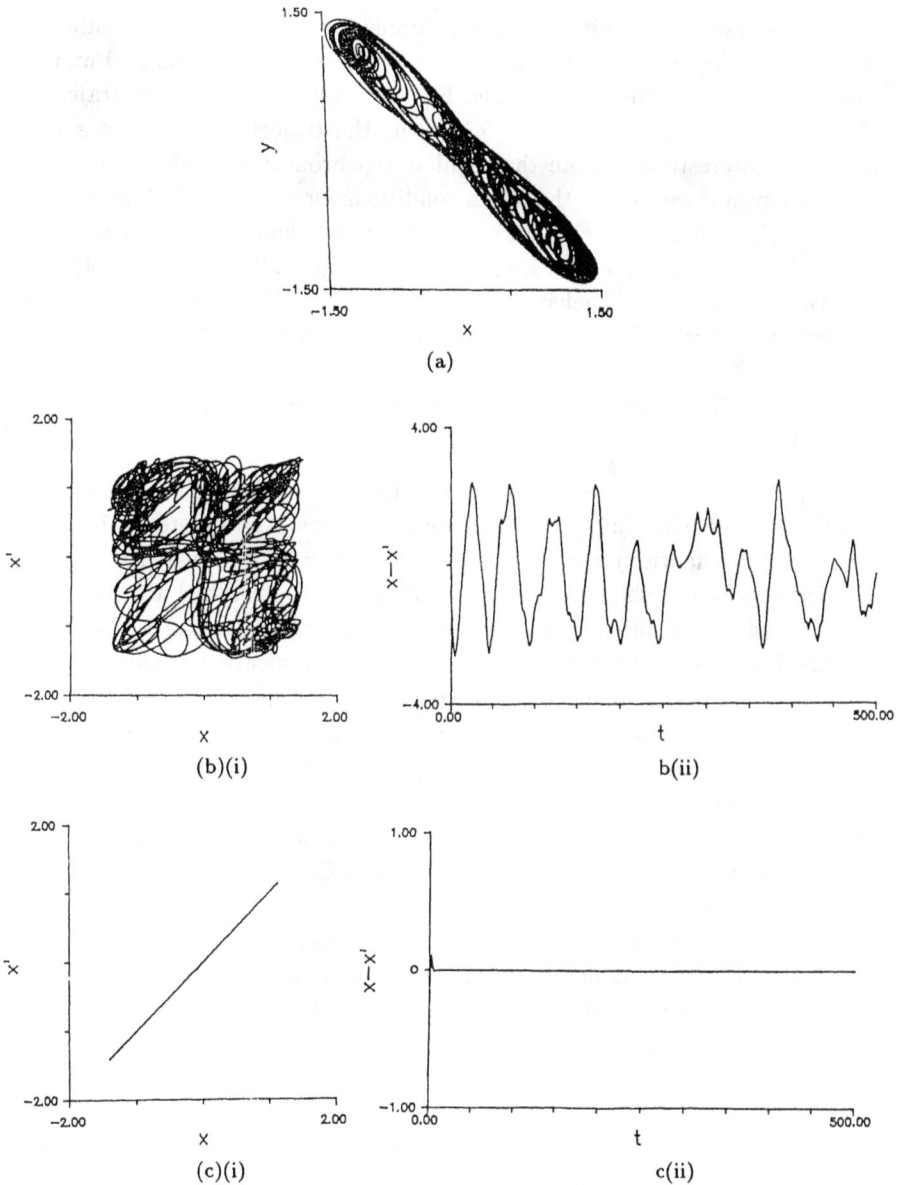

Fig. 9.14. (a) Double-band chaotic attractor of Eqs. (9.22a–b) for $f = 0.15$ and $\omega = 0.75$. (b) Unsynchronized motion of Eq. (9.22) for $\varepsilon = 0, f = 0.15$ and $\omega = 0.75$: (i) Trajectory plot in $(x–x')$ plane; (ii) Waveform $x^*(t)$ for $\varepsilon = 0$. (c) Synchronized motion of Eq. (9.22) for $\varepsilon = 1, f = 0.15$, and $\omega = 0.75$: (i) Trajectory plot in $(x–x')$ plane; (ii) Waveform $x^*(t)$ for $\varepsilon = 1$.

these two systems exhibit perfect synchronization among their variables as indicated in Fig. 9.13(d) even though the two systems are integrated numerically with different initial conditions. Figures 9.13(e)(i) depicts the trajectory of $x^* = (x - x')$ for $\varepsilon = 0$ and 9.13(e)(ii) the trajectory of x^* for $\varepsilon = 1$. Also it is interesting to note that similar synchronization behaviour can be observed even if we change the initial conditions for x, y and x', y' as shown in Fig. 9.13(f). Finally, for $F = 0.15$, a double-band chaotic attractor as shown in Fig. 9.14(a) is observed for Eqs. (9.12a–b). Figures 9.14(b) and 9.14(c) depict the unsynchronized behaviour for $\varepsilon = 0$, $F = 0.15$ and the synchronized behaviour for $\varepsilon = 1, F = 0.15$ respectively for the systems (9.22a–d).

9.3. Secure Transmission of Signals by Synchronized Chaotic Systems

In the previous section, we have discussed the aspects of chaos synchronization and the ways of achieving it in typical nonlinear circuits and systems. For some synchronizing chaotic systems the ability to synchronize is robust. As discussed in references (Refs. [89, 150, 198, 206, 207, 213–216, 220, 221] a combination of synchronization and unpredictability from purely deterministic nonlinear systems leads to some potentially interesting communication applications.

The earliest attempts to use random signals in secure *communications* was probably due to Vernam (Ref. [238]), which dates back to 1926. Since then, such usage has assumed considerable significance in fields like cryptography. The importance of the concept of chaos synchronization in secure communication has been quickly highlighted by signal processing, and circuits, and systems communities. In particular, Cuomo and Oppenheim (Refs. [213, 214]) have reported some important applications, in the form of "*chaotic masking and modulation*" and "*chaotic switching*". Specifically, these authors have shown how the concept of synchronization can be used to mask information by adding a chaotic signal to a speech signal which is to be transmitted. The chaotic signal used as a noise-like signal is recovered by the receiver using the synchronization effect. The speech signal is simply obtained from the received signal by subtraction of the original signal generated at the receiver. More recently, Kocarev *et al.* (Ref. [215]) have also applied the ideas of Pecora and Carroll, and Cuomo and Oppenheim, in the context of chaotic masking (secure communications). Using Chua's circuit as a simple generator of chaotic signals, they showed experimentally the synchronization effect in an application very similar to the one described in Ref. [213] in which the information signal is buried in the chaotic signal.

A variation of the above signal transmission is also studied in Ref. [216] and in Refs. [207, 213], where the message signal is a binary signal causing a parameter in the transmitter chaotic circuit to take on one of two possible values, thereby producing a modulated chaotic output. The receiver replica has the respective parameter fixed to one of the possible values in its counterpart. As a result, it tracks the transmitter anytime the binary input is in one state, and falls out of synchronization at times corresponding to the other input state. Lock and unlock conditions are easily detected, resulting in proper demodulation.

Some of the most complex issues offered by secure communication has been described in a paper by Halle *et al.* (Ref. [220]). The proposed idea is to multiply the information signal by a broad-band, noise-like chaotic signal. Simultaneously, a series of works (Refs. [150, 206, 207]) reported by Murali and Lakshmanan utilizes chaos synchronization for secure communication studies in the ADVP and Duffing oscillator models and after that in the MLC circuit model (Ref. [239]). In all these approaches, the power level of the information signal to be transmitted in a secure way must be kept significantly lower than the power level of the chaotic signal in order that synchronization is possible. We describe the salient features of these studies in the next two sections and show how both analog and digital signals can be transmitted in a secure way. We make use of both the methods of chaos synchronization discussed in the previous sections.

9.4. Chaotic Signal Masking and Transmission of Analog Signals

In this section, we discuss and demonstrate the concept of chaotic signal masking in the various nonlinear oscillator systems and illustrate the fact that synchronization of chaotic systems offer potential opportunity for novel applications for secure communications. The point is that when a chaotic signal is transmitted it cannot be deciphered in general at the receiving end unless full information about the transmitting (chaotic nonlinear) system is available, so that it can be coupled appropriately as discussed in Sec. 9.1 for synchronization. From this point of view the signal which is to be transmitted is masked by the noise-like chaotic signal by adding it at the transmitter to the information-bearing signal $s(t)$. Then, at the receiver the masking is removed. The basic idea is to use the received signal to regenerate the information-bearing signal by subtracting the masking chaotic signal (regenerated separately through chaos synchronization) to obtain $s(t)$. This task is feasible with the synchronizing receiver system since the ability to synchronize is robust, that is, it is not highly sensitive to perturbations in the drive signal (Refs. [213, 215]). It is assumed that, for masking, the power level of $s(t)$ is significantly lower than that of the chaotic signal to be used for masking (at the transmitter). While there are many possible variations, one can consider for example a transmitted signal of the form $r(t) = x(t) + s(t)$, where $x(t)$ is the chaotic signal of the transmitter. Then one can exploit the robustness of synchronization using $r(t)$ as the synchronizing drive signal at the receiver. If the receiver or response has synchronized with $r(t)$ as the drive signal, then $x_r(t) \approx x(t)$ and consequently $s(t)$ is recovered as $s^1(t) = r(t) - x_r(t)$.

The above procedure can be demonstrated easily with the aid of the ADVP oscillator, the Duffing oscillator, and the MLC circuit through numerical simulation. Especially, we use the two basic approaches of chaos synchronization, namely, cascading synchronization and the one-way coupling technique discussed previously.

9.4.1. *Signal Transmission Through Cascading Synchronization*

Already using this approach Cuomo *et al.* (Ref. [213]) and Kocarev *et al.* (Ref. [215]) have reported the secure transmission of analog signals in the Lorenz oscillator and Chua's circuit, respectively, using both experimental and numerical means. Presently, we show how the chaotic ADVP oscillator can be effectively utilized as a vehicle to transmit analog signals in a secure way. Following the scheme adopted by Cuomo *et al.* and Kocarev *et al.* we first consider the cascading synchronized chaotic ADVP oscillator. The total system of rescaled equations is represented in the following manner (Ref. [150]):

Drive:

$$\dot{x} = -\nu[x^3 - \alpha x - y]\,, \tag{9.27a}$$

$$\dot{y} = x - y - z\,, \tag{9.27b}$$

$$\dot{z} = \beta y\,; \tag{9.27c}$$

response 1:

$$\dot{y}' = x - y' - z'\,, \tag{9.27d}$$

$$\dot{z}' = \beta y'\,; \tag{9.27e}$$

response 2:

$$\dot{x}'' = -\nu[(x'')^3 - \alpha x'' - y']\,. \tag{9.27f}$$

(a)

(b)

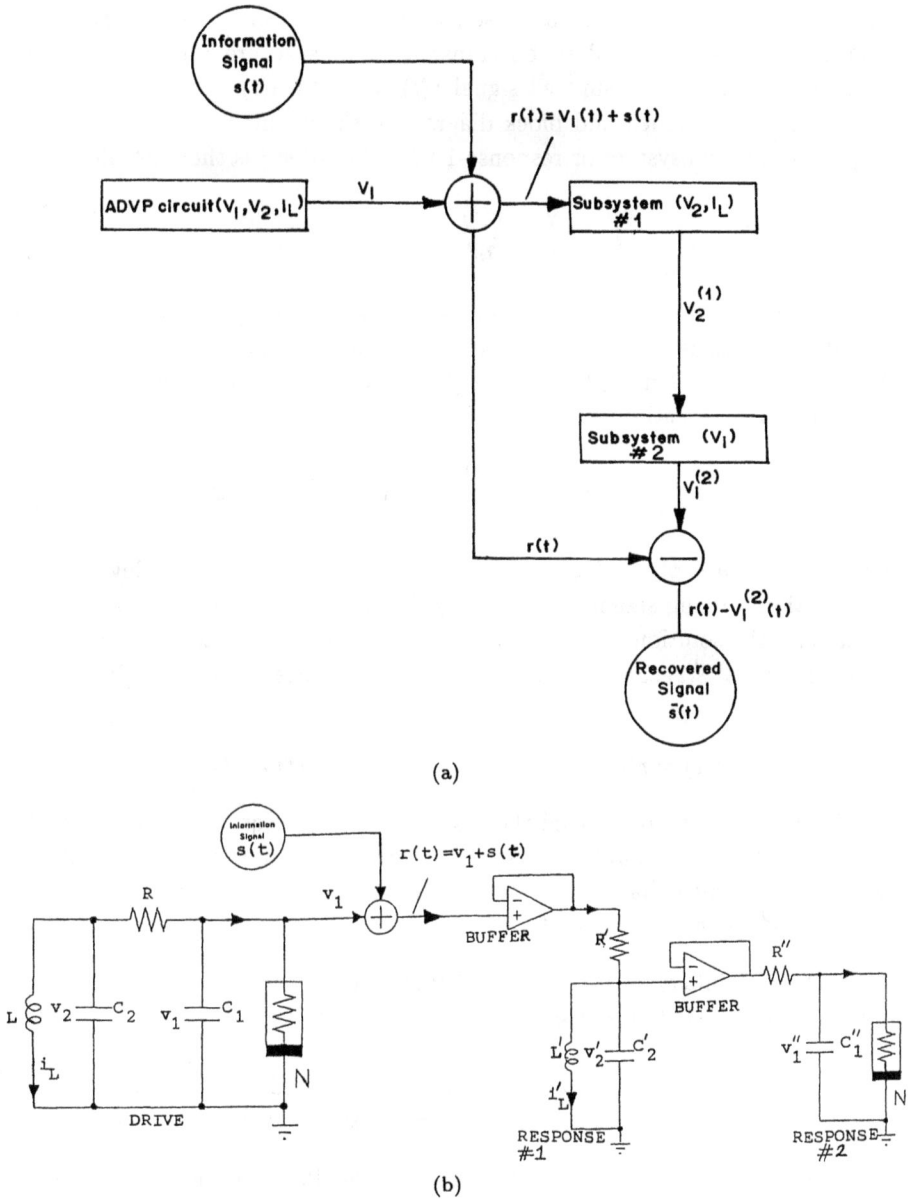

Fig. 9.15. (a) Block diagram of signal masking method for analog signal transmission. The voltages v_1, v_2 and the current i_L are rescaled as the variables x, y, and z, respectively, in Eqs. (9.27). (b) Practical circuit realization of (a).

For our subsequent numerical analysis, we use the $x(t)$ signal of the drive system [Eqs. (9.27a–c)] as a noise-like "masking signal" and $s(t)$ as an information-bearing signal to be transmitted in a secure way. Now let us consider the actual transmitted signal $r(t) = x(t) + s(t)$. The schematic diagram of this approach and block diagram of the circuit model are shown in Fig. 9.15. The subsystem or response-1 [Eqs. (9.27d–e)] is then modified as

$$\dot{y}' = r(t) - y' - z', \tag{9.27d'}$$
$$\dot{z}' = \beta y'. \tag{9.27e'}$$

The second response system (response-2) is the x'' subsystem driven by the signal y', which is the same as that represented by Eq. (9.27f). Now from Eqs. (9.27a–c), (9.27d'), (9.27e') and (9.27f) we have the following inhomogeneous linear differential equations:

$$\dot{y}^* = s(t) - y^* - z^*, \tag{9.28a}$$
$$\dot{z}^* = \beta y^*. \tag{9.28b}$$

where $y^* = (y - y')$ and $z^* = (z - z')$. Assuming the power level of the information-bearing signal $s(t)$ to be significantly lower than that of the $x(t)$ signal and the solution $x^* = (x - x'')$ to be significantly small with respect to $s(t)$, we see that $s(t)$ can be recovered from the response system-2 as (Refs. [150, 213, 215]):

$$\tilde{s}(t) = r(t) - x''(t) = x(t) + s(t) - x''(t) \approx s^1(t). \tag{9.29}$$

We have numerically solved the cascade system of Eqs. (9.27a–c), (9.27d'), (9.27e') and (9.27f) simultaneously with parameters $\alpha = 0.35, \nu = 100$ and $\beta = 300$ (for which chaos is observed). The information-bearing signal $s(t)$ is assumed to be any one of the following type:

(i) $s(t) = F\sin(\omega t)$ [single-tone, $F = 0.02, \omega = 1$],
(ii) $s(t) = F\sin(\omega t)[1 + f\sin(\Omega t)]$ [amplitude-modulated wave, $F = 0.02$,
                                    $\omega = 1, f = 1 \,\&\, \Omega = 0.2$]
(iii) $s(t) = F\sin[\omega t + f\sin(\Omega t)]$ [phase-modulated wave, $F = 0.02, \omega = 1$,
                                    $f = 0.2 \,\&\, \Omega = 0.2$]

From the numerical simulation results, the information signal $s^1(t)$ is recovered at the response system by adopting Eq. (9.29). Figures 9.16(a)–(c) depict the power spectrum of the information signal $s(t)$, the actual transmitted signal $r(t)[= s(t) + x(t)]$, and the recovered signal $s^1(t)$ for the above three

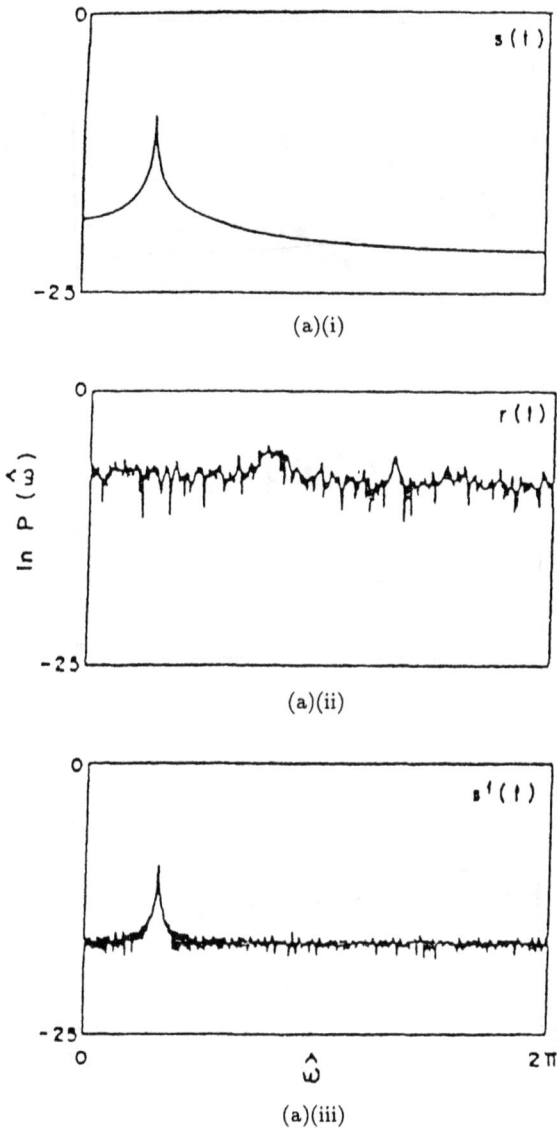

(a)(i)

(a)(ii)

(a)(iii)

Fig. 9.16. (a) Power spectra of signals: (i) $s(t) = F \sin \omega t$ [single-tone, $F = 0.02, \omega = 1$]; (ii) $r(t)$; (iii) $s^1(t)$. (b) Power spectra of signals: (i) $s(t) = F \sin \omega t(1 + f \sin \Omega t)$ [amplitude-modulated wave, $F = 0.02, \omega = 1, f = 1, \Omega = 0.2$]; (ii) $r(t)$; (iii) $s^1(t)$. (c) Power spectra of signals: (i) $s(t) = F \sin(\omega t + f \sin \Omega t)$ [phase-modulated wave, $F = 0.02, \omega = 1, f = 0.2, \Omega = 0.2$]; (ii) $r(t)$; (iii) $s^1(t)$.

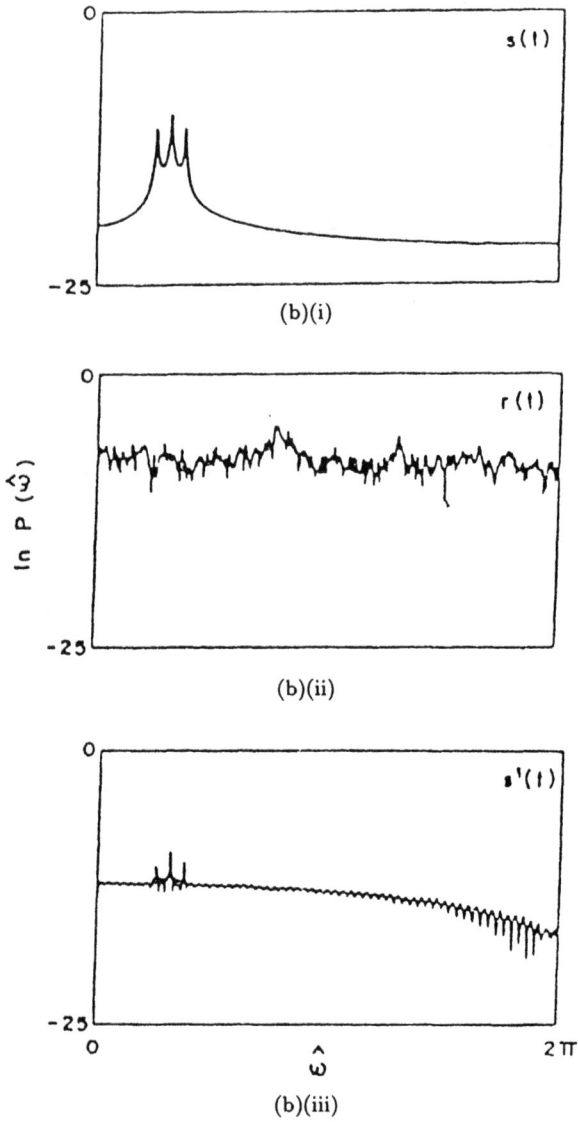

(b)(i)

(b)(ii)

(b)(iii)

Fig. 9.16. (*Continued*)

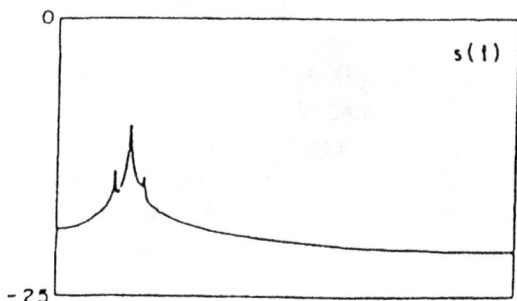

(c)(i)

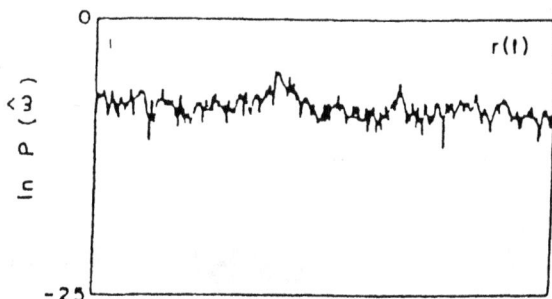

(c)(ii)

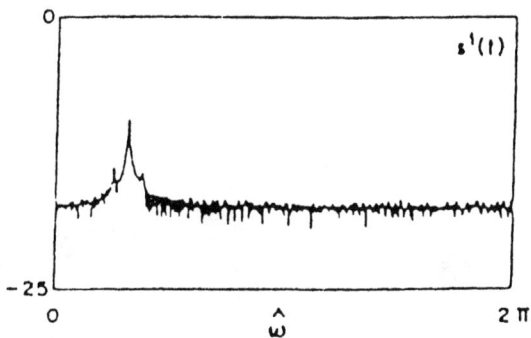

(c)(iii)

Fig. 9.16. (*Continued*)

different cases respectively. As the power level of $s(t)$ is significantly lower than that of the $x(t)$ signal, the frequency component of $s(t)$ is not discernible or detectable in Figs. 9.16(a)(ii)–9.16(c)(ii) due to the chaotic (broad-band) nature of the actual transmitted signal $r(t)$. However, as we observe from Figs. 9.16(a)(iii)–9.16(c)(iii), the quality of the recovered signal $s^1(t)$ is significantly comparable to that of the original signal $s(t)$. Using this chaotic-signal masking technique and Pecora & Carroll cascading synchronization concept, Cuomo and Oppenheim have for the first time demonstrated the performance of the synchronized Lorenz system as a model to transmit a segment of speech from the sentence *"He has the bluest eyes"* (Ref. [213]). Figures 9.17(a) and (b) show the original speech signal and the recovered speech signal at the receiver respectively. Also indicated in Fig. 9.18 is the power spectra of the chaotic masking signal and the speech signal.

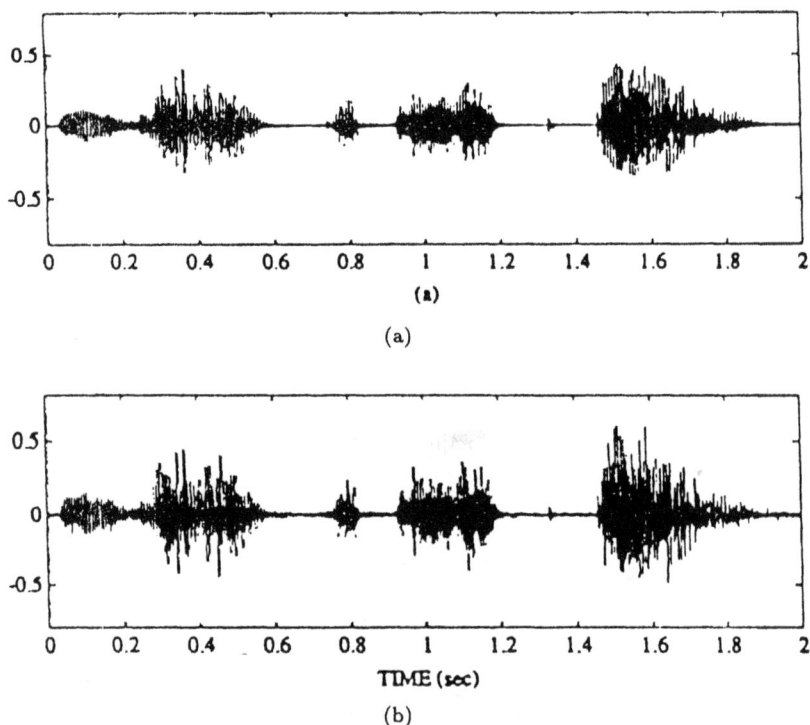

(a)

(b)

Fig. 9.17. Circuit data: speech waveforms (a) original; (b) recovered (Ref. [213]).

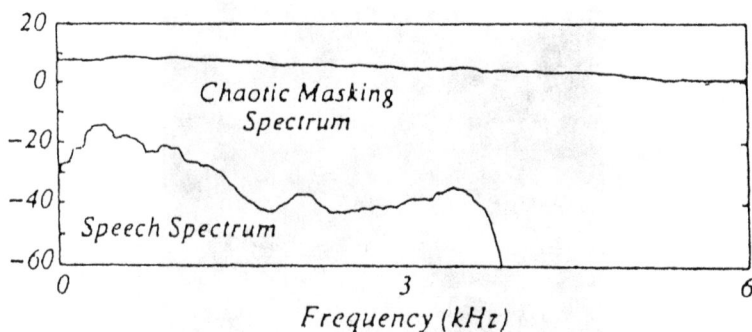

Fig. 9.18. Circuit data: power spectra of chaotic masking and speech signals (Ref. [213]).

Further, Kocarev *et al.* (Ref. [215]) have reported the first experimental demonstration of secure communications via chaotic synchronization using Chua's circuit as a universal chaotic building block. They have used as the information-bearing signal a single-tone(sine wave) of frequency 720Hz and a spiral Chua's attractor as a masking signal. Figures 9.19(a)–(c) show the actual transmitted signal and the recovered signal at the receiver.

9.4.2. *Method of One-way Coupling*

As discussed in Sec. 9.2, synchronization of chaos can be observed in certain nonlinear systems not only through cascading synchronization but also with the aid of one-way coupling of two identical systems. In the following, we use this approach of chaos synchronization to demonstrate the signal masking technique for transmitting analog signals. First, let us consider the ADVP oscillator with drive-response configuration and one-way coupling(see Sec. 9.2). The drive system is represented by Eqs. (9.18a–c). As discussed in the previous sub-section, let us consider now the actual transmitted signal $r(t) = x(t) + s(t)$. Then the response system is written as (Ref. [206])

response:

$$\dot{x}' = -\nu[(x')^3 - \alpha x' - y'] + \nu\varepsilon[r(t) - x'], \tag{9.30a}$$

$$\dot{y}' = x' - y' - z', \tag{9.30b}$$

$$\dot{z}' = \beta y'. \tag{9.30c}$$

Here, ε is the coupling parameter. In the absence of the signal $s(t)$, a synchronized chaotic behaviour of Eqs. (9.27a–c) and (9.30a–c) between x and x' variables for $\varepsilon = 1$ or 1.35, $\alpha = 0.35$, $\nu = 100$, and $\beta = 300$ is observed. Now by

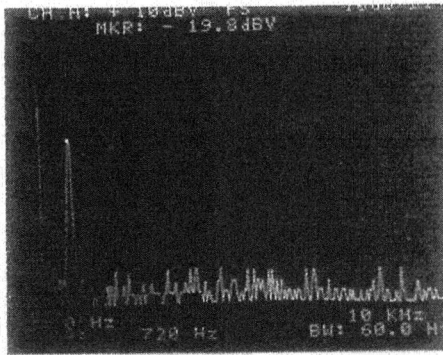

(a)

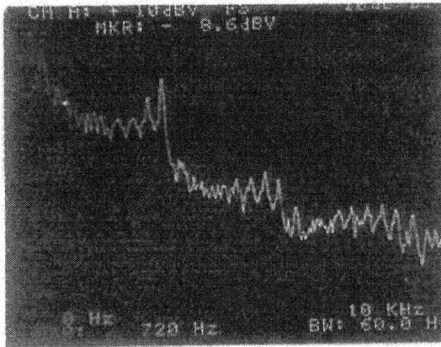

(b)

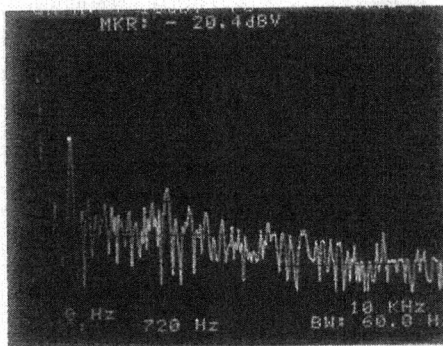

(c)

Fig. 9.19. (a) Power spectrum of the input (information-bearing) signal $s(t)$, which consists of a single-tone (sine wave) of frequency 720 Hz. The noise floor is at -65 dBV (Ref. [215]). (b) Power spectrum of the transmitted signal $r(t)$. Note that the component signal frequency at 720 Hz is not discernible (Ref. [215]). (c) Power spectrum of the recovered signal $s^1(t)$, highlighted by the bright spot at 720 Hz (Ref. [215]).

assuming that the signal $s(t)$ is present and its power level is significantly lower than that of the $x(t)$ signal and that the difference solution $x^*(t) = x(t) - x'(t)$ is significantly small with respect to $s(t)$, then we see that $s(t)$ can be recovered from the response system as (Ref. [206]) (see Fig. 9.20)

$$\tilde{s}(t) = r(t) - x'(t) = x(t) + s(t) - x'(t) \approx s^1(t). \qquad (9.31)$$

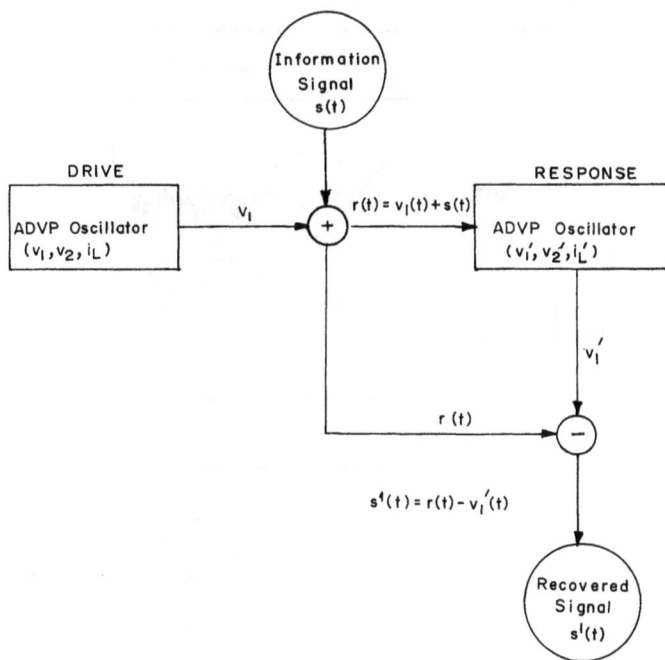

Fig. 9.20. Schematic of "signal masking technique" for analog signal transmission with one-way coupling. In Eqs. (9.30) rescaled variables are used.

Numerical solution of the equations (9.27a–c) and 9.30(a–c) with $s(t) = F\sin(\omega t)$ (single-tone, $F = 0.02$, $\omega = 1$) gives rise to the information signal $s^1(t)$ as recovered at the response system by adopting Eq. (9.31). Figure 9.21 depicts the power spectra of the signal $s(t)$, the actual transmitted signal $r(t) (= s(t) + x(t))$, and the recovered signal $s^1(t)$. As noted previously, the component of signal frequency $s(t)$ is not discernible or detectable in Fig. 9.21(b). For other signals like FM and PM also, a similar analysis can be performed.

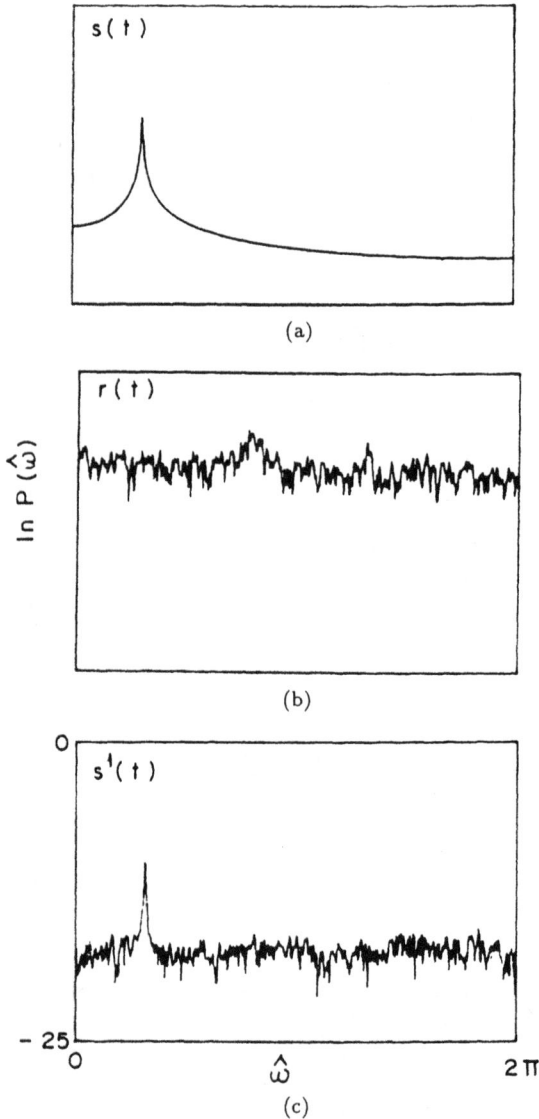

Fig. 9.21. Power spectra of the signals: (a) $s(t) = 0.02 \sin(t)$; (b) $r(t) = x(t) + s(t)$; (c) $s^1(t)$.

The applicability of this method of chaos synchronization and signal masking is not restricted to third-order autonomous systems alone, but can be equally well adopted for second-order non-autonomous systems. For example, for the Duffing oscillator with the drive system (Ref. [206]) we have

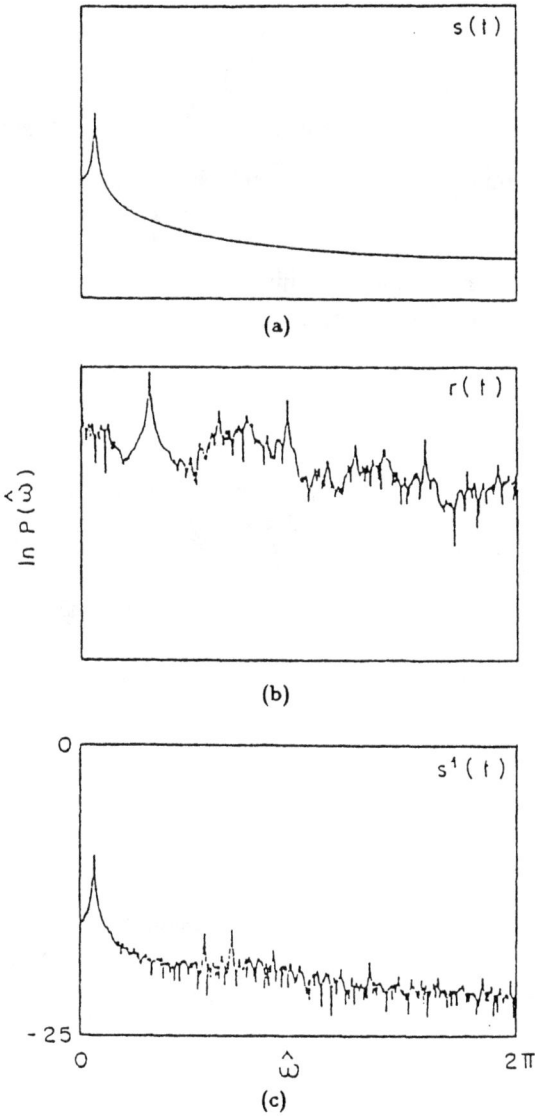

Fig. 9.22. Power spectra of the signals: (a) $s(t) = 0.02\sin(0.2t)$; (b) $r(t) = x(t) + s(t)$; (c) $s^1(t)$.

drive:

$$\dot{x} = y \,, \tag{9.32a}$$

$$\dot{y} = -py - x^3 + F\cos(\omega t) \,, \tag{9.32b}$$

and the response system with added information signal $s(t)$ to $x(t)$,

response:

$$\dot{x}' = y' + \varepsilon(r(t) - x'),\tag{9.32c}$$

$$\dot{y}' = -py' - (x')^3 + F\cos(\omega t), \qquad [r(t) = x(t) + s(t)].\tag{9.32d}$$

Assuming the information signal $s(t) = 0.02\sin(0.2t)$ (single-tone), Fig. 9.22 shows the power spectra of the signals $s(t), r(t)$ and $s^1(t)$ for $p = 0.05, F = 7.5, \omega = 1$ and $\varepsilon = 1$. Also, for the experimental feasibility of investigating the secure transmission of signals, a suitable analog simulation circuit can be employed for the Duffing oscillator.

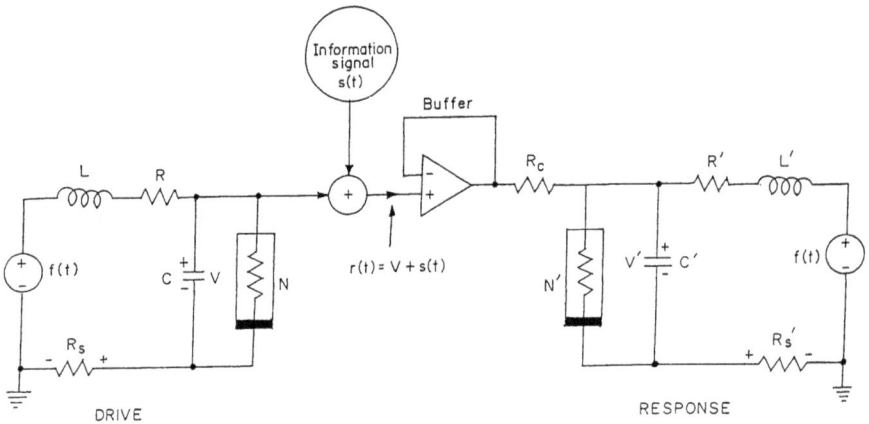

Fig. 9.23. Circuit realization of two MLC circuits with one-way coupling for secure transmission of analog signals.

Next we consider the MLC circuit as a vehicle to transmit analog signals in a secure way. The schematic circuit realization of this model is shown in Fig. 9.23. The synchronization of chaos behaviour of this circuit in the absence of $s(t)$ signal is discussed in Sec. 9.2. Now the normalized state equations of Fig. 9.23 with $r(t) = s(t) + x(t)$ are written as

drive:

$$\dot{x} = y - h(x),\tag{9.33a}$$

$$\dot{y} = -\beta(1 + v)y - \beta x + F\sin(\omega t),\tag{9.33b}$$

response:

$$\dot{x}' = y' - h(x') + \varepsilon(r(t) - x'),\tag{9.33c}$$

$$\dot{y}' = -\beta(1 + v)y' - \beta x' + F\sin(\omega t).\tag{9.33d}$$

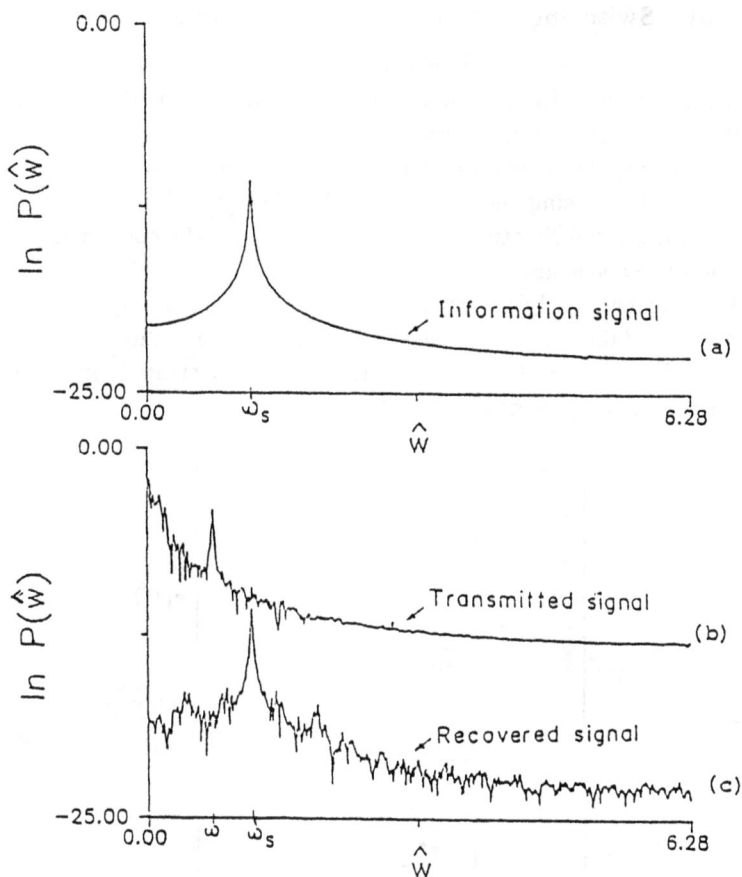

Fig. 9.24. Power spectra of the signals: (a) $s(t) = 0.01 \sin(1.2t)$; (b) $r(t) = x(t) + s(t)$; (c) $s^1(t)$.

By assuming the information signal $s(t) = 0.01 \sin(1.2t)$ and for the parameter $\beta = 1, \nu = 0.015, \varepsilon = 1, F = 0.15$ and $\omega = 0.75$, numerical simulation of Eq. (9.33) gives the power spectra of the signals $s(t)$, $r(t)$ and $s^1(t)$ as in Fig. 9.24.

The above studies clearly indicate that in view of the typical broad band spectrum, the chaotic signal $x(t)$ becomes an ideal candidate for *spread-spectrum secure communication applications* (Refs. [89, 213, 215, 220, 221]).

9.5. Chaotic Switching and Digital Signal Communication

It is not only the analog signals which can be securely transmitted through chaos synchronization. Binary-valued bit signals can be equally well transmitted by the use of synchronized chaotic systems (Refs. [207, 213, 216]). Here, the idea is essentially to modulate a control parameter associated with the transmitter or drive using the information-bearing digital wave form and accordingly transmit the chaotic signal. At the receiver, the coefficient modulation will produce a synchronization error between the received drive signal and the receiver's regenerated drive signal, with an error-signal amplitude that depends on the modulation. Using the synchronization error the modulation can be detected. This method has also been recently demonstrated experimentally for typical nonlinear systems (Refs. [207, 213, 216]).

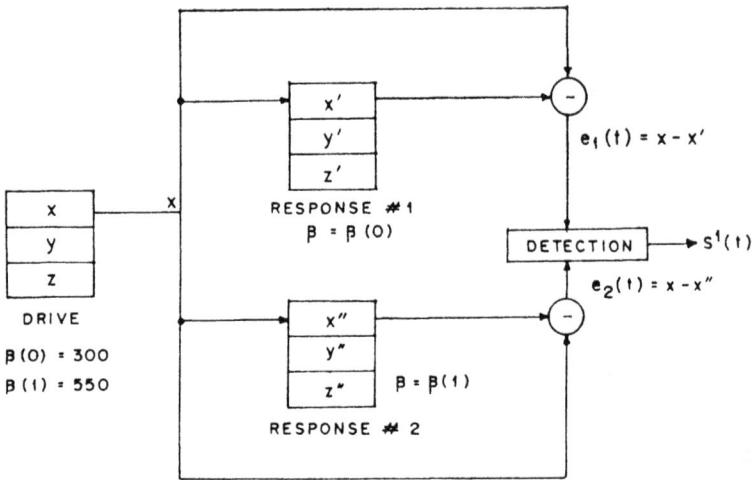

Fig. 9.25. Schematic of digital signal transmission technique. Here, #1 and #2 are two separate ADVP oscillators.

Now we explain the method for the ADVP oscillator, which is illustrated schematically in Fig. 9.25. Here the coefficient β in Eq. (9.27a–c) is modulated by the information waveform $s(t)$, which is now a binary-coded signal. The information is carried over the channel by the chaotic signal $x(t)$, which serves as the driving input to the receivers #1 and #2. Here, two identical ADVP oscillators with one-way coupling element are used. At the two receivers the modulation is detected by forming the difference between $x(t)$ and the reproduced signals $x'(t)$ of #1 and $x''(t)$ of #2. Then the synchronization error $e_1(t) = x(t) - x'(t)$ will be relatively large in amplitude during the time period when "1" value of binary information signal is transmitted and small in amplitude during the "0" value transmission. Also, the synchronization error $e_2(t) = x(t) - x''(t)$ will have an opposite nature to the previous one. Thus the synchronization receivers can be recognized as a form of matched filters for the chaotic transmitted signal $x(t)$.

To illustrate this technique numerically, we use a square wave for $s(t)$ as shown in Fig. 9.26(a). The square wave produces a variation in the transmitter (drive) coefficient β with zero-bit and one-bit coefficients corresponding to $\beta(0) = 300$ and $\beta(1) = 550$, respectively. Figure 9.26(b) shows the actual transmitted chaotic signal $x(t)$ from the transmitter (drive). Figure 9.26(c) depicts the synchronization error signal, $e_1^2(t) = (x - x')^2$, at the output of the response #1 (receiver). The coefficient modulation produces significant synchronization error during a "1" transmission (since $\beta(1) \neq \beta$ of response #1) and very little error during a "0" transmission (since $\beta(0) = \beta$ of the response #1). Also, Fig. 9.26(d) shows the synchronization error $e_2^2(t) = (x - x'')^2$, which has an opposite nature to the previous one (since $\beta(1) = \beta$ of the response #2) (see Fig. 9.25). Figures 9.26(e) and 9.26(f) depict the low-pass filtered signals of $e_1^2(t)$ and $e_2^2(t)$ respectively. Then by applying a threshold test to these low-pass filtered signals 9.26(e) and 9.26(f), the square wave modulation or information signal can be reliably recovered as shown in Fig. 9.26(g). The allowable data rate of $s(t)$ is, of course, dependent on the synchronization response time of the receiver system. Although we have used a low bit rate to demonstrate the technique numerically, the circuit time scale can be easily adjusted to allow much faster bit rates during experimental implementation.

We also note that the ability to communicate digital bit streams using this method does not depend on the periodic nature of the square-wave used to demonstrate the technique. The results apply to aperiodic or random bit streams as well. Using Chua's circuit and Pecora & Carroll method of cascading approach, the above procedure has also been shown to work well by Parlitz *et al.* (Ref. [216]).

Also, Halle *et al.* [220] have proposed an alternative idea of secure communication. Their idea is to multiply the information signal by a broad-spectrum noise-like chaotic signal. This kind of chaotic signal modulation offers several advantages over the parameter modulation or simple masking techniques. First, the whole range of the chaotic signal spectrum is used for hiding the information. Second, the sensitivity to parameter variation is increased, thus offering increased security. There is also an interesting approach based on

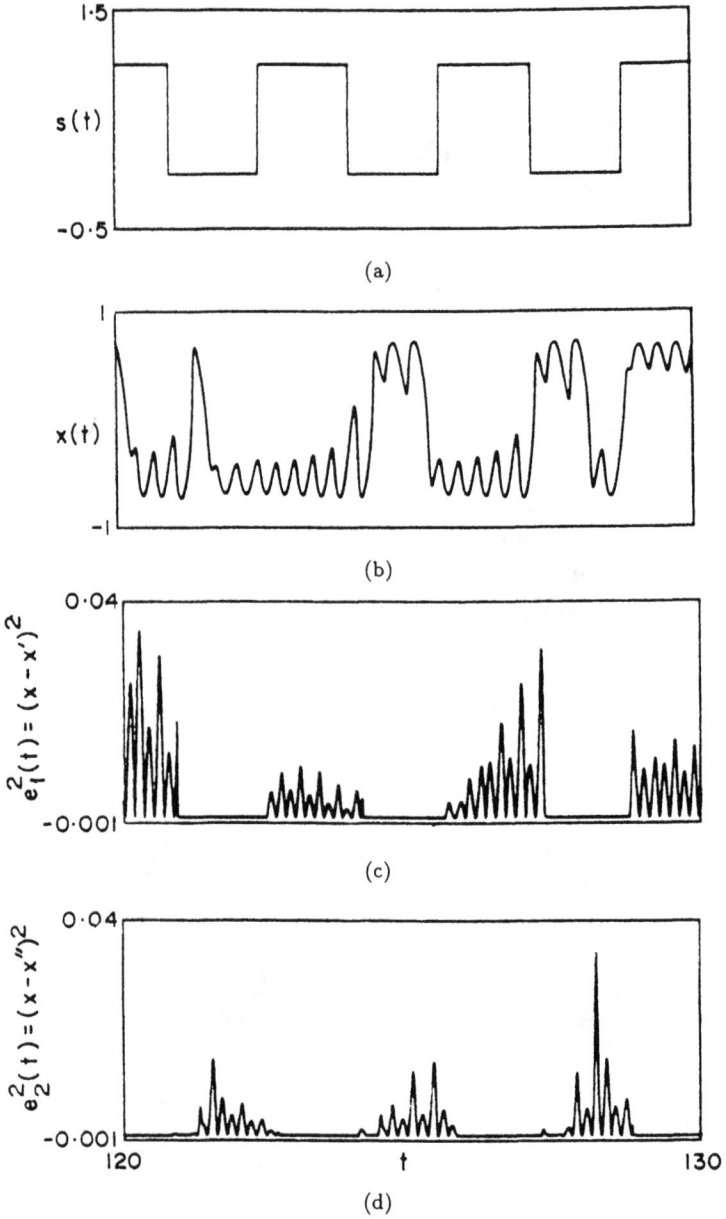

Fig. 9.26. Secure digital signal transmission: (a) $s(t)$-digital information signal; (b) actual transmitted signal; (c) error signal $e_1^2(t)$; (d) error signal $e_2^2(t)$; (e) low-pass filtered signal of (c); (f) low-pass filtered signal of (d); (g) recovered information signal $s^1(t)$.

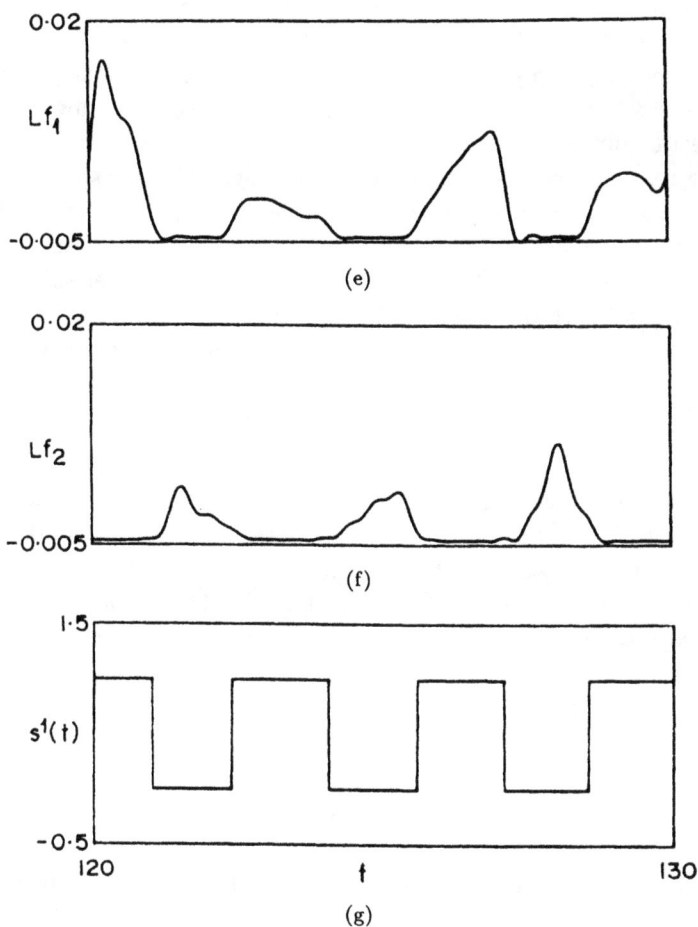

(e)

(f)

(g)

Fig. 9.26. (Continued)

the information theoretic formalism of chaos reported recently by Hayes *et al.* (Ref. [240]). Further, a number of other works on secure communications through chaos synchronization has appeared recently (Refs. [89, 221]) and practical technological realization appears imminent.

9.6. Recent Developments in Synchronization of Chaos

Finally, we wish to include a short discussion on some further developments in the understanding of synchronized chaotic systems.

(a) *Hyper-chaotic attractors of unidirectionally coupled systems*
(i) *Chua's Circuit*: Based on the approach of one-way coupling (as discussed in Sec. 9.4.2), recently Kapitaniak *et al.* (Ref. [228]) have studied the properties of hyperchaotic attractors (having more than one positive Lyapunov exponents) of unidirectionally coupled identical Chua's circuits. They have shown that for certain ranges of values of one-way coupling strength, the spectrum of the Lyapunov exponents is characterized by two maximal positive exponents and the "composite" coupled system will evolve on a higher-dimensional manifold on which the hyperchaotic attractor exists. However, for higher values of one-way coupling strength only one maximal Lyapunov exponent of the coupled "composite" system is positive, then the system evolves on the same manifold on which both systems evolve and synchronization between the two one-way coupled Chua's circuits occurs (Ref. [228]).

(ii) *MLC Circuit*: Now we discuss the observation of hyperchaotic attractors in the one-way coupled MLC circuits represented by Eqs. (9.22a–d). For numerical investigation we fixed the parameters of Eq. (9.22) at $\beta = 1, \nu = 0.015, \omega = 0.75, a = -1.02, b = -0.55$ and $F = 0.15$. In the case of $\varepsilon = 0$ (no coupling), as previously discussed, both the MLC circuits evolve along the double-band chaotic attractor. Let the spectrum of the Lyapunov exponents of the coupled system (9.22) be divided into two subsets $\lambda^{(1)}$ and $\lambda^{(2)}$ associated with the first (Eq. (9.22a–b)) and the second (Eq. (9.22c–d)) MLC circuits respectively.

For $F = 0.15$ the system admits double-band chaos and for $\varepsilon = 0$ the Lyapunov exponents of the drive system (first MLC circuit) Eqs. (9.22a–b) are $0.0869, -0.2317$ and 0. They are also the same for the response system (second MLC circuit) Eqs. (9.22c–d) for this specific choice of $\varepsilon = 0$. However, for $\varepsilon > 0$ the $\lambda^{(2)}$ Lyapunov exponents of the response system are equivalent to the conditional or sub-Lyapunov exponents. In Fig. 9.27 we present a plot of the maximal Lyapunov exponent of the response system (Eqs. (9.22 c–d) versus the coupling stiffness ε.

For smaller values of ε the chaotic trajectories of system (9.22) are characteried by two positive maximal Lyapunov exponents; one in the $\lambda^{(1)}$ -subset and the other in the $\lambda^{(2)}$-subset (see Fig. (9.27)) so that in this case the two MLC circuits cannot synchronize. In Fig. 9.28(a) we have shown a projection of the system trajectories on the $(x - x')$ plane for $F = 0.15$ and $\varepsilon = 0.015$, where no synchronization occurs between the two MLC circuits. In Fig. 9.28(b) the trajectory plot in the $(x - x')$ plane for $F = 0.15$ and $\varepsilon = 1$ is shown. In this

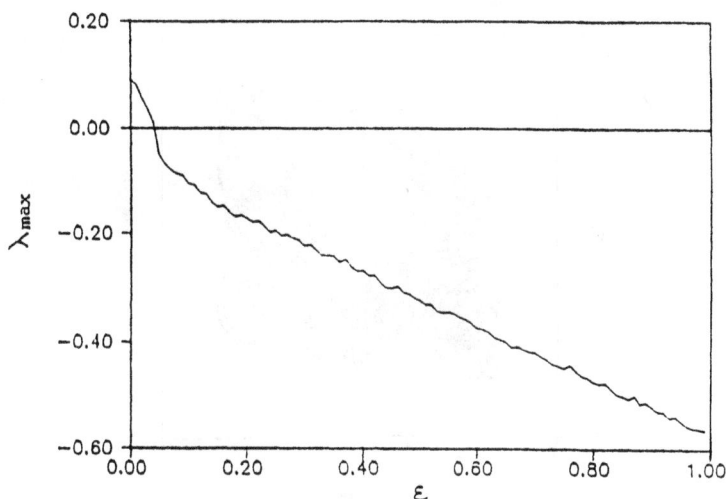

Fig. 9.27. Plot of maximal Lyapunov exponent λ_{max} versus ε for Eqs. (9.22c–d).

figure we observe a single-line characteristic in the synchronization regime. The simplicity of the $(x - x')$ projection of the attractors in these cases allows us to see the qualitative difference between chaotic and hyperchaotic attractors.

(b) *Controlling of chaos through synchronization*: Recently using the approach of chaos synchronization through one-way coupling (as discussed in Sec. 9.3) Kittel *et al.* (Ref. [209]) have demonstrated that synchronization of the current state of a chaotic system with its prerecorded history is possible. The one-way coupling perturbation (which acts as a small self-controlling feedback) transforms an unpredictable chaotic behaviour into a predictable chaotic or a periodic motion via stabilization of unstable, aperiodic or periodic orbits of the strange attractor. Furthermore, this method does not require any analytical knowledge of the system dynamics and can be simply implemented in an experiment by a purely analog technique. The above authors have used an electronic autonomous chaos oscillator shown in Fig. 9.29, as suggested by Shinriki *et al.* (Ref. [241]), as a model and demonstrated that both the unstable aperiodic orbits and unstable periodic orbits can be stabilized.

In the experiments of Ref. [209], initially the unstable aperiodic and periodic orbits of the circuit are stored (prerecorded) in a memory. Then by activating the one-way coupling of appropriate strength, the current chaotic state of the circuit behaviour has been shown to stabilize to the chosen prerecorded time traces in memory. Synchronization between the prerecorded history signal with the current state of the circuit is possible only in appropriate intervals

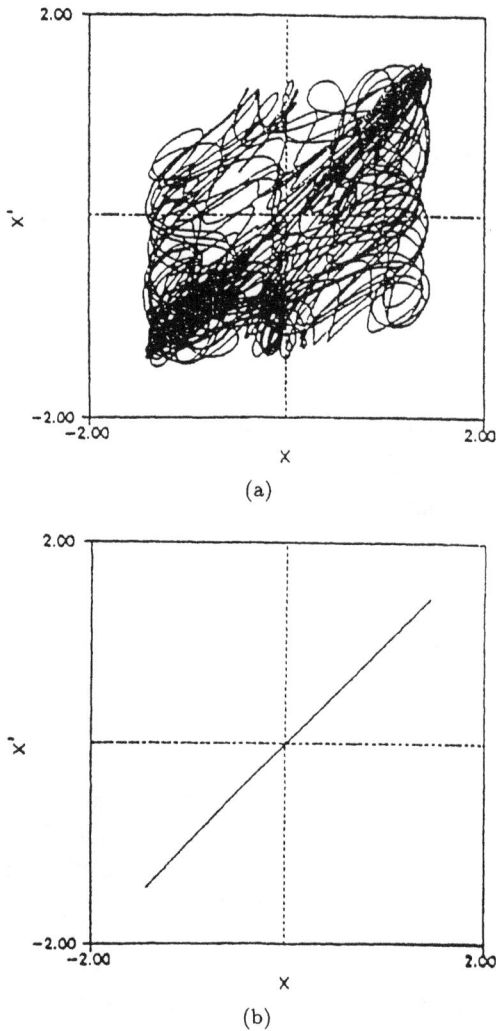

Fig. 9.28. Trajectory plot in $(x-x')$ plane of Eqs. (9.22): (a) Hyperchaotic behaviour for $f = 0.15$ and $\varepsilon = 0.015$. (b) Synchronized behaviour for $f = 0.15$ and $\varepsilon = 1$.

of the one-way coupling parameter where the maximal conditional Lyapunov exponent of the system is negative (Ref. [209]). The sample plot of stabilization of some of the unstable aperiodic orbits and periodic orbits are shown in Fig. 9.30. Stabilization of chaotic orbits for the MLC circuit, Eqs. (9.22), can be achieved through this method as described above, and the results are shown in Fig. 9.31.

NONLINEAR OSCILLATOR   CONTROL CIRCUIT

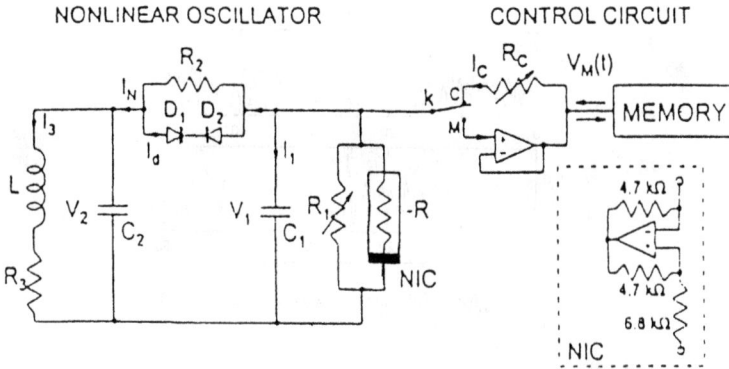

Fig. 9.29. The scheme of the nonlinear oscillator and the control circuit. At the lower right a scheme of the NIC is plotted. The variable resistor R_1 is a precise potentiometer fixed by hand. The op-amp is of the type TL071 biased with ± 15 V. D_1 and D_2 are 3.3 V Zener diodes BZX55C 3V3 (Ref. [209]).

(c) *Synchronizing chaotic systems using filtered signals*: Another interesting development in the field of synchronization of chaotic systems has recently been discussed by Carroll (Ref. [231]). It was reported that the concept of synchronization of cascaded chaotic systems may be extended to cases where the driving signal has been altered by a *filter* and reconstructed at the response system. Certain frequency components are subtracted from the driving signal at the transmitter, and added back at the receiver (response) in a process using a feedback loop at the receiver. The drive and response systems are not identical for this experiment, although they are effectively identical when they are synchronized. This type of synchronization is demonstrated in both numerical simulations and circuit experiments for a specific non-autonomous system. However, this approach is applicable equally well to autonomous systems also (Ref. [231]).

(d) *Synchronization of chaos in Hamiltonian systems*: Recently, Heagy and Carroll have shown, using the standard map (period-one return map for a periodically kicked pendulum) (Ref. [242]) as an example, that chaotic synchronization is indeed possible in this Hamiltonian system (Ref. [230]). They have discussed the analytic conditions for synchronization of two standard maps and numerical studies of the degree of chaotic synchronization as a function of the standard map parameter. Further, by employing suitable electronic circuit they have studied the chaos synchronization properties of a piecewise-linear standard map.

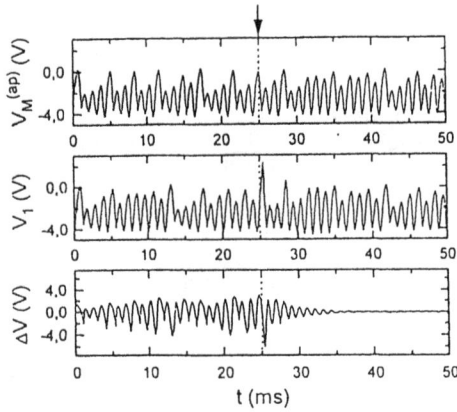

(a)

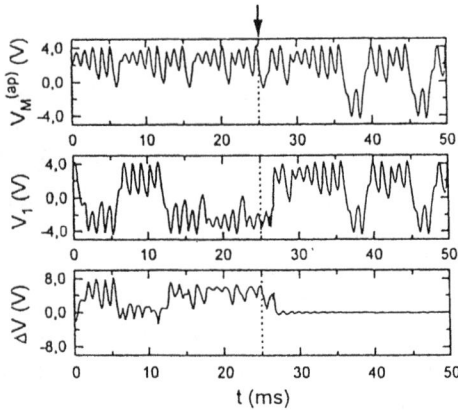

(b)

Fig. 9.30. Time traces of the recorded aperiodic output signal $V_M^{(ap)}(t)$, the dynamics of the output signal V_1, and the difference $\Delta V = V_1 - V_M^{(ap)}$ of Fig. 9.29 for mono-scroll regime (a) $[R_1 = 33.8\text{k}\Omega, R_c = 10\text{k}\Omega]$ and double-scroll chaos regime (b) $[R_1 = 39.8\text{k}\Omega, R_c = 10\text{k}\Omega]$. The arrows and the dashed lines mark the moment of switching onto the control (Ref. [209]).

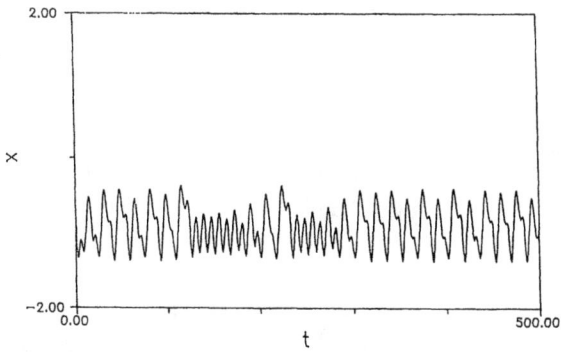

(a)(i)

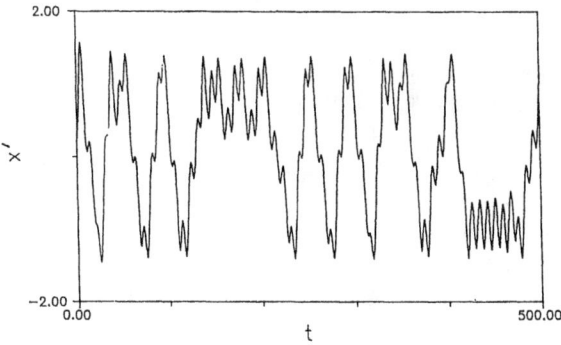

(a)(ii)

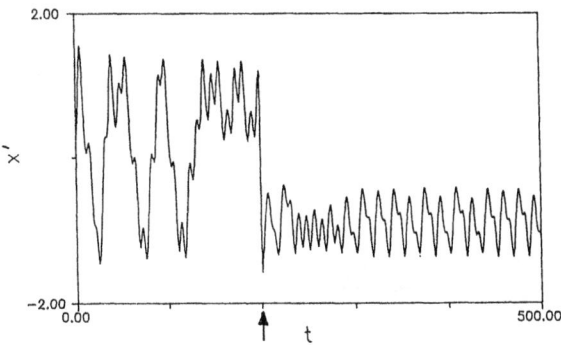

(a)(iii)

Fig. 9.31. (a) Trajectories of Eq. (9.22): (i) One-band chaotic signal x ($f = 0.1$). (ii) Double-band chaotic signal x' ($f = 0.15, \varepsilon = 0$ of Eq. (9.22c–d)); (iii) Stabilized one-band chaotic signal from Eqs. (9.22c–d) for $\varepsilon = 3.5$. The arrow indicates the moment of switching onto the control.

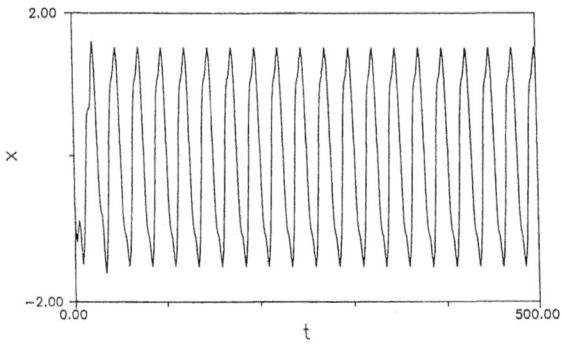

(b)(i)

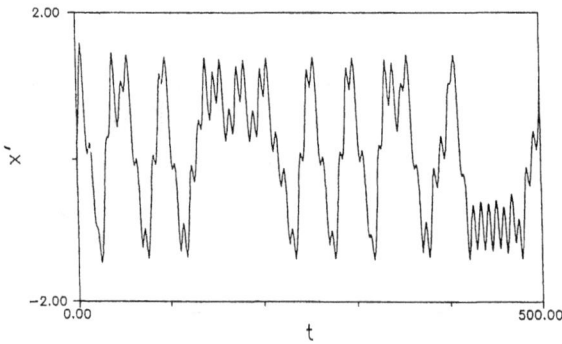

(b)(ii)

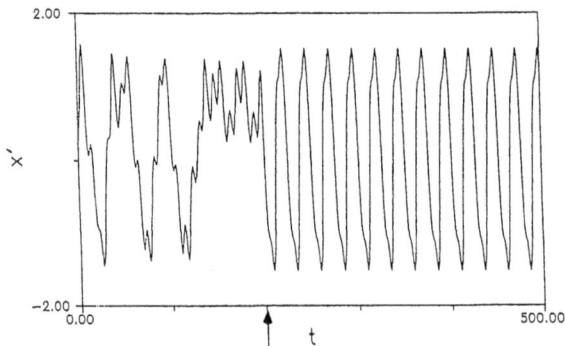

(b)(iii)

Fig. 9.31. (*Continued*) (b) Trajectories of Eq. (9.22): (i) Period-3 signal x of Eq. (9.22a–b) for $f = 0.2$; (ii) Double-band chaotic signal x' ($f = 0.15, \varepsilon = 0$ of Eq. (9.22c–d)); (iii) Stabilized period-3 signal from Eqs. (9.22c–d) for $\varepsilon = 3.5$. The arrow indicates the moment of switching onto the control.

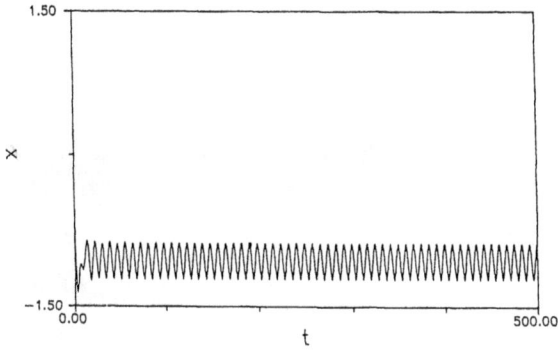

(c)(i)

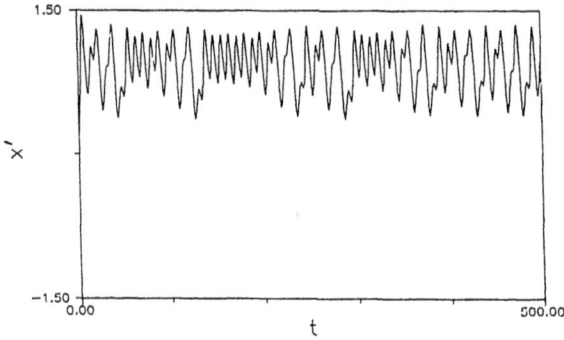

(c)(ii)

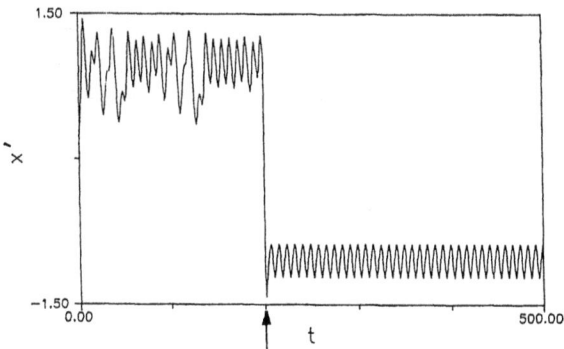

(c)(iii)

Fig. 9.31. (*Continued*) (c) Trajectories of the Eq. (9.22): (i) Period-1 signal x of Eq. (9.22a–b) for $f = 0.065$; (ii) One-band chaotic signal x' ($f = 0.1, \varepsilon = 0$ of Eq. (9.22c–d)); (iii) Stabilized period-1 signal from Eqs. (9.22c–d) for $\varepsilon = 3.5$. The arrow indicates the moment of switching onto the control.

APPENDIX A

PERTURBATION AND RELATED
APPROXIMATION METHODS

Considering the nonlinear oscillator system governed by the equation of motion

$$\ddot{x} + \omega_0^2 x = \varepsilon f(x, \dot{x}, t), \qquad (\varepsilon \ll 1), \ (\cdot = d/dt), \qquad \text{(A.1)}$$

where the function g contains the nonlinearity, several perturbative methods can be developed (Refs. [36, 243]) to obtain approximate periodic solutions valid for small strengths of the parameter $\varepsilon (\ll 1)$. These include

(i) Lindstedt–Poincaré perturbation method
(ii) Multiple-scale perturbation method
(iii) Harmonic balance method
(iv) Averaging methods

and so on. Each one of these procedures has its own advantages and disadvantages as well as range of validity and is suitable to deal with specific situations. Salient features of these methods can be summarized as follows.

(i) *Lindstedt–Poincaré perturbation method*: This well-known method essentially develops a series expansion to the solution $x(t)$ of Eq. (A.1) in the small parameter ε, but avoids the notorious secular terms present in the standard series expansion through a renormalization procedure. One looks for a solution of the form

$$x(t) = x(\tau, \varepsilon) = \sum_{n=0} \varepsilon^n x_n(\tau), \qquad \tau = \omega t, \qquad \text{(A.2)}$$

where the new independent variable τ is an unspecified function of ε. The system of new governing recursive set of differential equations for the $x_n(\tau)$'s, which are linear in nature, will contain ω in the coefficient of the second derivative, and this permits the frequency and amplitude to interact. Then one can choose the function ω in such a way as to eliminate secular terms.

(ii) *Multiple-scale perturbation method*: The uniformly valid expansion given in Eq. (5.8) may be considered as a function of two independent variables $T_0 = \varepsilon^0 t = t$ and $T_1 = \varepsilon^1 t \equiv \varepsilon t$, rather than a function of t alone. Extending this idea, one can look for solutions where the response is a function of multiple independent variables or scales, $T_n = \varepsilon^n t, n = 0, 1, 2 \ldots$. Rewriting the defining equation in terms of these multiple variables, one can set up a recursive set of linear differential equations, which can be solved systematically. Though the method is a little more involved, it has definite advantages in that different orders of contributions can be treated through different scales.

(iii) *Harmonic balance method*: Here the idea is to express the periodic solution of an equation of the form (A.1) as a finite Fourier series,

$$x(t) = \sum_{m=0}^{M} A_m \cos m(\omega t + \delta) . \qquad (A.3)$$

Substituting the series (A.3) in the given ode and equating each of the lowest $(M + 1)$ harmonics to zero, one can obtain a system of $(M + 1)$ algebraic equations relating ω and A_m. Usually these equations are solved for $A_0, A_2, A_3 \ldots A_m$ and ω in terms of A_1. The accuracy of the resulting solution will then depend on the value of A_1 and the number of harmonics in the assumed solution (A.3).

(iv) *Averaging methods*: Another useful technique to deal with nonlinear oscillator equations of the type given by Eq. (A.1) is the method of averaging. There are various versions of it, including the Krylov–Bogoliubov method, the Krylov–Bogoliubov–Mitropolsky technique, the generalized method of averaging, averaging using canonical variables, averaging using Lie series, and transforms and averaging using Lagrangians (Ref. [36]).

Most of the averaging procedures start with the method of variation of parameters to transform the dependent variable from x to $a(t)$ and $\beta(t)$, where

$$x = a(t) \cos(\omega_0 t + \beta(t)) , \qquad (A.4a)$$

$$\dot{x} = -a(t)\omega_0 \sin(\omega_0 t + \beta(t)) . \qquad (A.4b)$$

Using Eqs. (A.4) in (A.1), one obtains a set of differential equations for the slowly changing amplitude $a(t)$ and phase $\beta(t)$ in the form

$$\dot{a} = -\frac{\varepsilon}{\omega_0} f(a\cos(\omega_0 t + \beta), \ -a\omega_0 \sin(\omega_0 t + \beta), t) \sin(\omega_0 t + \beta), \quad \text{(A.5a)}$$

$$\dot{\beta} = -\frac{\varepsilon}{\omega_0 a} f(a\cos(\omega_0 t + \beta), \ -a\omega_0 \sin(\omega_0 t + \beta), t) \cos(\omega_0 t + \beta). \quad \text{(A.5b)}$$

Now we can assume that $a(t)$ and $\beta(t)$ are 'slowly' varying so that they remain nearly constant during a time interval of duration $T_0 = 2\pi/\omega_0$. As a result one obtains the autonomous system of 'averaged' equations

$$\dot{a} = -\frac{\varepsilon}{\omega_0 2\pi} \int_0^{2\pi} f\left(a\cos(\Theta + \beta), \ -a\omega_0 \sin(\Theta + \beta), \frac{\Theta}{\omega_0}\right) \sin(\Theta + \beta) d\Theta,$$

$$\dot{\beta} = -\frac{\varepsilon}{\omega_0 a 2\pi} \int_0^{2\pi} f\left(a\cos(\Theta + \beta), \ -a\omega_0 \sin(\Theta + \beta), \frac{\Theta}{\omega_0}\right) \cos(\Theta + \beta) d\Theta.$$

$$\text{(A.6a)}$$

Equations (A.6) are equivalent to those obtained from the asymptotic method of Bogoliubov and Mitropolsky.

APPENDIX B

VAN DER POL OSCILLATOR AND CHAOS

As a mathematical model to describe the oscillations in a vacuum tube circuit, van der Pol introduced (Ref. [244]) the second-order nonlinear differential equation with linear restoring force and nonlinear damping

$$\ddot{x} - \varepsilon(1 - x^2)\dot{x} + x = 0. \qquad (\cdot = d/dt) \qquad (\text{B}.1)$$

It also describes self-excited oscillations in several problems in electronics, physics, biology, neurology and many other disciplines. Equation (B.1) essentially admits limit-cycle oscillations for small ε in the region $(x, \dot{x} = -2, 2)$ and relaxation oscillations for large ε (Refs. [245, 246]). Typical forms of the limit cycle and relaxation oscillations are given in Fig. B.1.

When driven by external periodic forcing, Eq. (B.1) exhibits interesting bifurcation structures. Considering the driven van der Pol oscillator

$$\ddot{x} - \varepsilon(1 - x^2)\dot{x} + x = f\cos\omega t, \qquad (\text{B}.2)$$

Parlitz and Lauterborn (Ref. [118]) have made a detailed investigation of its dynamics and brought out the existence of various mode-locking responses and period-doubling cascades. Other relevant references include (Refs. [247–251]). The results can be summarized (with $\varepsilon = 5.0$) as in Table B.1.

Table B.1.

Value of f	Region of ω	Type of orbits	Figures
1.0	(0.0, 1.5)	mode-lockings and quasiperiodicity	B.2 and B.3
2.5	(0.0, 6.0)	mode-lockings, large period oscillations and quasiperiodicity	B.4
5.0	(2.424, 2.502)	periodic windows, period-doubling bifurcations to chaos	B.5, B.6 and B.7

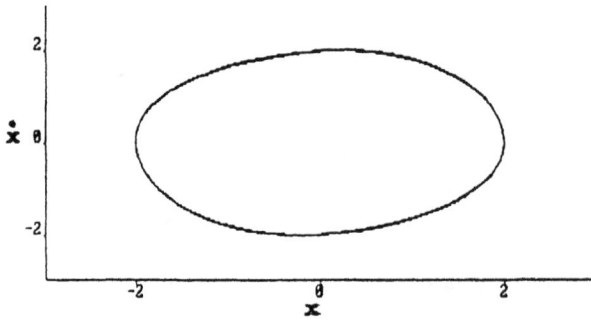

(a)

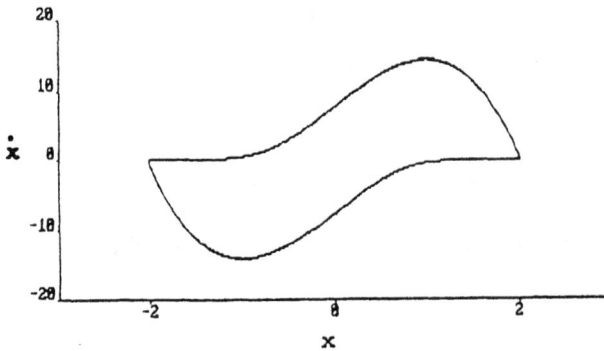

(b)

Fig. B.1. (a) Stable limit cycle oscillation in $(x - \dot{x})$ plane for $\varepsilon = 0.1$ of the van der Pol oscillator Eq. (B.1). (b) Stable relaxation oscillation in $(x - \dot{x})$ plane for $\varepsilon = 10$ of Eq. (B.1).

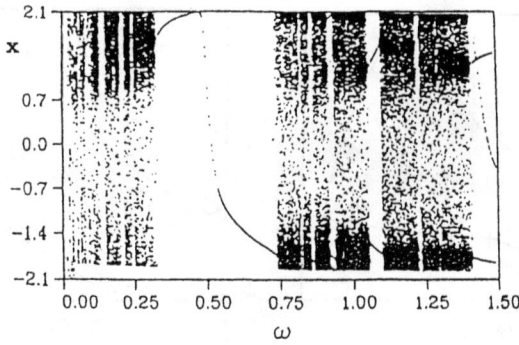

Fig. B.2. Bifurcation diagram for $f = 1.0$ and $\varepsilon = 5$ of Eq. (B.2) (Ref. [118]), showing the presence of mode-locking and quasiperiodic responses.

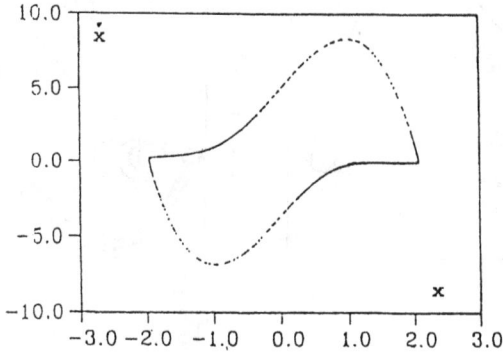

Fig. B.3. Poincaré map of an attractor lying on an invariant torus in phase space for $\varepsilon = 5.0, f = 1$, and $\omega = 1$ (Ref. [118]).

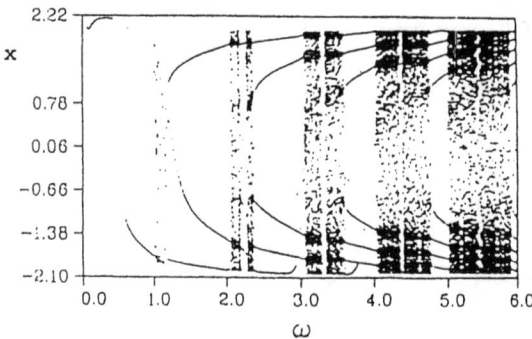

Fig. B.4. Bifurcation diagram for $f = 2.5$ (c.f. Fig.B.2). Between the extended locking regions of period $1, 3, 5, 7, \ldots$ parameter intervals with large-period oscillations occur, each of them undergoing the same bifurcation scenario when the excitation amplitude f is increased (Ref. [118]).

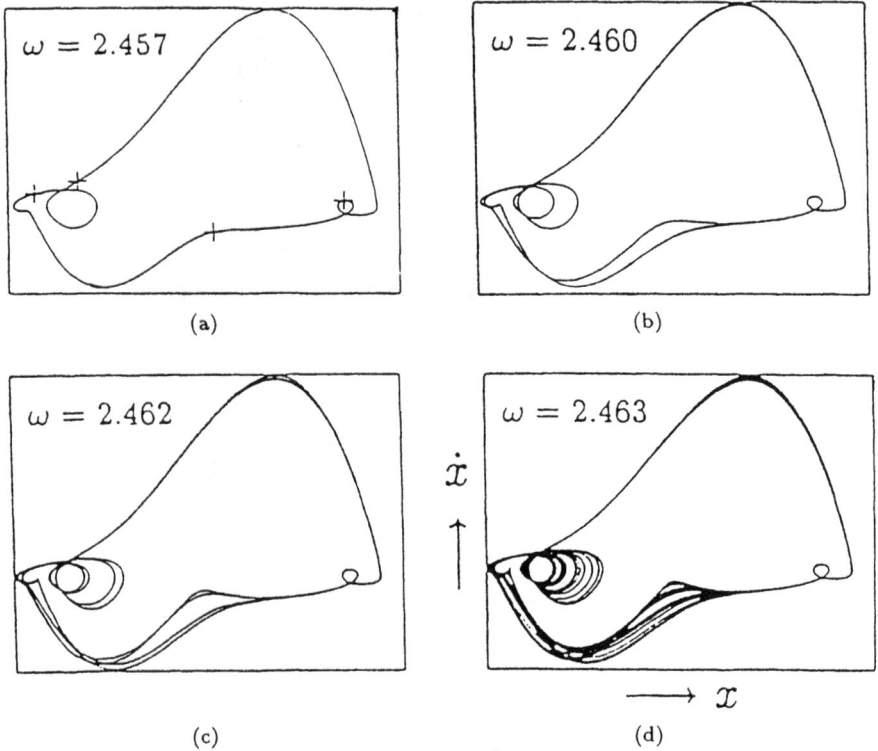

Fig. B.5. Period-doubling sequence of a (basic) period-4 attractor. Here, $\varepsilon = f = 5$ (Ref. [118]). (a) Period-4 attractor; $\omega = 2.457$. (b) Period-4 $\times$ 2^1 attractor; $\omega = 2.460$. (c) Period-4 $\times$ 2^2 attractor; $\omega = 2.462$. (d) Chaotic attractor; $\omega = 2.463$.

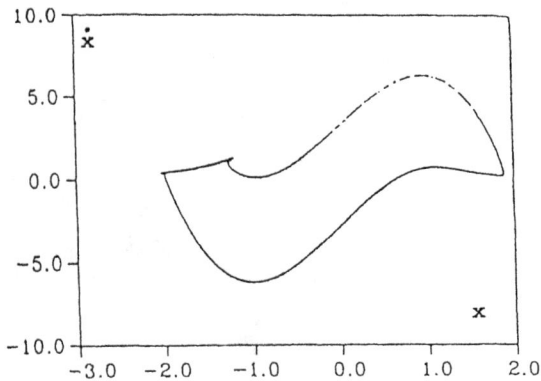

Fig. B.6. Poincaré map of the strange attractor for $\varepsilon = 5, f = 5$, and $\omega = 2.466$ (Ref. [118]).

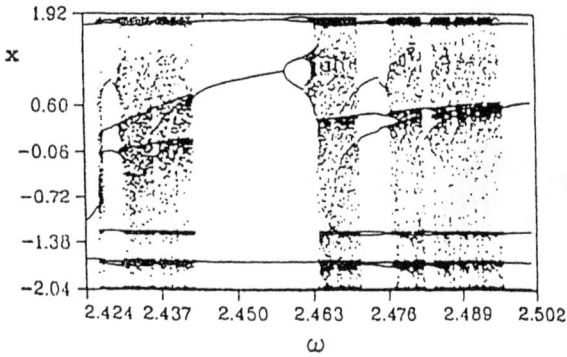

Fig. B.7. Bifurcation diagram of the period-3 to period-5 interval for $f = 5$ showing complete period-doubling cascades into chaos. Owing to the symmetry of the system for each period-doubling cascade a counterpart exists(which is reached from other initial conditions) that is not included here (Ref. [118]).

APPENDIX C

SOME FURTHER UBIQUITOUS
CHAOTIC OSCILLATORS

In this appendix, we wish to give a brief sketch of the basic features associated with some of the other important nonlinear oscillators, of both autonomous and nonautonomous types, not considered in the main body of the book. Among these, the Lorenz system of equations and damped and driven pendulum have played historically a paradigmic role in the development of the field of chaotic dynamics (Refs. [4, 8, 40, 252–255]). On the other hand, the Rössler system (Ref. [256]) has been one of the pioneering models studied to understand the nature of the strange attractor, while the Brusselator system (Ref. [257]) played a crucial role for understanding autocatalytic chemical reactions and nonlinear dynamics of open systems. The damped and driven Morse oscillator (Ref. [258]) has important applications in atomic physics and nonlinear optics.

C.1. Lorenz Equations

Historically, the Lorenz system of equations is perhaps the first of the nonlinear dynamical systems found to exhibit sensitive dependence on initial conditions and chaos. Lorenz (Ref. [4]) deduced these equations as a simple model for thermally induced fluid convection in the atmosphere. Fluid heated from below becomes lighter and rises, whereas heavier fluid falls under gravity. Such motions often produce convection rolls similar to the motion of fluid in a circular torus. In such a model three state variables (x, y, z) are used. The variable x is proportional to the amplitude of the fluid velocity circulating in the fluid ring, while y and z measure the distribution of temperature around the ring.

The nondimensional forms of Lorenz's equations are

$$
\begin{aligned}
\dot{x} &= \sigma(y - x)\,, \\
\dot{y} &= rx - y - xz\,, \\
\dot{z} &= xy - bz\,.
\end{aligned}
\qquad\qquad (\text{C.1})
$$

Here, the parameters σ and r are related to the Prandtl number and Rayleigh number respectively and the third parameter b is a geometric factor. For $\sigma = 10$ and $b = 8/3$, there are three equilibrium states for $r > 1$, for which the origin is an unstable saddle. When $r > 25$, the other two equilibria become unstable spirals and a complex chaotic trajectory moves between all the three equilibria as shown in Fig. C.1 ($r = 28$) (Refs. [4, 40]). The Lorenz equations are also similar to those that model the chaotic behaviour of laser devices (Ref. [252]).

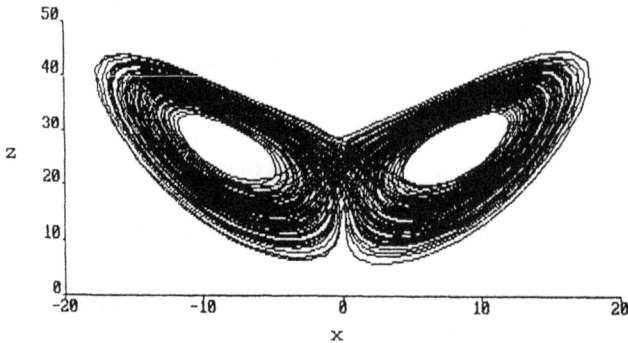

Fig. C.1. Phase portrait in $(z - x)$ plane of Lorenz equations (C.1) for $\sigma = 10.0, b = 8/3$, and $r = 28$.

C.2. Rössler Equations

A simple model motivated by the dynamics of chemical reactions in a stirred tank is the following set of equations proposed by Rössler (Refs. [256, 259]):

$$
\begin{aligned}
\dot{x} &= -(y + z)\,, \\
\dot{y} &= x + ay\,, \\
\dot{z} &= b + z(x - c)\,.
\end{aligned}
\qquad\qquad (\text{C.2})
$$

If $x < c$, $z(t)$ settles down near to the value $b/(c - x)$ (and rapidly, if $(c - x)$ is large), whereas if $x > c$, $z(t)$ increases exponentially. Thus the only nonlinearity in this dynamics involves this lifting and reinjection of the

$z(t)$ dynamics relative to the 'linear' (x, y) motion. From another point of view, the motion in the (x, y) plane provides a "stretching" operation, whereas the $z(t)$ motion produces a "folding" of the dynamics back to the region near the centre of the spiral motion and a z-compression (squeezing) when $x < c$ (Ref. [259]). The system often studied is the case $a = b = 1/5$. Period-1, 2 and 4 motions may be found to occur for $c = 2.6, 3.5$ and 4.1 respectively. Chaotic motions may be found for $c > 4.23$ (Ref. [259]). Figure C.2 shows one such Rössler chaotic attractor for $c = 5.7$.

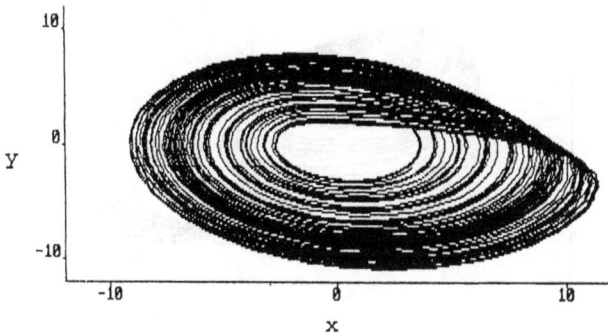

Fig. C.2. Phase portrait in $(y - x)$ plane of Rössler equations (C.2) for $a = b = 1/5$ and $c = 5.7$.

C.3. Brusselator Model

The reaction scheme of the original Brusselator (Ref. [257]) is given by

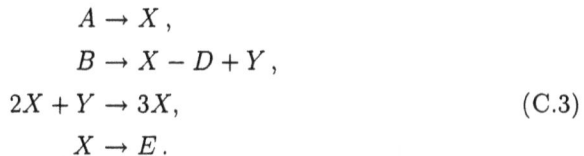

$$A \rightarrow X \, ,$$
$$B \rightarrow X - D + Y \, ,$$
$$2X + Y \rightarrow 3X, \tag{C.3}$$
$$X \rightarrow E \, .$$

This reaction system is open with respect to the parameters A, B, D, and E, which are held constant. System (C.3) describes a hypothetical three-molecular chemical reaction with autocatalytic step under far-from-equilibrium conditions. It displays a limit-cycle oscillation if $B \geq A^2 + 1$. Kai and Tomita (Ref. [260]) have applied a sinusoidal perturbation to the Brusselator which yields the following set of differential equations for the kinetics:

$$dX/dt = A + X^2Y - BX - X + f \cos \omega t \, ,$$
$$dY/dt = BX - X^2Y \, , \tag{C.4}$$

where f is the amplitude and ω is the angular frequency of the perturbation and all rate constants are set equal to unity. The response patterns in the $[f,\omega]$ phase plane have been thoroughly investigated by Kai and Tomita (Ref. [260]) and by Hao and Zhang (Ref. [261]). Its behaviour includes several bifurcation scenarios discussed earlier in the present book such as entrainment, quasiperiodicity, period-doubling, and chaos (see Figs. C.3 and C.4). Figure C.3 depicts a typical strange attractor exhibited by this system for $A = 0.4, B = 1.2, f = 0.08$ and $\omega = 0.86$ (Refs. [260, 261]).

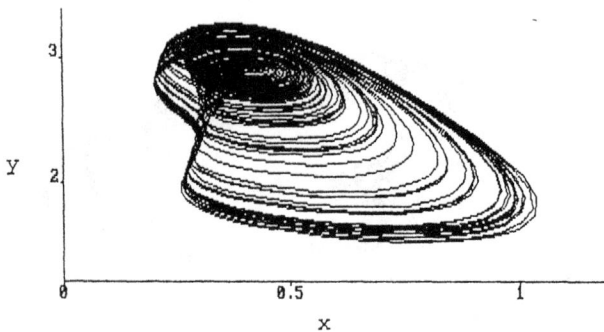

Fig. C.3. Phase portrait in $(y - x)$ plane of Brusselator model (C.4) for $A = 0.4, B = 1.2, f = 0.08$, and $\omega = 0.86$.

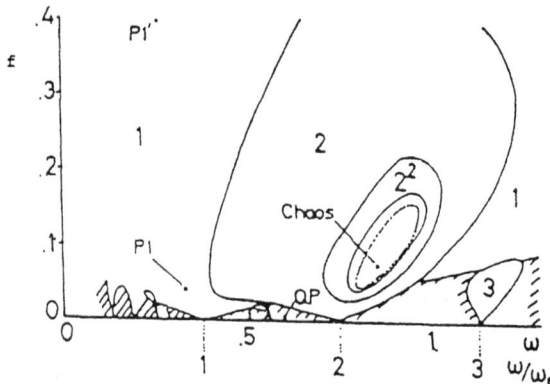

Fig. C.4. Phase diagram of the Brusselator according to Tomita and Kai (Ref. [260]). ω_0 is the frequency of the free running oscillator, ω is the perturbation frequency. The numbers in the diagram denote the periodicity of the response; quasiperiodic motion occurs in the dashed regions. P1: periodic motion close to a bifurcation to quasiperiodic motion. $f = 0.040; \omega = 0.29$ rad/s. P1': periodic motion far away from any bifurcation. $f = 0.4; \omega = 0.25$ rad/s. QP: quasiperiodic motion. $f = 0.0072; \omega = 0.6$ rad/s. Chaos: chaotic motion. $f = 0.08; \omega = 0.852$ rad/s.

C.4. Damped Driven Pendulum

The simple pendulum that is damped as well as driven by an oscillatory torque has been widely investigated because the differential equation also models the time dependence of the quantum phase difference of a Josephson junction and occurs in a wide variety of physical phenomena (Ref. [8]). Also, the

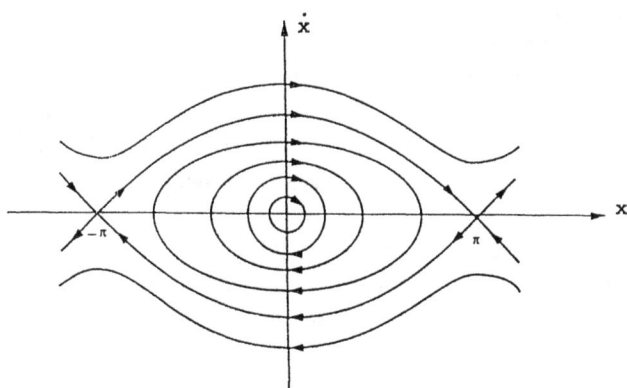

(a)

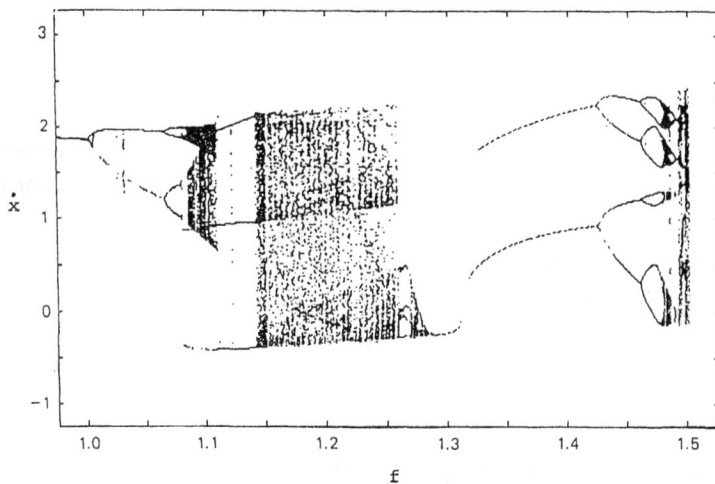

(b)

Fig. C.5. (a) Phase-space trajectories for a simple pendulum (C.5) for $\alpha = f = 0$. (b) Bifurcation diagram in $(\dot{x} - f)$ plane of Eq. (C.5) for $\alpha = 0.5, \Omega = 1$, and $\omega = 2/3$ (Ref. [8]).

equation of this model has been simulated with an analog simulation circuit to explore extensively the dynamical behaviour (Refs. [253–255]). The differential equation for the damped, driven pendulum is given by

$$\ddot{x} + \Omega^2 \sin x = -\alpha\dot{x} + f\cos\omega t\,, \tag{C.5}$$

where Ω represents the natural frequency of the pendulum for small oscillations, α, f and ω measure the strength of damping, the forcing amplitude and the forcing frequency respectively. With α and f both zero the phase curves for the simple pendulum are familiar and given in Fig. C.5(a). If $\alpha > 0$ and $f = 0$, due to damping almost all orbits approach the origin. If f is included and varied then the system admits familiar period-doubling bifurcation sequences to chaos. A typical bifurcation diagram in the $(\dot{x}$–$f)$ plane for $\alpha = 0.5, \omega = 2/3, \Omega = 1.0$, and f in the range $(1.0, 1.5)$ is shown in Fig. C.5(b) (Ref. [8]). Also, Pederson and Davidson (Ref. [255]) have classified the dynamical behaviour (both chaotic and mode-locked states) for a region of parameter $(f$–$\omega)$ space with specific damping value.

C.5. Morse Oscillator

The equation of motion of the damped and driven Morse oscillator (Refs. [258, 263–265]) is represented in terms of dimensionless variables as

$$\ddot{x} + \alpha\dot{x} + \beta e^{-x}(1 - e^{-x}) = f\cos\omega t\,. \tag{C.6}$$

For over half a century the Morse potential (Ref. [266]) has provided a useful model for the interatomic potential and for fitting the vibrational spectra of diatomic molecules (Ref. [267]). The damped and driven Morse oscillator (C.6) has been widely used:

(i) as models for infrared multiphoton excitation and dissociation of molecules,
(ii) for the problems of laser isotope separation,
(iii) for explanation of the anomalous gains observed in the stimulated Raman emissions, and
(iv) for dissociation of van der Walls complexes.

Several authors have studied (Refs. [258, 263, 265, 268, 269]) the classical dynamics of the sinusoidally driven Morse oscillator with damping (Refs. [258, 263]) or without damping (Refs. [264, 268]), and found that the system exhibits chaotic phenomenon through Feigenbaum's period-doubling route or Chirikov's resonance overlap criterion, respectively. Typical period-doubling route to chaos of this oscillator is shown in Fig. C.6 in the form of bifurcation diagram in the $(\dot{x}-\omega)$ plane for $\alpha = 0.8, \beta = 8.0, f = 2.5$, and ω in the range $(0, 3.0)$ (Ref. [258]).

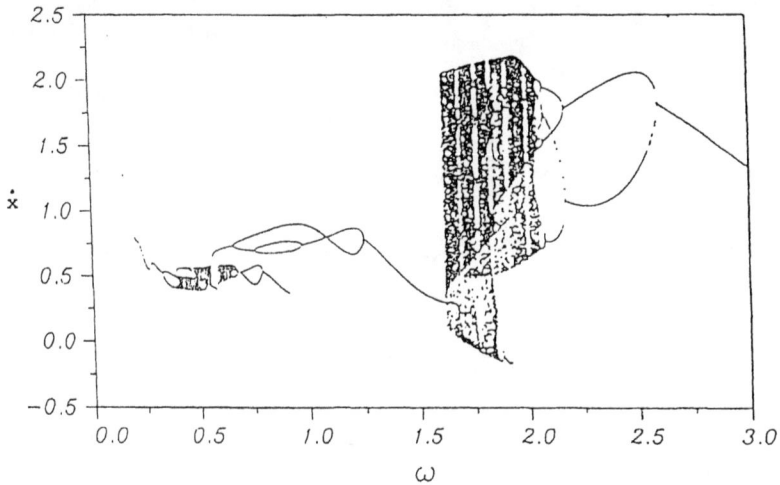

Fig. C.6. Bifurcation diagram in the $(\dot{x} - \omega)$ plane of Eq. (C.6) for $\alpha = 0.8, \beta = 8$, and $f = 2.5$ (Ref. [258]).

GLOSSARY

Antimonotonicity: Creation or destruction of periodic orbits near a homo-clinic tangency value when the control parameter is monotonically varied.

Attractor: The bounded region or set of points of phase space (with zero volume) towards which orbits/trajectories originating from nearby points (basins) are attracted during long-time evolution in the case of dissipative systems. Point attractors (such as stable focus, centre, stable node), limit cycles, torus and chaotic (strange) attractors all are examples.

(Autonomous) Duffing–van der Pol (ADVP) oscillator: An oscillator represented by a system of three coupled first-order odes:

$$\dot{x} = -\nu(x^3 - \alpha x - y), \quad \dot{y} = x - y - z, \quad \dot{z} = \beta y \,.$$

Bifurcation: Sudden/abrupt qualitative change in the system dynamics at critical control parameter values when the parameter is varied.

Bonhoeffer–van der Pol (BVP) oscillator: It is the set of two first-order coupled nonlinear odes $\dot{V} = V - (V^3/3) - R + I(t)$, $\dot{R} = c(V + a - bR)$ for the voltage pulse $V(t)$ and recovery variable $R(t)$ of an axon of the nerve fibre.

Brusselator: It describes a hypothetical three-molecular chemical reaction with autocatalytic step under far-from-equilibrium conditions. It has the form $\dot{X} = A + X^2Y - BX - X$, $\dot{Y} = BX - X^2Y$.

Chaos: Complicated and aperiodic motions which are highly sensitively dependent upon initial conditions in deterministic nonlinear systems. Correspondingly, in the phase space two nearby chaotic trajectories diverge exponentially, though still confined to a bounded domain (in the case of dissipative systems).

Chua's circuit: A simple, third-order, autonomous electronic circuit consisting of two linear capacitors, a linear inductor, a linear resistor, and only one nonlinear element, namely, Chua's diode.

Chua's diode: A five-segment piecewise-linear (nonlinear) resistor, whose effective nontrivial operating region is the three middle segments (negative-resistance region) with two break points and two slopes. The current–voltage (V_R–i_R) characteristic for Chua's diode is represented as

$$i_R = f(V_R) = G_b V_R + 0.5(G_a - G_b)[|V_R + B_p| - |V_R - B_p|].$$

Controlling chaos: Conversion of chaotic motion to desired periodic/regular motion of a nonlinear system through minimal perturbations/preassigned changes.

Damped and driven pendulum: A very important model oscillator system described by the second-order nonlinear ode

$$\ddot{x} + \Omega^2 \sin x = -\alpha \dot{x} + f \cos \omega t.$$

Devil's staircase: The plot of the winding number versus frequency indicating the frequency-locked intervals. Typically it shows flat steps over certain frequency values.

Duffing oscillator: The second-order damped and driven nonlinear oscillator with cubic restoring force satisfying the ode $\ddot{x} + \alpha \dot{x} + \omega_0^2 x + \beta x^3 = f \cos \omega t$. It may be of single-well, double-well or double-hump type for $(\omega_0^2 > 0, \beta > 0), (\omega_0^2 < 0, \beta > 0)$ or $(\omega_0^2 > 0, \beta < 0)$ respectively.

Duffing–van der Pol (DVP) oscillator: A combination of Duffing and van der Pol equations: $\ddot{x} + p(x^2 - 1)\dot{x} + \omega_0^2 x + \beta x^3 = f \cos \omega t$.

Feigenbaum constant: A universal property of certain chaotic dynamical systems related to the period-doubling bifurcation sequence. The ratio of successive differences between period-doubling bifurcation control parameter approaches the universal number 4.6692... for dissipative systems.

Fractal dimension: A quantitative measure of a set of points in an n-dimensional space that characterizes its space-filling properties.

Hopf bifurcation: The onset of a limit cycle from a fixed or an equilibrium point as the control parameter of the system is varied.

Hyperchaos: Chaotic motion of a system for which more than one positive Lyapunov exponents exist.

Hysteresis: Residual motion when a control parameter completes a circuit in the parameter space.

Intermittency: A type of chaotic motion in which long-time intervals of regular or periodic motion are interrupted by chaotic bursts. The time interval between bursts is not fixed but is unpredictable.

Linear oscillators: Dynamical systems exhibiting amplitude-independent periodic oscillations and modelled by linear odes.

Lorenz system: The paradigmic nonlinear chaotic system originally introduced by E. Lorenz in 1963: $\dot{x} = \sigma(y - x)$, $\dot{y} = rx - y - xz$, $\dot{z} = xy - bz$.

Lyapunov criterion for asymptotic stability: It is one of the practical ways to establish the asymptotic stability of a response system or sub-system by finding the appropriate Lyapunov function.

Lyapunov exponents: Numbers providing a quantitative average measure of the divergence of nearby trajectories in phase space. All negative exponents represent regular and periodic orbits, while at least one positive exponent signals the presence of chaotic motion.

Melnikov criterion: A lower threshold criterion for the onset of chaotic motion for periodically perturbed systems. It gives a measure for the typical ways in which a fixed point of an unperturbed system breaks and complicated motion sets in under weak perturbation.

Morse oscillator: The Morse potential has provided a useful model for interatomic potentials and for fitting the vibrational spectra of diatomic molecules. The damped and driven Morse oscillator is $\ddot{x} + \alpha\dot{x} + \beta e^{-x}(1 - e^{-x}) = f\cos\omega t$.

Murali–Lakshmanan–Chua (MLC) circuit: The simplest second-order non-autonomous, nonlinear circuit consisting of a linear resistor, a linear inductor, a linear capacitor, a sinusoidal driving force and only one nonlinear element, namely, Chua's diode.

Nonlinear oscillators: Oscillatory systems modelled by nonlinear differential equations.

Painlevé property: If a nonlinear differential equation is free from movable (integration constant/initial condition dependent) critical singular points (branch points and essential singularities) it is said to possess the Painlevé property. In this case the system is expected to be integrable.

Period-doubling: Denotes the bifurcation sequence of periodic motions for a nonlinear dynamical system in which the period doubles at each bifurcation as the control parameter is varied. Beyond a critical accumulation parameter value, chaotic motions occur.

Phase diagram: A plot of the different types of dynamical behaviour as a function of two or more control parameters.

Phase- or mode-locking: As the frequency of the forcing of a driven nonlinear oscillator approaches the natural frequency, beating disappears and the latter gets locked or entrained by the driving frequency.

Phase space/Phase portrait: The abstract space of the dynamical variables is the phase space in which the state point moves along the phase trajectory constituting the phase portrait.

Poincaré map: Any suitable hyperplane of the phase space is the Poincaré surface of section. The relation between the successive intersections of the phase trajectory with this section in a single direction constitutes the Poincaré map. For periodically driven systems, it is simply the stroboscopic map taken at every period of the external force.

Power spectrum: The distribution of power in a signal $x(t)$ is most commonly quantified by means of the power density spectrum or simply the power spectrum. It is the magnitude-square of the Fourier transform of the signal $x(t)$. It can detect the presence of chaos when the spectrum is broad-banded.

Quasiperiodicity: In phase space, this corresponds to a torus. Quasiperiodic behaviour occurs in a dynamical system when two incommensurate frequencies are present.

Rössler system: A simple model equation related to chemical reactions in a stirred tank: $\dot{x} = -(y + z)$, $\dot{y} = x + ay$, $\dot{z} = b + z(x - c)$.

Routes to chaos: The specific ways of instability by which a dissipative nonlinear system dynamics changes from regular to chaotic behaviour. The routes include period-doubling, intermittency and quasiperiodic.

Secure communications (through chaos): Method of transmitting information signals in a secure way through synchronization using chaotic signals as masking signals.

Synchronization (of chaotic systems): The possibility of driving two identical chaotic systems to have the same in phase and amplitude through appropriate coupling.

REFERENCES

[1] H. Goldstein, *Classical Mechanics* (Narosa Publishing House, New Delhi, 1990).

[2] J. Gleick, *Chaos: Making A New Science* (Viking, New York, 1987).

[3] J. P. Crutchfield, J. D. Farmer, N. H. Packard and R. S. Shaw, *Sci. Amer.* **255**, (1986) 38.

[4] E. N. Lorenz, *J. Atmos. Science* **20**, (1963) 130.

[5] J. M. T. Thompson and H. B. Stewart, *Nonlinear Dynamics and Chaos* (John-Wiley, Singapore, 1988).

[6] I. Prigogine and I. Stengers, *Order Out of Chaos* (Flamingo, London, 1985).

[7] F. C. Moon, *Chaotic and Fractal Dynamics* (John-Wiley, New York, 1992).

[8] G. L. Baker and J. P. Gollub, *Chaotic Dynamics: An Introduction* (Cambridge Univ. Press, Cambridge, 1990).

[9] H. G. Schuster, *Deterministic Chaos: An Introduction* (Physik-Verlag, Weinheim, 1988).

[10] J. Guckenhimer and P. Holmes, *Nonlinear Oscillations, Dynamical Systems and Bifurcations of Vector Fields*, (Springer-Verlag, New York, 1983).

[11] R. L. Devaney, *Chaotic Dynamical Systems* (Benjamin/Cummings, Menlo Park, CA, 1986).

[12] A. J. Lichtenberg and M. A. Lieberman, *Regular and Stochastic Motion* (Springer-Verlag, New York, 1983).

[13] P. Cvitanovic, *Universality in Chaos* (Adam Hilger, Bristol, 1984).

[14] Hao Bai Lin (ed.), *Chaos* (World Scientific, Singapore, 1984).

[15] S. Neil Rasband, *Chaotic Dynamics of Nonlinear Systems* (John-Wiley, New York, 1990).

[16] P. G. Drazin, *Nonlinear Systems* (Cambridge Univ. Press, Cambridge, 1992).

[17] S. Rajasekar and M. Lakshmanan, *Physica* **D32**, (1988) 146.

[18] K. Murali and M. Lakshmanan, *Int. J. Bifurcation and Chaos* **1**, (1991) 369.

[19] L. O. Chua, *IEICE Trans. Fundamentals Electron. Commun. Sci.* E76-A, (1993) 704.

[20] T. Matsumoto, M. Komuro, H. Kokubu and R. Tokunaga, *Bifurcations: Sights, Sounds and Mathematics*, (Springer-Verlag, Tokyo, 1993).

[21] M. C. Gutzwiller, *Chaos in Classical and Quantum Mechanics* (Springer-Verlag, Berlin, 1990).

[22] L. E. Reichl, *The Transition to Chaos in Conservative Classical Systems: Quantum Manifestations* (Springer-Verlag, New York, 1992).

[23] F. Haake, *Quantum Signatures of Chaos* (Springer-Verlag, Berlin, 1991).

[24] K. Nakamura, *Quantum Chaos: A New Paradigm of Nonlinear Dynamics* (Cambridge Univ. Press, New York, 1993).

[25] M. J. Ablowitz and P. A. Clarkson, *Solitons, Nonlinear Evolution Equations and Inverse Scattering* (Cambridge Univ. Press, Cambridge, 1991).

[26] M. Lakshmanan (ed.), *Solitons: Introduction and Applications* (Springer-Verlag, Berlin, 1988).

[27] A. Ramani, B. Grammaticos and T. Bountis, *Phys. Rep.* **180**, (1989) 160.

[28] M. Lakshmanan and R. Sahadevan, *Phys. Rep.* **224**, (1993) 1.

[29] M. Lakshmanan and R. Sahadevan, *J. Math. Phys.* **32**, (1991) 75.

[30] M. Lakshmanan and M. Senthil Velan, *J. Math. Phys.* **25**, (1992) 1259.

[31] P. J. Olver, *Applications of Lie Groups to Differential Equations* (Springer-Verlag, New York, 1986).

[32] S. Wiggins, *Introduction to Applied Nonlinear Dynamical Systems and Chaos* (Springer-Verlag, New York, 1990).

[33] I. G. Main, *Vibrations and Waves in Physics*, (III Edition) (Cambridge Univ. Press, Cambridge, 1993).

[34] G. Duffing, *Erzwungene Schwingungen bei Veränderlicher Eigen Frequenz und ihre Technische Bedeutung* (Vieweg, Brauschweig, 1918).

[35] S. Parthasarathy and M. Lakshmanan, *J.Sound & Vib.* **137**, (1990) 523.

[36] A. H. Nayfeh and D. T. Mook, *Nonlinear Oscillations* (John Wiley, New York, 1979).

[37] F. C. Moon, *Chaotic Vibrations: An Introduction to Applied Scientists and Engineers* (John-Wiley, New York, 1987).

[38] Y. Ueda, *J. Stat. Phys.* **20**, (1979) 181.

[39] V. Englisch and W. Lauterborn, *Phys. Rev.* **A44**, (1991) 916.

[40] C. Sparrow, *The Lorenz Equations: Bifurcation, Chaos and Strange Attractor* (Springer-Verlag, New York, 1982).

[41] A. V. Holden (ed.), *Chaos* (Manchester Univ. Press, Manchester, 1986).

[42] J. P. Eckmann and D. Ruelle, *Rev. Mod. Phys.* **57**, (1985) 617.

[43] R. Seydel, *From Equilibrium to Chaos: Practical Bifurcation and Stability Analysis* (Elsevier, New York, 1988).

[44] P. R. Fenstermacher, H. L. Swinney and J. P. Gollub, *J. Fluid Mech.* **94**, (1979) 103.

[45] Y. Pomeau and P. Manneville, *Comm. Math. Phys.* **74**, (1980) 189.

[46] K. Kaneko, *Collapse of Tori and Genesis of Chaos in Dissipative Systems* (World Scientific, Singapore, 1986).

[47] G. Qin, R. Li, D. Gong and L. Jiang, *Phys. Lett.* **A137**, (1989) 255.

[48] M. Hasler and J. Neirynck, *Nonlinear Circuits* (Artech House Inc., Massachusetts, 1986).

[49] L. O. Chua, C. A. Desoer and E. S. Kuh, *Linear and Nonlinear Circuits* (McGraw-Hill, Singapore, 1987).

[50] M. Marek and I. Schreiber, *Chaotic Behaviour of Deterministic Dissipative*

Systems (Cambridge Univ. Press, Cambridge, 1991).
[51] D. R. He, W. J. Yeh and Y. H. Kao, *Phys. Rev.* **B31**, (1985) 1359.
[52] Y. H. Kao, Y. C. Huang and Y. S. Gou, *Phys. Rev.* **A35**, (1987) 5228.
[53] M. Lakshmanan and K. Murali, *Physics News* **24**, (1993) 3.
[54] Y. H. Kao and C. S. Wang, *Phys. Rev.* **E48**, (1993) 2514.
[55] A. N. Willson (Jr.), *Nonlinear Networks: Theory and Analysis* (IEEE, New York, 1975).
[56] L. O. Chua, *Introduction to Nonlinear Network Theory* (McGraw-Hill, New York, 1969).
[57] L. O. Chua and R. N. Madan, *IEEE Circuits and Devices Magazine* (1988) 3.
[58] P. S. Linsay, *Phys. Rev. Lett.* **47**, (1981) 1349.
[59] T. Matsumoto, L. O. Chua and S. Tanaka, *Phys. Rev.* **A30**, (1984) 1155.
[60] C. Jeffries and J. Perez, *Phys. Rev.* **A26**, (1982) 2117.
[61] Y. Nishio and S. Mori, *Trans. IEICE* J75-A, (1992) 754.
[62] Y. Nishio and S. Mori, *Trans. IEICE* J75-A, (1992) 1819.
[63] K. Akiyama, K. Araki and M. Morisue, *Tech. Rep. on Nonlinear Problem of IEICE* NLP-89-6 (1989) 35.
[64] A. E. A. Aranjo, A. C. Soudack and J. R. Marti, *IEE Proceedings*-C140 (1993) 237.
[65] B. van der Pol and J. van der Mark, *Nature* **120**, (1927) 363.
[66] T. S. Parker and L. O. Chua, *Proc. IEEE* **75**, (1987) 1081.
[67] Y. Yasuda and K. Hoh, *IEICE Trans. Fund.* E76-A, (1993) 1126.
[68] T. Saito, *Electronics and Comm. Japan* **72**, (1990) 58.
[69] M. P. Kennedy, K. R. Krieg and L. O. Chua, *IEEE Trans. Circ. Syst.* **36**, (1989) 1133.
[70] J. J. Healey, D. S. Broomhead, K. A. Cliffe, R. Jones and T. Mullin, *Physica* **D48**, (1991) 322.
[71] M. Shinriki, Y. Yamamoto and S. Mori, *Proc. IEEE* **69**, (1981) 394.
[72] S. Inaba, T. Saito and S. Mori, *Trans. IEICE* **E70**, (1987) 744.
[73] L. O. Chua, *Archiv. für Electronik und Übertrag. Tech.* **46**, (1992) 250.
[74] M. P. Kennedy, *IEEE Trans. Circ. Syst.-I* **40**, (1993) 640; 657.
[75] M. P. Kennedy, *Frequenz* **46**, (1992) 66.
[76] L. O. Chua, *IEICE Trans. Fund.* **E76-A**, (1993) 704.
[77] K. Murali, M. Lakshmanan and L. O. Chua, *IEEE Trans. Circ. Syst.-I* **41**, (1994) 462.
[78] K. Murali, M. Lakshmanan and L. O. Chua, *Int. J. Bifurcation and Chaos* **4**, (1994) 1511.
[79] K. Murali, M. Lakshmanan and L. O. Chua, *Int. J. Bifurcation Chaos* **5**, (1995) 563.
[80] J. Luprano and M. Hasler, *IEEE Trans. Circ. Syst.* **36**, (1989) 146.
[81] A. Azzouz, R. Duhr and M. Hasler, *IEEE Trans. Circ. Syst.* **30**, (1983) 913.
[82] M. Hasler, *Proc. IEEE* **75**, (1987) 1009.
[83] A. Rodriguez-Vazquez, J. L. Huertas, A. Rueda, B. Perez-Verdu and L. O. Chua, *Proc. IEEE* **75**, (1987) 1090.
[84] T. P. Weldon, *Am. J. Phys.* **58**, (1990) 936.

[85] M. De Sousa Viera, A. J. Lichtenberg and M. A. Lieberman, *Phys. Rev.* **A46**, (1992) R7359.

[86] T. Endo and L. O. Chua, *IEEE Trans. Circ. Syst.* **35**, (1988) 987.

[87] L. O. Chua and T. Lin, *IEEE Trans. Circ. Syst.* **35**, (1988) 648.

[88] F. Zou and J. A. Nossek, *IEEE Trans. Circ. Syst.-I* **40**, (1993) 166.

[89] M. J. Ogorzalek, *IEEE Trans. Circ. Syst.* **36**, (1989) 1221.

[90] J. H. B. Deane and D. C. Hamill, *Electronics Lett.* **27**, (1991) 1172.

[91] F. Moss and P. V. E. McClintock (eds.), *Noise in Nonlinear Dynamical Systems* (Vol. 3) (Cambridge Univ. Press, Cambridge, 1989).

[92] Y. Ueda, *Nonlinear Science Today* 2, (1992) 1.

[93] F. C. Moon and P. J. Holmes, *J. Sound & Vib.* **65**, (1979) 275.

[94] P. J. Holmes, *Phil. Trans. R. Soc. London* **292**, (1979) 420.

[95] U. Parlitz and W. Lauterborn, *Phys. Lett.* **A107**, (1985) 351.

[96] W. Szemplinska-Stupnicka and J. Rudowksi, *Chaos* 3 (1993) 375.

[97] E. Kreuzer, M. Kleczka and S. Schaub, *Chaos, Solitons & Fractals* **1**, (1991) 439.

[98] I. Kan and J. A. Yorke, *Bull. American Math. Soc.* **23**, (1990) 469.

[99] Lj. Kocarev, K. S. Halle, K. Eckert and L. O. Chua, *Int. J. Bifurcation and Chaos* **3**, (1993) 1051.

[100] L. Perko, *Differential Equations and Dynamical Systems* (Springer-Verlag, New York, 1991).

[101] C. L. Olson and M. G. Olsson, *American J. Phys.* **59**, (1991) 907.

[102] B. A. Huberman and J. P. Crutchfield, *Phys. Rev. Lett.* **43**, (1979) 1743.

[103] R. Raty, J. Von Boehm and H. M. Isomaki, *Phys. Rev.* **A34**, (1986) 4310.

[104] W. Szemplinska-Stupnicka, *The Behaviour of Nonlinear Vibrating Systems* (Kluwer–Academic, Dordrecht, 1990).

[105] J. D. Fournier, G. Levine and M. Tabor, *J. Phys.* **A21**, (1988) 33.

[106] T. Bountis, V. Papageorgiou and T. Bier, *Physica* **24D**, (1987) 292.

[107] S. Parthasarathy and M. Lakshmanan, *J. Phys.* **A23**, (1990) L1223; *Phys. Lett.* **157**, (1991) 365.

[108] T. Bountis, *Int. J. Bifurcation and Chaos* 2, (1992) 217.

[109] E. L. Ince, *Ordinary Differential Equations* (Dover, New York, 1956).

[110] M. Lakshmanan and P. Kaliappan, *J. Math. Phys.* **24**, (1983) 795.

[111] Y. F. Chang and G. Corliss, *J. Inst. Math. Appl.* **25**, (1980) 349.

[112] M. R. Guevara, L. Glass, M. C. Mackey and A. Shrier, *IEEE Trans. Syst. Man & Cybern.* **13**, (1983) 790.

[113] N. Minorsky, *Nonlinear Oscillations* (Van Nostrand, Princeton, 1962).

[114] S. Rajasekar and M. Lakshmanan, *J. Theor. Biology* **133**, (1988) 473.

[115] A. C. Scott, *Neurophysics* (John-Wiley, New York, 1977).

[116] S. Yasin, M. Friedman, S. Goshen, A. Rabinovitch and R. Thieberger, *J. Theor. Biol.* **160**, (1993) 179.

[117] M. H. Jensen, P. Bak and T. Bohr, *Phys. Rev. Lett.* **50**, (1983) 1637.

[118] U. Parlitz and W. Lauterborn, *Phys. Rev.* **A36**, (1987) 1428.

[119] S. Rajasekar, S. Parthasarathy and M. Lakshmanan, *Chaos, Solitons & Fractals* **2**, (1992) 271.

[120] T. Kapitaniak and W. H. Steeb, *J. Sound & Vib.* **143**, (1990) 167.

[121] T. C. Bountis, L. B. Drossos, M. Lakshmanan and S. Parthasarathy, *J. Phys.* **A24**, (1993) 6927.

[122] T. C. Bountis, L. B. Drossos and I. C. Percival, *J. Phys.* **A24**, (1991) 3217.

[123] L. O. Chua, C. W. Wu, A. Huang and G. Q. Zhong, *IEEE Trans. Circ. Syst.-I* **40**, (1993) 732; 745.

[124] M. P. Kennedy, *Proc. 1st Experimental Chaos Conference* eds., S. Vohra, M. Spano, M. Shlesinger, L. Pecora and W. Ditto (World Scientific, Singapore, 1992).

[125] L. P. Shil'nikov, *Int. J. Bifurcation and Chaos* **4**, (1994) 489.

[126] P. Kevorkian, *IEEE Trans. Circ. Syst.-I* **40**, (1993) 762.

[127] R. Madan (ed.) Special issue on Chua's circuit: *A Paradigm for Chaos, Part I & II, of J. Circ. Syst. Comput.* **3**, (1993).

[128] J. M. Cruz and L. O. Chua, *IEEE Tran. Circ. Syst.* **39**, (1992) 985.

[129] J. M. Cruz and L. O. Chua, *Int. J. Circ. Theory Appl.* **21**, (1993) 309.

[130] K. Murali and M. Lakshmanan, *IEEE Trans. Circ. Syst.-I* **39**, (1992) 264.

[131] K. Murali and M. Lakshmanan, *IEEE Trans. Circ. Syst.-I* **40**, (1993) 836; *Int. J. Bifurcation and Chaos* **2**, (1992) 621.

[132] L. O. Chua, *Proc. IEEE* **56**, (1968) 1325.

[133] G. Q. Zhong and F. Ayrom, *Int. J. Circ. Theory Appl.* **13**, (1985) 93.

[134] T. Matsumoto, *IEEE Trans. Circ. Syst. CAS* **31**, (1984) 1055.

[135] T. Matsumoto, L. O. Chua, and M. Komuro, *IEEE Trans. Circ. Syst. CAS-32*, (1985) 797.

[136] T. Matsumoto, L. O. Chua, and K. Tokumasu, *IEEE Trans. Circ. Syst. CAS-33*, (1986) 828.

[137] L. O. Chua, M. Komuro and T. Matsumoto, *IEEE Trans. Circ. Syst.* **33**, (1986) 1073 (Part I & II).

[138] L. O. Chua, Y. Yao and Q. Yang, *Int. J. Circ. Theory Appl.* **14**, (1986) 315.

[139] K. Fukushima and T. Yamada, *J. Phys. Soc. Japan* **57**, (1988) 4055.

[140] Z. Su, R. W. Rollins and E. R. Hunt, *Phys. Rev.* **A40**, (1989) 2689.

[141] G. R. Qin, R. Li, D. C. Gong and J. Jiang, *Phys. Lett.* **A137**, (1989) 255; *Phys. Lett.* **A141**, (1989) 412.

[142] O. Maldonao, M. Markus and B. Hess, *Phys. Lett.* **A144**, (1990) 153.

[143] S. Rajasekar and M. Lakshmanan, *Phys. Lett.* **A147**, (1990) 264.

[144] E. Lindberg, *J. Circ. Syst. Computers* **3**, (1993) 537.

[145] H. J. Reich, *Functional Circuits and Oscillators* (D. Van Nostrand Co., Princeton, NJ, 1961).

[146] Y. Ueda and N. Akamatsu, *IEEE Trans. Circ. Syst.* **28**, (1981) 217.

[147] C. T. Sparrow, *J. Math. Anal. Appl.* **83**, (1981) 275.

[148] G. P. King and S. T. Gaito, *Phys. Rev.* **A46**, (1993) 3092.

[149] P. Ashwin, G. P. King and J. W. Swift, *Nonlinearity* **3**, (1990) 585.

[150] K. Murali and M. Lakshmanan, *Phys. Rev.* **E48**, (1993) R1624.

[151] T. B. Fowler, *IEEE Tran. Auto. Control* **34**, (1989) 201.

[152] E. Ott, C. Grebogi and J. Yorke, *Phys. Rev. Lett.* **64**, (1990) 1196.

[153] B. A. Huberman and E. Lumer, *IEEE Trans. Circ. Syst.* **37**, (1990) 547.

[154] S. Sinha, R. Ramaswamy and J. Subba Rao, *Physica* **D43**, (1990) 118.

[155] R. Lima and M. Pettini, *Phys. Rev.* **A41**, (1990) 726.

[156] J. Singer, Y. Wang and H. H. Bau, *Phys. Rev. Lett.* **66**, (1991) 1123.

[157] Y. Braiman and I. Goldhirsch, *Phys. Rev. Lett.* **66**, (1991) 2545.

[158] L. Fronzoni, M. Giocondo and M. Pettini, *Phys. Rev.* **A43**, (1991) 6483.

[159] G. A. Johnson and E. R. Hunt, *J. Circ. Syst. Computers* **3**, (1993) 109; 119; E. R. Hunt, *Phys. Rev. Lett.* **67**, (1991) 1953.

[160] W. L. Ditto, S. N. Rauseo and M. L. Spano, *Phys. Rev. Lett.* **65**, (1990) 3211.

[161] A. Garfinkel, M. L. Spano, W. L. Ditto and J. N. Weiss, *Science* **257**, (1992) 1230.

[162] V. V. Aleixeev and A. Y. Loskutov, *Sov. Phys. Dokl.* **32**, (1987) 1346.

[163] A. S. Pikovsky, *Radiophys. Quantum Elect.* **27**, (1984) 390.

[164] N. J. Mehta and R. M. Henderson, *Phys. Rev.* **A44**, (1991) 4861.

[165] G. Nitsche and U. Dressler, *Physica* **D58**, (1992) 153; U. Dressler and G. Nitsche, *Phys. Rev. Lett.* **68**, (1992) 1.

[166] R. Genesio and A. Tesi, *J. Circ. Syst. Computers* **3**, (1993) 151.

[167] T. T. Hartley and F. Mossayebi, *Int. J. Bifurcation and Chaos* **2**, (1992) 881; *J. Circ. Syst. Computers* **3**, (1993) 173.

[168] T. Hogg and B. A. Huberman, *IEEE Trans. Syst. Man. Cyber.* **21**, (1991) 1325.

[169] E. A. Jackson, *Phys. Lett.* **A151**, (1990) 478; *Physica* **D50**, (1991) 341; *Phys. Rev.* **A44**, (1991) 4839.

[170] E. A. Jackson and A. Hubler, *Physica* **D44**, (1990) 407.

[171] E. A. Jackson and A. Kodogeorgiou, *Physica* **D54**, (1991) 253.

[172] A. W. Hubler, *Helv. Phys. Acta* **62**, (1989) 343.

[173] A. W. Hubler and E. Luscher, *Naturwissenschaft* **76**, (1989) 67.

[174] G. Chen and X. Dong, *Int. J. Bifurcation and Chaos* **2**, (1992) 407; *J. Circ. Syst. Computers* **3**, (1993) 139; *Int. J. Bifurcation and Chaos* **3**, (1993) 1363.

[175] T. Kapitaniak, Lj. Kocarev and L. O. Chua, *Int. J. Bifurcation and Chaos* **3**, (1993) 459; T. Kapitaniak, *Chaos, Solitons & Fractals* **2**, (1992) 519.

[176] S. Rajasekar and M. Lakshmanan, *Physica* **D67**, (1993) 282.

[177] K. Murali and M. Lakshmanan, *J. Circ. Syst. Computers* **3**, (1993) 125.

[178] T. Shinbrot, E. Ott, C. Grebogi and J. A. Yorke, *Phys. Rev. Lett.* **65**, (1990) 3215.

[179] T. Shinbrot, C. Grebogi, E. Ott and J. A. Yorke, *Nature* **363**, (1993) 411.

[180] Y. C. Lai, T. Tel and C. Grebogi, *Phys. Rev.* **E48**, (1993) 709.

[181] Y. C. Lai, *Computers in Phys.* **8**, (1994) 62.

[182] R. Roy, T. W. Murphy, T. D. Maier, Z. Gills and E. R. Hunt, *Phys. Rev. Lett.* **68**, (1992) 1259.

[183] R. Sinha, *Phys. Lett.* **A156**, (1991) 475.

[184] E. J. Romeiras, C. Grebogi, E. Ott and W. P. Dayawansa, *Physica* **D58**, (1992) 165.

[185] Y. C. Lai, M. Ding and C. Grebogi, *Phys. Rev.* **E47**, (1993) 86.

[186] T. Tel, *J. Phys.* **A24**, (1991) L1359.

[187] K. Wiesenfeld and B. McNamara, *Phys. Rev.* **A33**, (1986) 629.

[188] P. Bryant and K. Wiesenfeld, *Phys. Rev.* **A33**, (1986) 2525.

[189] H. Svensmark and M. R. Samuelsen, *Phys. Rev.* **A36**, (1987) 2413.

[190] D. Auerbach, P. Cvitanovic, J. P. Eckmann, G. Gunaratne and I. Procaccia, *Phys. Rev. Lett.* **58**, (1987) 2387.

[191] D. P. Lathrop and E. J. Kostelich, *Phys. Rev.* **A40**, (1989) 4028.

[192] F. Romeiras and E. Ott, *Phys. Rev.* **A35**, (1987) 4404.

[193] J. Brindley and T. Kapitaniak, *Chaos, Solitons & Fractals* **1**, (1991) 327.

[194] C. Grebogi, E. Ott, S. Pelikan and J. A. Yorke, *Physica* **D13**, (1984) 261.

[195] J. Heagy and W. L. Ditto, *J. Nonlinear Sci.* **1**, (1991) 423.

[196] S. Rajasekar and M. Lakshmanan, *Physica* **A167**, (1990) 793.

[197] S. P. Dawson and C. Grebogi, *Chaos, Solitons & Fractals* **1**, (1991) 137.

[198] L. M. Pecora and T. L. Carroll, *Phys. Rev. Lett.* **64**, (1990) 821; *Phys. Rev.* **A44**, (1991) 2374; T. L. Carroll and L. M. Pecora, *IEEE Trans. Circ. Syst.* **38**, (1991) 453; *Physica* **D67**, (1993) 126; *Int. J. Bifurcation and Chaos* **2**, (1992) 659; *IEEE Trans. Circ. Syst.* **40**, (1993) 646.

[199] R. He and P. G. Vaidya, *Phys. Rev.* **A46**, (1992) 7387.

[200] L. O. Chua, Lj. Kocarev, K. Eckert and M. Itoh, *Int. J. Bifurcation and Chaos* **2**, (1992) 705.

[201] T. Endo and L. O. Chua, *Int. J. Bifurcation and Chaos* **1**, (1991) 701.

[202] M. de Sousa Vieira, A. J. Lichtenberg and M. A. Lieberman, *Int. J. Bifurcation and Chaos* **1**, (1991) 691; *Phys. Rev.* **A46**, (1992) 7359.

[203] J. F. Heagy, T. L. Carroll and L. M. Pecora, *Phys. Rev.* **E50**, (1994) 1874.

[204] N. Gupte and R. E. Amritkar, *Phys. Rev.* **E48**, (1993) 1620.

[205] K. Murali and M. Lakshmanan, *Int. J. Bifurcation and Chaos* **3**, (1993) 1057.

[206] K. Murali and M. Lakshmanan, *Phys. Rev.* **E49**, (1994) 4882.

[207] M. Lakshmanan and K. Murali, *Current Science* **67**, (1994) 989.

[208] K. Pyragas, *Phys. Lett.* **A170**, (1992) 421; *Z. Naturforsch,* **A48**, (1993) 629; *Phys. Lett.* **A181**, (1993) 203.

[209] A. Kittel, K. Pyragas and R. Richter, *Phys. Rev.* **E50**, (1994) 262.

[210] N. F. Rulkov, A. R. Volkovskii, A. Rodriguez-Lozano, E. Del Rio and M. G. Velarde, *Int. J. Bifurcation and Chaos* **2**, (1992) 669; *Chaos, Solitons & Fractals* **4**, (1994) 201.

[211] V. S. Afraimovich, N. N. Verichev and M. I. Rabinovich, *Radiophysics & Quantum Electr.* **29**, (1986) 795.

[212] V. Anishchenko, T. E. Vadivasova, D. E. Postnov and M. A. Safonova, *Radioengn. & Electr.* **36**, (1991) 338.

[213] K. M. Cuomo and A. V. Oppenheim, *Phys. Rev. Lett.* **71**, (1993) 65; K. M. Cuomo, A. V. Oppenheim and S. H. Strogatz, *IEEE Trans. Circ. Syst.* **40**, (1993) 626.

[214] K. M. Cuomo, *Int. J. Bifurcation and Chaos* **3**, (1993) 1327; *Int. J. Bifurcation and Chaos* **4**, (1994) 727.

[215] Lj. Kocarev, K. S. Halle, K. Eckert, L. O. Chua and U. Parlitz, *Int. J. Bifurcation and Chaos* **2**, (1992) 709.

[216] U. Parlitz, L. O. Chua, Lj. Kocarev, K. S. Halle and A. Shang, *Int. J. Bifurcation and Chaos* **2**, (1992) 973.

[217] L. O. Chua, M. Itoh, Lj. Kocarev and K. Eckert, *J. Circ. Syst. Computers* **3**,

(1993) 93.

[218] Lj. Kocarev, A. Shang and L. O. Chua, *Int. J. Bifurcation and Chaos* **3**, (1993) 479.

[219] C. W. Wu and L. O. Chua, *Int. J. Bifurcation and Chaos* **3**, (1993) 1619; **4**, (1994) 979.

[220] K. S. Halle, C. W. Wu, M. Itoh and L. O. Chua, *Int. J. Bifurcation and Chaos* **3**, (1993) 469.

[221] H. Dedieu, M. P. Kennedy and M. Hasler, *IEEE Trans. Circ. Syst.-II* **40**, (1993) 634.

[222] Y. C. Lai and C. Grebogi, *Phys. Rev.* **E47**, (1993) 2357; *Phys. Rev.* **E50**, (1994) 1894.

[223] Y. H. Yu, K. Kwak and T. K. Lim, *Phys. Lett.* **A191**, (1994) 233.

[224] T. C. Newell, P. M. Alsing, A. Gavrielides and K. Kovanis, *Phys. Rev. Lett.* **72**, (1994) 1647.

[225] M. Ding and E. Ott, *Phys. Rev.* **E49**, (1994) R945.

[226] E. D. Rio, M. G. Velarde, A. Rodriguez–Lozano, N. F. Rulkov and A. R. Volkovskii, *Int. J. Bifurcation and Chaos* **4**, (1994) 1003.

[227] V. S. Anishchenko, T. Kapitaniak, M. A. Safonova and O. V. Sosnovzeva, *Phys. Lett.* **A192**, (1994) 207.

[228] T. Kapitaniak and L. O. Chua, *Int. J. Bifurcation and Chaos* **4**, (1994) 477; T. Kapitaniak, L. O. Chua and G. Q. Zhong, *Int. J. Bifurc. Chaos* **4**, (1994) 483.

[229] N. F. Rulkov, L. S. Tsimring and H. D. I. Abarbanel, *Phys. Rev.* **E50**, (1994) 314.

[230] J. F. Heagy and T. L. Carroll, *Chaos* **4**, (1994) 385.

[231] T. L. Carroll, *Phys. Rev.* **E50**, (1994) 2580.

[232] A. M. Maritan and J. R. Banavar, *Phys. Rev. Lett.* **72**, (1994) 1451; **73**, (1994) 2932.

[233] A. S. Pikovsky, *Phys. Rev. Lett.* **73**, (1994) 2931.

[234] T. S. Parker and L. O. Chua, *Practical Numerical Algorithms for Chaotic Systems* (Springer-Verlag, New York, 1989).

[235] D. W. Jordon and P. Smith, *Nonlinear Ordinary Differential Equations* (Oxford Univ. Press, New York, 1987).

[236] K. Ogata, *Modern Control Engineering* (Prentice-Hall, New Delhi, 1991).

[237] M. Lakshmanan and K. Murali, *Phil. Trans. Roy. Soc. London, Trans.* **A353**, (1995) 33.

[238] G. S. Vernam, *J. Amer. Inst. Elec. Engn.* **55**, (1926) 109.

[239] K. Murali, *Bifurcation, Controlling and Synchronization of Certain Chaotic Nonlinear Electronic Circuits* (1994), Ph. D. Thesis, Bharathidasan University.

[240] S. Hayes, C. Grebogi and E. Ott, *Phys. Rev. Lett.* **70**, (1993) 3031.

[241] M. Shinriki, M. Yamamoto and S. Mori, *Proc. IEEE* **69**, (1981) 394.

[242] B. V. Chirikov, *Phys. Rep.* **52**, (1979) 265.

[243] P. Hagedron, *Nonlinear Oscillations* (Clarendon Press, Oxford, 1988).

[244] B. van der Pol, *Phil. Mag.* **43**, (1927) 700.

[245] M. L. Cartwright and J. E. Littlewood, *J. London Math. Soc.* **20**, (1945) 180.

[246] N. Levinson, *Ann. Math.* **50**, (1949) 127.

[247] P. J. Holmes and D. A. Rand, *Q. Appl. Math.* **35**, (1978) 495.

[248] J. Guckenheimer, *Physica* **D1**, (1980) 227.

[249] M. P. Kennedy and L. O. Chua, *IEEE Trans. Circ. Syst. CAS-33*, (1986) 974.

[250] J. W. Swift and K. Wisenfeld, *Phys. Rev. Lett.* **52**, (1984) 705.

[251] G. Qin, D. C. Gong, R. Li and X. D. Wen, *Phys. Lett.* **A141**, (1989) 412.

[252] H. Haken, Light, Vol. 2: *Laser Light Dynamics*, (North-Holland, Amsterdam, 1985).

[253] E. G. Gwinn and R. M. Westervelt, *Phys. Rev. Lett.* **54**, (1985) 1613; *Phys. Rev.* **A33**, (1986) 4143.

[254] B. A. Huberman, J. P. Crutchfield and N. H. Packard, *Appl. Phys. Lett.* **37**, (1980) 750.

[255] N. F. Pedersen and A. Davidson, *Appl. Phys. Lett.* **39**, (1981) 830.

[256] O. E. Rössler, *Phys. Lett.* **71A**, (1979) 155.

[257] I. Prigogine and R. Lefever, *J. Chem. Phys.* **48**, (1968) 1695.

[258] W. Knop and W. Lauterborn, *J. Chem. Phys.* **93**, (1990) 3950.

[259] J. Peinke, J. Parisi, O. E. Rössler and R. Stoop, *Encounter with Chaos*, (Springer-Verlag, Berlin, 1992).

[260] T. Kai and K. Tomita, *Progr. Theor. Phys.* **61**, (1979) **54**; K. Tomita and T. Kai, *J. Stat. Phys.* **21**, (1979) 65.

[261] B. L. Hao and S. Y. Zhang, *J. Stat. Phys.* **28**, (1982) 769.

[262] Th. M. Kruel, A. Freund and F. W. Schneider, *J. Chem. Phys.* **93**, (1990) 416.

[263] C. G. Lie and J. M. Yuan, *J. Chem. Phys.* **84**, (1986) 5486.

[264] M. E. Goggin and P. W. Milonni, *Phys. Rev.* **A37**, (1988) 796.

[265] J. Heagy and J. M. Yuan, *Phys. Rev.* **A41**, (1990) 796.

[266] P. M. Morse, *Phys. Rev.* **34**, (1929) 57.

[267] G. Herzberg, *Spectra of Diatomic Molecules* (Van Nostrand, New York, 1950).

[268] J. R. Ackerhalt and P. W. Milonni, *Phys. Rev.* **A34**, (1986) 1211; 5137.

[269] R. B. Walker and R. K. Preston, *J. Chem. Phys.* **67**, (1977) 2017.

INDEX

Brusselator, 3, 18, 295, 297, 298, 303

capacitor, 22, 25, 26, 28–30, 32, 133, 161, 162, 305
center, 10
chaos, 3, 18, 19, 31, 42, 43, 48, 51, 52, 57, 67, 77–80, 94, 101, 105–107, 113,
        115, 132, 133, 140, 146, 147, 151, 153, 155, 160, 161, 164, 165, 172, 180,
        181, 183–187, 194, 196, 197, 199, 207–209, 216–218, 220, 226, 227, 232,
        235, 236, 238, 239, 242, 248, 251, 256–258, 265, 270, 275–277, 279, 290,
        295, 298, 300, 306, 307
chaotic, 3, 4, 15, 16, 18, 19, 21, 29, 31, 32, 34, 36, 42, 44, 48, 49, 51, 59–61, 67,
        75, 78, 80, 82, 91, 101, 105, 107, 109–111, 114, 115, 127, 132, 133, 139,
        142, 147, 148, 151, 153, 159, 162, 169, 172, 175, 181, 183–187, 190, 193,
        199–201, 205, 207–210, 213, 215, 216, 220, 221, 226, 227, 230, 231, 235,
        236, 245–248, 253, 254, 256–258, 264, 265, 271–273, 275–279, 295, 296,
        298, 300, 303–305, 307
chaotic switching, 256
chemical reactions, 295, 296, 297, 303, 306
Chua's attractor, 133, 142, 172
Chua's circuit, 31, 127, 132, 133, 135, 138, 139, 142, 147, 159, 185, 209, 210,
        212, 213, 215, 216, 238, 239, 246, 258, 265, 276, 304
Chua's diode, 30, 31, 127–133, 135, 141, 142, 161–163, 175, 212, 227, 304, 305
Chua's oscillator, 135, 139
circuits, 3, 5, 6, 21, 22, 25, 28–30, 32, 87, 88, 127, 138, 161, 176–179, 183, 226,
        248, 256, 276
coexistence of attractors, 159
conservative systems, 3
correlation, 16, 107, 207
crisis, 42, 48, 60, 107, 112, 116, 147, 154, 170
criteria for chaos, 67
criterion for Lyapunov's asymptotic stability, 305
critical points, 81, 84
current, 22, 23, 28, 33, 34, 62, 92, 93, 95–97, 101, 133, 142, 230, 277

deterministic system, 303
Devil's staircase, 101, 159, 304
digital signal transmission, 236, 272, 274
dimension, 42, 103, 104, 207, 304
dissipative systems, 3, 4, 303, 304
double-band, 42, 170, 181, 216, 231, 249, 256, 276

torus, 133, 291, 295, 303, 306
trajectory plot, 63, 149, 153–155, 276

universal constants, 18, 147

v–i characteristic, 22, 128, 131, 134, 136
van der Pol oscillator, 3, 17, 90, 91, 104, 111, 120, 176–179, 289
voltage, 22, 25, 26, 28, 34, 62, 92, 112, 129, 130, 133, 142, 146, 147, 161, 215, 227, 230–232, 303

wave, 63, 218, 219, 260, 261, 265, 266, 272, 273
winding number, 91, 101, 102, 116, 304
window, 39, 44, 49, 63, 142, 148, 151, 153, 155, 156, 159